Fundamental Theories of Physics

Volume 208

The international monograph series "Fundamental Theories of Physics" aims to stretch the boundaries of mainstream physics by clarifying and developing the theoretical and conceptual framework of physics and by applying it to a wide range of interdisciplinary scientific fields. Original contributions in well-established fields such as Quantum Physics, Relativity Theory, Cosmology, Quantum Field Theory, Statistical Mechanics and Nonlinear Dynamics are welcome. The series also provides a forum for non-conventional approaches to these fields. Publications should present new and promising ideas, with prospects for their further development, and carefully show how they connect to conventional views of the topic. Although the aim of this series is to go beyond established mainstream physics, a high profile and open-minded Editorial Board will evaluate all contributions carefully to ensure a high scientific standard.

Léon Brenig · Nikolai Brilliantov · Mustapha Tlidi
Editors

Nonequilibrium Thermodynamics and Fluctuation Kinetics

Modern Trends and Open Questions

Springer

Editors
Léon Brenig
Physics Department
Université Libre de Bruxelles (U.L.B.)
Bruxelles, Belgium

Mustapha Tlidi
Physics Department
Université Libre de Bruxelles (U.L.B)
Bruxelles, Belgium

Nikolai Brilliantov
School of Computing and Mathematical
Science
University of Leicester
Leicester, UK

ISSN 0168-1222 ISSN 2365-6425 (electronic)
Fundamental Theories of Physics
ISBN 978-3-031-04457-1 ISBN 978-3-031-04458-8 (eBook)
https://doi.org/10.1007/978-3-031-04458-8

This Springer imprint is published by the registered company Springer Nature Switzerland AG
The registered company address is: Gewerbestrasse 11, 6330 Cham, Switzerland

Preface

While equilibrium statistical mechanics and thermodynamics belong to a well-settled area of the physical science, their non-equilibrium counterpart, and especially the fluctuation kinetics, are still the object of intense research activities and are in rapid development. Though the theory of small fluctuations out of equilibrium and of linear response is well established, large fluctuations and, more generally, far-from-equilibrium non-linear regimes lack a general theory. Moreover, even very basic concepts of the equilibrium theory, like temperature, entropy, and other thermodynamic potentials continue to be under debate. The application domain of the non-equilibrium thermodynamics and fluctuation kinetics permanently grows. It expands beyond systems and phenomena, traditional for statistical thermodynamics, such as fluid mechanics, physics of plasma, atmospheric science etc. and engulfs new areas of science. Granular and active matter, non-linear systems, dusty plasma, high-Q resonators, networks and even linguistic questions may be mentioned among the most prominent examples.

Fluctuation-dissipation theorem (FDT) and corresponding fluctuation-dissipation relations provide a common background to these diverse phenomena. FDT, initially discovered for equilibrium systems, is associated with such outstanding scientists as Einstein, Landau, Onsager, Kubo and other. Later FDT has been generalized for non-equilibrium processes. Professor Vjacheslav (Slava) Belyi greatly contributed to the theory of non-equilibrium fluctuations, including the application of the Callen-Welton fluctuation-dissipation theorem to non-equilibrium systems and its generalization. Slava Belyi was actively working on the elaboration of a framework of non-equilibrium thermodynamics, thus providing the basis for the future development of the field.

This book comprising the collection of articles by leading experts in non-equilibrium thermodynamics and fluctuation kinetics is dedicated to the memory of Slava Belyi, who passed away unexpectedly on May 20, 2020. It contains review papers on hot topics from the field with some personal dedications from Slava's colleagues.

The book starts with a short biographic in Chap. 1 describing the main stages of Slava's scientific career. The notes of this section have been written by his son,

Andrey Belyi, and friends with whom Slava collaborated during his life. They reflect both the scientific and personal life of Slava and may be interesting for a "western" reader as they enlighten some peculiar aspects of the scientific research.

Chapter 2 of the book is devoted to the general problems of non-equilibrium thermodynamics and fluctuation kinetics. It offers to the reader a review of fluctuation-dissipation relations (FDR), from the first formulations due to Einstein and Onsager, to the recent developments in the framework of stochastic thermodynamics of non-equilibrium systems. The general theory is demonstrated in the context of spin models, granular media and active matter. The nonlinear flux-force relations for the systems, out of Onsager's region, that respect the existing thermodynamic theorems for far-from-equilibrium systems are also considered. The reader will, then, find a discussion about the relationship between deterministic dynamical systems and their stochastic description when noise is included. The equivalence relationship between the resulting stochastic differential systems and the Fokker-Planck equations is compared to a recent result establishing an equivalence between the deterministic descriptions of dynamical systems and urn random processes. Next, the book explores the dynamics of an automaton on spatial lattices with spins at their nodes. The interactions between the automaton and the spins modify the latter and generate a complex trajectory that is equivalent to the functioning of a controller in a Turing machine. This illustrates an interesting link between non-equilibrium statistical mechanics and computer science. The most recent theoretical techniques in the field are also illustrated in this chapter. For instance, the application of the non-linear eigenvalue problem of an effective Hamiltonian in the complex energy plane allowing to explore unstable dynamical systems.

Then follows Chap. 3 which addresses the kinetic theory of far-from-equilibrium processes. Here the very popular theoretical tool of this area—the Boltzmann equation—is discussed in detail, as well as its application to the ballistic aggregation phenomena and to transport properties of granular mixtures. The derivation of Navier–Stokes equations and the corresponding transport coefficients is given. Different forms of the collision integral of the Boltzmann equations are considered. These range from the collision integral for granular gases, to that describing electrostatic interactions in dusty plasma. For the latter systems the characteristics of large-scale non-equilibrium fluctuations are derived from the first principles. In this section, non-equilibrium phase transitions occurring in active matter are also considered. Scaling phenomena and the similarity between the phase transitions in equilibrium and far-from-equilibrium systems are discussed.

The last Chap. 4 focuses on fluctuations and kinetics in non-linear and non-equilibrium systems. Here important for applications systems are considered such as dissipative solitons and frequency comb generation in both scalar and vectorial high-Q resonators. In particular, those simple devices have revolutionized many fields of science and technology, such as high-precision spectroscopy, metrology and photonic analog-to-digital conversion. Then the generalized entropy-production principle and the FDT relations are analysed in the context of their application to nanomaterials, such as carbon nanotubes and graphene films. Furthermore, a stochastic equation that describes the dynamics of optical vortices formation in liquid crystals cells is discussed. The statistical properties of the emerging vortices are determined by the properties of non-equilibrium fluctuations. The reported experimental results are in fair agreement with the theoretical findings. The thermodynamic instability of the atmospheric boundary layer is analysed in this chapter in terms of the non-equilibrium thermodynamics of open systems. This problem is very important for application, as it is related to earthquakes forecast. Finally, kinetic methods are applied to such a problem as information spread in co-evolutionary networks. Real networks from everyday life are addressed and the efficiency of the kinetic method is demonstrated.

We believe that this book would be useful both to the experts in the field and for newcomers. Moreover, we believe that with this book we express our gratitude for Slava Belyi's contribution to the area of science, to which he devotedly served all his life.

Bruxelles, Belgium — Léon Brenig
Leicester, UK — Nikolai Brilliantov
Bruxelles, Belgium — Mustapha Tlidi

Contents

Kinetic Theory of Far-from-Equilibrium Processes

Fluctuations and Kinetics in Non-linear Non-equilibrium Systems

Memorials

In Memoriam Professor Viacheslav (Slava) Belyi (1945–2020)

I. Veretennicoff, L. Brenig, L. Bindler, J. P. Boon, C. De Mol, R. Gerold, J. Wallenborn, M. Mareschal, A. G. Zagorodny, and N. Brilliantov

Abstract This chapter gathers contributions of numerous Slava's friends and colleagues from various institutions.

I. Veretennicoff (✉)
Department of Applied Physics, Vrije Universiteit Brussels, Pleinlaan 2, 1050 Brussels, Belgium
e-mail: ivereten@vub.be

L. Brenig · J. P. Boon · C. De Mol · J. Wallenborn · M. Mareschal
Faculté des Sciences, Université Libre de Bruxelles (ULB), Campus Plaine, CP. 231, 1050 Bruxelles, Belgium
e-mail: leon.brenig@ulb.be

J. P. Boon
e-mail: jean-pierre.boon@ulb.be

C. De Mol
e-mail: christine.de.mol@ulb.be

J. Wallenborn
e-mail: jean.wallenborn@ulb.be

L. Bindler
Brussels, Belgium

R. Gerold
Bonn, Germany

A. G. Zagorodny
M.M. Bogolubov Institute for Theoretical Physics, The National Academy of Science of Ukraine, 14b, Metrologichna Str., 03680 Kyiv, Ukraine
e-mail: azagorodny@bitp.kiev.ua

N. Brilliantov
Center for Computational and Data-Intensive Science and Engineering, Skolkovo Institute of Science and Technology, 30 Bolshoi Boulevard, bld. 1, Moscow 121205, Russia
e-mail: nb144@leicester.ac.uk

School of Computing and Mathematical Science, University of Leicester, University Road, Leicester, LE1 7RH, UK

L. Brenig et al. (eds.), *Nonequilibrium Thermodynamics and Fluctuation Kinetics*, Fundamental Theories of Physics 208,
https://doi.org/10.1007/978-3-031-04458-8_1

1 I. Veretennicoff: Towards a Productive Scientific Collaboration Between Izmiran and the Vrije Universiteit Brussel (1980–1991)

1.1 Yuri Lvovich Klimontovich and Radu Balescu

When Yuri Lvovich Klimontovich, professor at the Physics Department of Moscow State University (MSU) meets Radu Balescu professor at the Université Libre de Bruxelles (ULB) and his group in 1979–1980, Yuri Lvovich Klimontovich was a highly respected theoretical physicist, not only in the Soviet Union but also abroad. One of his research topics was related to the transport theory of plasma's, i.e. ionized gases such as the ionosphere. One of his young co-workers, Slava Belyi, was working hard towards a Doctorate in Physics, the highest title that could be obtained at a Russian University. Klimontovich had met Radu Balescu on several occasions and they had become friends. Radu Balescu belonged to Ilya Prigogine's research group and was esteemed both for the results of his research and his kind and efficient communication skills. It was Radu Balescu who invited Klimontovich to pay a visit to his group, who counted new doctoral and post-doctoral students (like me!) and new topics to explore. All the co-workers were invited to present "their" topic to Klimontovich and have a serious discussion with him. And so I met him for the first time in person: quite an impressive encounter! He obviously was interested in our results. And some months later—bureaucracy is always slow—I received an official invitation to come and visit the Academy of Science of the USSR, deliver a talk and discover Moscow. As I was the mother of two young girls, the wife of a busy researcher himself and a full-time academic member of staff, the only possible week available for me was starting at the end of October 1980, the week where the children had a holiday!

1.2 My Own Roots Are to Be Found in St. Petersburg, Russia

My name could have been Irina Vladimirovna Veretennicova. But I was born in Antwerp in 1944 during WWII and it was not allowed to give a newborn baby a name belonging to the enemy. And so officially I became Irène Veretennicoff. My father wanted me to become an authentic Belgian citizen, not a refugee. I learned to speak French, Dutch and English, no Russian! However, once in a while, nostalgia suddenly was invading Vladimir. Together with his friends he would empty a bottle (or two ...) of real Russian vodka, grasp his guitar and sing romantic songs from before the Soviet Revolution. A typical Russian soul, Slava's type! Vladimir was born in St. Petersburg, 1901. His mother, born Vera von Ratch, was a descendent of the Serbian teacher who educated Peter the Great in martial arts. The Tsar appreciated it so much that he ennobled him, gave him the title of Prince and made him rich. My grandmother was a real princess!

My grandfather Alexis was educated as a military civil engineer and was the chief of staff (Chef d'Etat Major) on the top of the Tsaristic Army, before and during WWI. Vera and Alexis had five children together, the youngest one was my father Vladimir. They settled in Belgium after a long and dangerous journey. They had lost all their belongings and privileges. Alexis found a job as inspector for the construction of houses devoted to the working class in "Small Russia", situated in Zelzate near Ghent. Both my grandparents lived, suffered and passed away there ....

Vladimir was more fortunate: he got one of the fellowships attributed by Cardinal Mercier to study engineering at the University of Leuven, together with a lot of other young (and not so young!) Russian immigrants. At the end of his studies, Vladimir was immediately engaged to work as a mechanical engineer for the brand-new General Motors Continental factory in Antwerp. He stayed loyal to this firm until he retired, as managing director of the reliability and quality control in the 1960s.

1.3 My Arrival in Moscow

I arrived by plane in the capital of the USSR on October 20, 1980. The contrast with Brussels airport was impressive. Impressive were the officers at the border with their impressive uniforms and unfriendly way to welcome the travelers. Smiling and joking were not allowed! We had to queue and be patient. Then I identified the long row of people expecting the travelers. I looked for Professor Klimontovich. Without success. He was not there. However, I spotted a man in his 30-ties carrying some flowers and waving at me. It was Slava. What a relief!

I discovered that the guests of the Academy of Sciences were treated according to the rules of the Russian hospitality: we had a car and a chauffeur at our disposal for the time of my stay. My room had been booked in the hotel for the guests of the USSR Academy of Sciences. Slava had brought everything necessary to provide me with not only a comfortable but also a festive welcome. As restaurants and shops were rare and mostly not so attractive, we picked-nicked on the spot and made plans for the week.

The knowledge of Slava's French and English was sufficient to understand one another. We visited the most touristic places in and around the town and visited Zagorsk, a place where future orthodox priests got their education. I was impressed by their stature and the religious beauty of the environment. We went to a ballet performance and an opera. All of this was wonderful and I still vividly remember the atmosphere existing in the Moscow in the late Soviet times.

But what impressed me the most was the kindness of all the people that I met during that week. The director of IZMIRAN, Professor Miguline invited us for lunch. His appreciation for Slava's work and personality was obvious. He seemed happy to have me as a guest together with Slava. According to the terms of his invitation, a talk was planned during my stay.

In those times—the very early 1980—most of the modern equipment for presentations consisted in transparent slides and a retro-projector to project them. I remember arriving well on time to check the equipment and found out that it was out of order! I also noticed that the conference room was fully occupied by men and women, young and old, with profiles that were not necessarily matching those of scientists. Given the circumstances, I proposed to deliver the talk on the blackboard, a suggestion that seemed to please the organizers and the audience. I could take the necessary time for my presentation. I did my best.

Then the chairman thanked me—in Russian—for the "interesting seminar" and asked the audience whether there were any questions or remarks. To my surprise nobody reacted! I concluded that I did not properly do my job.... Fortunately, Slava told me not to worry: the problem was that the vast majority of the audience did not understand English! In the times of the Soviet Union orders were orders; there was no escape when a visitor was giving a talk, whatever the subject and the language. Even today I still feel very sorry for this misunderstanding.

1.4 Collaboration

Slava and I had an excellent scientific collaboration between 1980 and 1990 despite the distance between our institutions and our heavy agendas. We became very good friends. Together we produced 15 original papers largely devoted to the anomalous transport and fluctuations in strongly inhomogeneous systems. Some of them have been published in well-known international journals such as the "Physics of Fluids". In 1991, Klimontovich, Veretennicoff and Belyi received the very first and prestigious Russian State Price. Those were the times when Slava developed an impressive international network of friends and colleges that continued to expand until he sadly passed away, a victim of the Coronavirus on May 20, 2020.

As to us, we lost the common scientific track in 1991. The year when Michael Gorbachov was replaced by Boris Yeltsin. At the VUB, I had been promoted to full professor and director of the Department of Applied Physics and Photonics not only with a lot of interesting science but also a lot of administration and a growing number of (foreign) students. And Slava became more and more interested in questions related to the fluctuations in economics and social sciences inspired by Ilya Prigogine. For instance, he was invited to join Prigogine in talks and conferences with Gorbatchev or workshops in his field. And he became a friend of the family!

1.5 Traveling with Slava

Slava could have embraced many professions! Among those was the organization for his friends and colleagues on short trips to the ex Soviet Republics. I will never forget our visits to Armenia, Crimea, Georgia, Uzbekistan and so many exotic others. Warmhearted, intelligent, generous and also worried scientists, all were Slava's friends. Despite all his work and social connections, Slava was a devoted member of the Belyi clan. The most striking example for me is the way how he supported his son Andrei during the many difficult years leading to a doctorate obtained at the ULB and a professorship at the University of Tallin.

He was a man of all seasons; he loved his family and friends. He loved science and arts, traveling, cooking, singing and dancing. We have been fortunate to count him among our friends. We shall always remember him.

2 L. Brenig

It is through Radu Balescu and Irina Veretennicoff that I met the first time Slava Belyi at the beginning of the year 1980. Radu Balescu, my former Ph.D. director at Brussels Free University (ULB), was a renowned expert in Plasma Statistical Physics and a friend of Yuri L. Klimontovitch, a well-known Russian physicist working in the same field. As it happens, Slava Belyi had been a Ph.D. student of Y. L. Klimontovitch. Irina Veretennicoff, who also made her Ph.D. thesis under the direction of Radu Balescu, met Slava Belyi in Moscow where she had been invited by Klimontovitch. In return, Slava Belyi was invited for a stay in our Department at the ULB. This is how the Universe conspired to make me meet Slava.

Though we frequently discussed about scientific questions, I did not properly collaborate with Slava. Our research focuses were slightly different. However, we had common interests in non-equilibrium phenomena and, more generally, in statistical physics. We also had vivid conversations about the political and social situation in the Soviet Union and elsewhere in the world. Slava was not a fan of the Soviet regime but he was realistic and managed to adapt to this situation, though, at the cost of a perpetual and exhausting effort for keeping a decent life for his family.

Slava Belyi was a multi-gifted person. Apart from being an outstanding scientist, he had a natural talent for singing Russian nostalgic or joyful songs that would create an isba-in-the-taiga atmosphere. Everybody present in the assembly would soon cry or laugh.

Slava also had a social rallying gift. Each time he was expected to come for a visit, here in Belgium or anywhere else, there would be a sudden buzz among his local friends. People who did not see each other for months would suddenly contact each other and organize a welcome party for his arrival.

One of our frequent conversation topics concerned the Jewish history and culture. Slava's interest in this domain came from his childhood in Tashkent. He had been living there with his parents near the Jewish quarter and, though he was from a Russian non-Jewish family, he spent a lot of time playing with the Jewish children of the neighborhood. He liked to say that they considered him as one of them. Along his life, he kept many contacts and good friends from the Jewish background.

Slava was a mixture of an enthusiastic and a fatalistic person. He spoke fluently French and his main motto was "A quoi bon?", an expression which could translate

Fig. 1 Viacheslav (Slava) Belyi and Nadine Galland. Photo taken in Belgium in September 1994

Fig. 2 Viacheslav (Slava) Belyi. Photo taken at Moscow University in January 2017

into English as "What's the point?". However, as soon as he pronounced these words he started laughing, thereby negating the first pessimistic impression (Figs. 1 and 2).

"A quoi bon?" certainly does not apply to his life. By his creativity and his great humanity and sense of friendship, Slava illuminated the life of many of us and will stay in our memories forever.

3 L. Bindler, Formerly at Belgonucleaire, Brussels, Belgium

I met Slava Belyi in May 1992 via a friend of mine, a scientist Georges Severne from the Vrije Universiteit Brussel (VUB). We found a common ground when I assisted his son, Andrei, still a student, in getting to Belgium via Rotary Exchange programme. We also collaborated professionally. In the past, I worked at Belgonucleair, a national nuclear fuel producer from Dessel in Belgium. Belgonucleaire was famous for techniques it had developed in the 1970s for obtaining civil nuclear fuel from plutonium. In 1972, the first mixed oxide fuel (MOX), using both uranium and plutonium, was used industrially.[1] In 1984, Belgonucleaire, together with its French partners, participated in the generation of a new type of MOX which could be used for converting military-grade plutonium for civil purposes. In the early 1990s, in the aftermath of the collapse of the Soviet Union, Belconucleaire's interest turned to Russia because the company was able to introduce techniques to convert Soviet weapon-graded plutonium into energy for civil use. I introduced Slava to representatives of Belgonucleaire, who had begun its activities in Russia. At that time, my colleagues needed a contact in Moscow who could introduce them to local industry representatives, and who had a broad understanding of the nuclear industry. Slava was the right person for that. In due time, the innovative MOX nuclear fuel was introduced in Russian nuclear power plants, and a couple of years later, Russia even generated MOX fuel itself and became an important international player in the civil nuclear industry. We have kept a close friendship and had common personal and professional contacts such as Andre Jaumotte, a former rector of the Free University of Brussels (ULB) who was also acting head of the Solvay Institutes from 2003 to 2004. In turn, Slava's collaboration with the Solvay Institutes commenced with his close friendship with Ilya Prigogine, who had directed the Solvay Institutes for four decades till his death in 2003. Slava Belyi was a remarkable person because of his ability to maintain a wide range of contacts and relationships. He also helped me to find my cousin in Moscow. My father was originally from Russia and happened to live in Antwerp during the Bolshevik revolution. He stayed there through the aftermath whilst his brother (my uncle) lived in Moscow. The two brothers broke contact in the 1930s because having a foreign relative could threaten the life of anyone who lived in the Soviet Union. I grew up knowing about the existence of a female cousin I had in Soviet Russia but had not had an opportunity to contact her. Slava just decided to help me to find my Russian cousin, conducting a brief investigation in Moscow. I met my cousin, Galina Bindler, thanks to Slava's efforts. After all, I keep bright memories about Slava as a dynamic, competent, jovial and seducing personality.

[1] J-V Vilet, A. Michel, L. Bindler 'Belgonucleaire 1990–2005' in Histoire du nucléaire en Belgique, 1990–2005, p. 185.

4 J. P. Boon

When Slava first came to Brussels to visit Prigogine's group in the 1980s, I didn't get to know him very well. But some years later, in 2001, I invited him to participate in a summer school that I organized at the Institute des Etudes Scientifiques in Cargese (Corsica). When Slava arrived in Cargese and discovered the landscape and beautiful beach, he expressed how delighted he was, saying, with his characteristic "slavish" accent "Comme Crimee, comme Crimee!", and as a present, he offered me a bottle of J. P. B.'s Vodka with a picture of mine on the label. I still have the bottle which I intended to open and share on the occasion of Slava's next visit. The bottle is still unopened on the shelve in my office ...

5 C. De Mol

I met Slava Belyi during one of his research visits in Belgium (I think the first one) in the group of Professor Ilya Prigogine at the 'Université libre de Bruxelles' (ULB). At the time, he mainly collaborated with Irina Veretennicoff.

We had over the years many interesting conversations about several aspects of life, but rarely about physics since our research fields were essentially disjoint. There was one exception: when Slava learned that I had become interested in economics, he tried to convince me to work on some problems in econophysics that were related to his expertise in statistical mechanics. I then bought and started studying the book by Stanley and Mantegna "An Introduction to Econophysics". I regret that these exchanges never got concretized into some real work and collaboration.

Our conversations, which were also a good opportunity to practice the Russian language I had learned at evening courses, developed into a genuine friendship. I must say that Slava had a real cult for friendship. He introduced me during festive events to some of his friends in Belgium, and he had many. He liked to gather them around memorable uzbek plovs (lamb and rice pilaf) he cooked with enthusiasm, using special mysterious flavouring berries which he called "berberis" and brought along with him from Russia, until he discovered that hedges in Belgium were full of them!

In January 1991, Slava organized for me a professional visit to his research institute "IZMIRAN" in Troitsk, south of Moscow. I discovered what a real Russian winter was like and also experienced the lifestyle in times of "deficit" of basic goods, including food! At the occasion of this visit, I could meet his family and I enjoyed very much their hospitality in his home in Troitsk, as well as in Andrei Slavnov's (his brother-in-law) in Moscow. Slava was very proud of his family and enjoyed talking with me about their kayaking holidays in Siberia and their stays in his datcha. One of the greatest satisfactions of his life was that he managed to arrange that his beloved son, Andrei, could complete his studies and then a Ph.D. in political science at ULB. We had the great pleasure to see him more often during that period.

During the pandemic of 2020, we regularly talked by Skype. Slava was very worried about the growing number of Covid contaminations in Russia. Rightly so, helas! Our last contact was by SMS during the beginning of his stay at the hospital where he passed away.

We miss him deeply.

6 R. Gerold

Slava Belyi was a charming person, an excellent scientist and communicator and a real friend of me for more than 20 years. He was a recognized physicist specialized in plasma physics working at IZMIRAN, a prestigious centre for space and geophysics. Already during the Cold War period, he had established his first contacts with fellow scientists in Belgium. With the collapse of the Soviet Union, restrictions on international cooperation in research were lifted and Slava became a regular and highly estimated scientific guest in Brussels. In order to give his son Andrei who had finished high school in Russia an international education, he looked for families who would host Andrei for an initial period. This way we met both in 1994 and agreed that Andrei would stay with us for a certain time. We made friends and Andrei became a member of our family.

At the time I was working at the Directorate General for Science, Research and Development of the European Commission and was in charge of international cooperation. It was a particularly thrilling time because after the fall of the Berlin wall I had the task to promote and manage the opening of the European Research Framework Programme to Eastern European countries and the Baltic States. Our aim was to boost scientific cooperation with these countries in view of their later membership in the European Union.

Strengthening scientific cooperation with Russia and the other "Newly Independent States of the Former Soviet Union" was also among the objectives pursued by the EU at the time. This included the highly political undertaking aiming at the conversion of military to civil research managed by the International Science and Technology Center, an intergovernmental organization including Russia, the United States, Japan, the EU and other countries. A more modest initiative was the setting up of INTAS, the "International Association for the promotion of cooperation with scientists from the independent states of the former Soviet Union". Its aim was to promote fundamental research based on a bottom-up approach. The financing came from the Research Framework Programme of the European Union and contributions from its associated countries. In the practical implementation of this cooperation Slava was of great support. He familiarized me with the structure of the Russian research system, which was dominated by the illustrious Academy of Sciences. Slava also helped us to establish contact with the newly established Russian Foundation of Basic Research which became a major partner of INTAS. Slava's role can be described as one of honorary ambassadors.

I recall in particular a mission in 1995 to St. Petersburg where I carried out some onsite visits to cooperation projects co-financed by INTAS. Slava arranged a meeting with Ludvig Faddeev, the president of the Academy of Sciences in St. Petersburg and a visit to the prestigious Joffe Institute. We also paid a visit to a university institute which kept historic records of dialects spoken in Siberia a century ago engraved in wax cylinders which started to deteriorate. Thanks to the funding of INTAS, these historic records were digitalized and preserved for future generations.

The friendship with Slava continued and every time he came to Brussels to meet his scientific partners, he also came to our home with Andrei. Many times he took over the kitchen to prepare his legendary dish "riz pilav" and my wife and family still remember his charming and heart-warming presence.

In 2015, we were Slava's guests in Moscow. He put us up in his new flat and together with his wife Ludmilla, his family and his friends we celebrated Slava's 70th birthday. We spent a delightful day at his datcha. A particular highlight was a very generous invitation by Andrei to a cruise of several days on the Wolga river together with Slava. We retain unforgettable impressions of the vast countryside and of the very friendly Russian families traveling with us on the boat.

Our last meeting with Slava was in 2017 when he spent a week with us in our datcha in Brandenburg together with his grandson. Despite suffering from health problems that he played down as usual, Slava was in a good mood and still very engaged in a dispute about a scientific publication. The corona pandemic interrupted all further personal contacts, but we remained in touch via the internet and occasionally received humorous WhatsApp messages often addressing political issues. We were completely shaken up when we heard in May 2020 about his Coronavirus infection and very soon thereafter his passing away.

Our family will keep Slava in lasting memory as an extraordinary and charming personality who became a close friend whom we sincerely miss.

Rainer Gerold, November 2021.

7 J. Wallenborn

More than a collaboration

Slava was not only a colleague, above all he was a friend.

I became acquainted with his family in the small apartment they inhabited in Troitsk, a small town near Moscow where is also situated IZMIRAN, the research institute at which he was affiliated. His wife Lucia was very welcoming and made efforts to make my stay pleasant despite the often difficult situation and the difficulties in finding food at that time. Their children, Nusha and Andrei were still at school. Also, I had the opportunity to meet Slava's parents who lived in Troitsk too. Later on, I went during weekends to the datcha he had just acquired, a piece of ground on which he build a traditional wooden house. He drove his family there, first in the gigantic Volga, after in the relatively small Golf, cars he bought in Belgium and Lucia cultivated a pleasant vegetable garden.

The first time I came to Troitsk was, with Léon Brenig, in 1989. At this occasion we visited Izmiran—of course—but Slava guided us to Moscow: Kremlin, Red Square, Moskwa and Arbat street, where the number of people was impressive due to the perestroika that was in full swing. There was an unexpected wind of freedom. In contrast, we went silently sightseeing in the surrounding country between Troitsk and Moscow, thanks to a car provided by the Academy of Sciences: Slava mistrusted the driver and was afraid of the words that Léon and I might say.

Slava was a very good guide. He was still a better travel agent. Several times, he took advantage of one of my work stays to organize a nice trip through the USSR. We were in St. Petersburg twice, including once with my wife and Nusha, Slava's daughter. Another time, we went to Kiev by train for a scientific meeting on statistical physics. The welcome music on our arrival at the destination station was impressive. But there I saw the difference in treatment between Soviet citizens and foreigners: at the hotel, Slava received a small room in the attic while I received a comfortable one.

The most extraordinary travel led us to Uzbekistan. We went by train—two days and three nights—with my wife and the young Andrei. In Samarcand we were received by the vice-rector of the university and his wife in traditional dress served us. After breakfast, starting with fruits and ending with meat and soup, the vice-rector went to cotton fields where students were working and his wife, no more servant, dressed herself accidentally to guide us to the university where she was a professor.

In Bukhara, the rector himself welcomed us. After a visit to the university, he said he wanted to invite the rector of the Brussels university to which I belong. When he learned out that this rector was a woman, he said, "Too bad, but it doesn't matter. She will be invited". Then we were the honored guests of a wedding. My wife was the only woman at the men's table; even the just married girl was confined to the women's room. We have given many toasts to the Uzbek-Belgian friendship. Finally, as we were a little late for the train, the rector drove us to the station platform with a sound of the horn to chase away the teeming travelers.

Slava repeatedly told me that I had nothing to prepare, so in Tashkent I gave without preparation a seminar on the transport properties of gases. Fortunately, the audience was not interested, because according to Slava, I said some nonsense. As Slava was born in Tashkent, we had the opportunity to stay with friends of his parents. In the garden were growing vegetables and good grapes were ripening. It was the sweetness of life, very different from the academic hotels in Moscow where the babushkas were afraid because our scientific discussions were a little too noisy, especially when we were collaborating with Yuri Kukharenko.

The Slava's knowledge of Uzbekistan was well appreciated in Belgium. He had become a specialist in plof, a typically Uzbek dish, made from mutton, onions, yellow carrots, rice and spices. On several occasions during the many parties organized by his Belgian friends and colleagues, he cooked this delicious dish, especially at my home. These meetings were very joyful. Slava, who was timid and quiet when he came for the first time to Belgium, became the principal animator of these assemblies. All participants enjoyed his bass voice and his loud laugh, and of course his jokes.

The scientific collaboration with Slava was therefore very pleasant, thanks to its ancillary activities. They are very pleasant memories.

8 M. Mareschal

There are several aspects which give my encounter and later friendship with Slava a particular flavour, a flavour made of a mixture of nostalgia for a time which has gone and for the extraordinary period we have lived in. And it is with those feelings, sadness for the loss of a friend and proudness of what we have lived and achieved that I write these lines.

I met Slava in 1982 when he came to visit our Department in Brussels. He was invited by Radu Balescu who had established contacts with Yuri Klimontovich. Klimontovich was a student of Bogoliubov, one of the founders of non-equilibrium statistical mechanics. While the names of Klimontovich and Bogoliubov were known to many of us, personally I was not familiar with books and articles coming from the Soviet Union (with the notable exception of the physics course of Landau and Lifshitz). Visits by Russian scientists were still exceptional at the time and the language barrier was a real obstacle. We have to be thankful to people like Balescu and Prigogine to have created those links which allowed a whole community to benefit from the results obtained by the important Russian School of Statistical Physics and Mathematics.

During my thesis in Prigogine's group, I had addressed a problem suggested by Alkis Grecos, my thesis advisor: can we compute how the sound modes are affected in a collisional plasma by the existence of the plasma oscillations? Alkis had found a simple model, a superposition of the Vlassov self-consistent field with an elementary collision operator. The model was elementary but it allowed a full treatment, showing indeed the predominance of the plasma frequency oscillations in the hydrodynamic regime. Slava had read the article and liked it so that we spent a few days discussing the meaning and importance of the result. Slava was very quick, both in understanding the physics and in learning the French language (including familiar expressions!).

This was the time when I shifted my interests to numerical simulations. Slava was curious about those techniques, but he remained faithful to the analytical approaches to kinetic theories so that we never really published together but remained closely related in discussing scientific problems concerning non-equilibrium systems. I later organized several workshops and conferences, in Brussels as well as in Lyon where I became director of a European center dedicated to atomic and molecular modeling (CECAM 1). I had the opportunity to invite Slava on many occasions, and he always took advantage of those possibilities to interact with the community of non-equilibrium molecular dynamics. It was amazing to witness how easily Slava could interact with a group, even when the techniques used were very much different.

Later, in Zaragoza, in 2012, I organized the 27th Rarefied Gas Dynamics conference. An important Russian community was attending the conference with many scientists using the Boltzmann equation, analytically or computationally. Slava participated and could present a paper to the conference. He was as much at ease with either group, helping many potential connections to be realized.

Slava was a giver. He would set no limits on spending time and energy with his friends. I remember once how he helped me in Moscow on a trip I was making to Novosibirsk, to attend a conference of the European Liquids group. That was back in the summer 1989. The whole trip was complicated to organize (I was also attending before that a conference in Poland). The Belgian travel agent had made reservations for me: I had a plane ticket for a flight from Prague to Sheremetyevo, the international Moscow airport, then a telegram was sent to the travel agent and confirmed the reservation for a flight from the other airport (Domodedovo), with national flights connecting to Novosibirsk. Slava told me that he would be there to help me in going from one airport to the other, and it would be difficult for me otherwise!

Slava was there when I came out of USSR entrance controls. We then went to taxis and there was a (rather hard) negotiation to fix a price to go to Domodedovo. That was my first visit to USSR and I was quite surprised to see the rather poor state of roads and taxis. The atmosphere at the other airport was also a bit special: people camping everywhere in the airport, waiting possibly a few days to get a seat for their flight.

I could see no signs in English. Slava was asking around and the only thing I could understand was intourist. At some point, we understood that we had to go to small construction reserved for foreigners. The easiest way was to go out and walk on the field for about 200 m and get to the construction. Slava convinced someone in change to let us go to the field and walk to the intourist building. Which we did. With planes moving around, we walked to the intourist building where they were quite surprised to see us coming from the field's side. The policemen in charge had a long discussion with Slava, after which he accepted that I could enter provided I would go through security screening. They also gave me a seat number. In all this negotiation, Slava remained calm, arguing and strong, finally getting the policemen in charge to solve the problem. I was impressed and very thankful to him. I will never forget!

Slava was a very generous person, and the critical views he could have would never lead him to aggressive statements. Rather he would use his (great) sense of humor and remain calm and strong in his views. He was an idealistic person in his thoughts, with a rather realistic view of humanity. A great personality, a friend, ...a man of value!

9 A. G. Zagorodny

The sad news of Slava Belyi's death was unexpected and extremely bitter for me. I have lost my old friend and colleague after over 40 years of friendship, warm meetings and numerous discussions of scientific problems, in particular those of our common research interests. We got acquainted with Slava in the early 1980s due to our common teacher, world-known scientist, a specialist in statistical physics and kinetic theory of many-particle systems professor of Moscow University Yurii Klimontovich. The acquaintance turned out to be highly fruitful not only because Slava was an amiable man and talented theorist but also because both of us dealt with the studies of large-scale fluctuations in non-equilibrium systems and hence could compare our experience that has laid a strong foundation to our friendly communications. Our contribution in this book concerns just this field of research. Slava has visited Kyiv many times, and he had talks both at the international conferences and seminars held in our institute—Bogolyubov Institute for Theoretical Physics. He had presented his doctorate in our institute and I tried my best to help him with various technical formalities. Slava, in turn, was most helpful for my report in the Institute of Terrestrial Magnetism, Ionosphere, and Radio Wave Propagation on the materials of the doctoral thesis to be successful and to obtain positive recommendations. We have met during various conferences in Moscow, Lviv, Tbilisi, Kiel etc. The meetings have always been warm and friendly. Slava has told me for many times that his family had been of Ukrainian descent. He loved Ukraine and supported our independence. He was aware of the Maidan events and delighted about our success. So it is nothing strange that his letters of that time ended with the slogan "Slava Ukraini (Glory to Ukraine)". A good scientist and friend of Ukraine have passed away. May his memory live forever.

10 N. Brilliantov

I met Slava Belyi for the first time when I was a Ph.D. student, while he was already a recognized scientist, an expert in kinetic theory. My Ph.D. supervisor recommended Slava as the external examiner for my thesis. Preparing a report on the thesis Slava gave me a lot of valuable comments about my work, which later developed into a few lines of research. We also discussed kinetic theory in general and the life of scientists. Slava had numerous scientific contacts in the Soviet Union and abroad. The respective stories about the lives immersed in science, which he narrated, sounded very attractive; this provided me with a strong motivation to remain in science after the viva. Hence, the meeting of a Ph.D. student and his thesis Referee resulted in a friendship that continued for decades till Slava's death. As an older friend and mature scientist, he supported me in my scientific career—sometimes giving good advice, or being a Referee for my Ph.D. students, sometimes inviting me as a speaker at an important conference, and sometimes directly, helping to get a job in an academic institute.

We have had multiple discussions about the Boltzmann equation in its various forms, about the fluctuation-dissipation theorem, both for equilibrium and non-equilibrium systems. The discussions were deep on the boundary of science and philosophy. Sometimes we addressed technical issues and Slava showed me nice theoretical tricks. I admired his skills and the ability to apply sophisticated mathematical constructions to physical problems. At the same time, he could clearly and simply explain complicated things. All such meetings were really enjoyable. After Slava has passed I feel a big gap in my scientific and human surroundings; however, disseminating his achievements would be good to honor his memory.

Biographic Note

Andrei Belyi

In this chapter, I will briefly describe Viacheslav (Slava) Belyi's biography. Being in a position of his son, I witnessed a large part of his scientific pathway, his profound curiosity about scientists' contributions to the real world. My father was convinced that scientific knowledge could help people understand the world around them in a broader sense, and could enable them to be successful in both politics and business. He was lucky enough to have frequently visited the families of the business leaders of the European industrial past, such as Schlumberger and Solvay. Both names connote successful people who wholeheartedly wanted to promote all aspects of science. Many witnessed his pleasant social manners and a remarkable ability to communicate with people from different cultures and his persistent attempt to bridge Moscow and Brussels in his personal life. His acquaintances were from different circles ranging from scholars to ambassadors and business leaders.

Belyi was a genuine academic who had a deep faith in scientific communities. He often helped his junior colleagues and friends in performing scientific results and he usually combined academic cooperation with a sense of friendship. He often told me: give to others more than you can receive from them! He was greatly inspired by Ilya Prigogine, the Nobel prize laureate of 1977, and was well-acquainted with his school of thought. In these lines of research, he achieved important results for the fluctuation–dissipation theorem in Plasma.

A. Belyi (✉)
Centre for Climate Change, Energy and Environmental Law, University of Eastern Finland, Yliopistonkatu, Joensuu, Finland
e-mail: andrey.belyy@uef.f

L. Brenig et al. (eds.), *Nonequilibrium Thermodynamics and Fluctuation Kinetics*, Fundamental Theories of Physics 208,
https://doi.org/10.1007/978-3-031-04458-8_2

1 Early Life

Slava Belyi grew up in Tashkent, the capital of the Uzbek Soviet Socialist Republic. At age of seventeen, in 1962, he moved to Moscow to study at the faculty of Physics of the Moscow State University (MSU). During Slava's studies at the university, courses on statistical mechanics were taught by Yuri Klimontovich, a relatively young professor at the university. Klimontovich specialized in the statistical mechanics of ionized gases and quickly gained an academic recognition with his book published in 1967.[1] Klimontovich liked both the conceptual approach and the differentiated formulas elaborated proposed by Slava and proposed him to arrange the findings into an article which they published together. Further, Klimontovich invited Belyi to continue with postgraduate studies to obtain a title of a 'candidate of science' (remotely equivalent to a doctoral title in the west). Scientific cooperation with Klimontovich continued throughout 1970s, as they visited together important academic symposiums from the West-Siberia to Lithuania's Baltic coast.

Once obtained a postgraduate degree in 1971, he was offered a job at the Soviet Institute of Terrestrial Magnetism, Ionosphere, and Radio Wave Propagation (abbreviated in Russian as IZMIRAN). By that time, IZMIRAN had become internationally recognized for scientific space physics and geophysics. This Institute was established in 1944 on the basis of a geomagnetic observatory planned back in 1939 on the eve of World War 2, next to a village called Troitskoe in Moscow's neighborhood. Later, this location was transformed into a town dedicated to researchers, a so-called *Akademgorodok* [literal translation: academic town]. A decade later, the Institute gained an international renommé, as it had become one of the world's largest centers for data collection on geomagnetic trends around the world. In 1958, IZMIRAN was among the co-organizers of an international symposium on geophysics. Research areas also included the ionosphere. Studies on the ionosphere's density have been important in defining the necessary conditions for radio wave propagation on which modern communication systems are based. This includes communication systems for aviation, shipping transport, and—later on—for the development of a global navigation system.[2] Belyi was involved in the topic of statistical mechanics of plasma throughout all his academic career.

2 Scientific Exchanges with Belgium

By the time Belyi joined IZMIRAN, the Institute was directed by Vladimir Migulin, a former Deputy to the Director General of the International Atomic Energy Agency (1955–1957). Twelve years later, in 1969, Migulin was nominated as IZMIRAN's director and on the same year as an agreement was struck between the Soviet

[1] Klimontovich [1].

[2] Klimenko et al. [2].

Union and the Union Belgium-Luxembourg on Scientific and Technical Cooperation. A supervisory board composed of representatives of the signatory countries was created, with annual meetings in Brussels and Moscow. This board was responsible for providing grounds for scientific exchange between members of scientific communities in the countries involved. Vladimir Migulin's political weight and reputation were critical to the inclusion of IZMIRAN in the scientific exchanges.

The very foreign first visitor to IZMIRAN was Irina Veretennikoff from the Free University of Brussels. Her former thesis supervisor, Radu Balescu, had met Yuri Klimontovich back in 1965 at a symposium on plasma studies organized near a Siberian city of Novosibirsk.[3] Contact with the Belgian scholar became significantly valuable for Klimontovich partly because Balescu published a seminal work on statistical mechanics with a focus on non-equilibrium theories, which were not studied in standard textbooks of that time.[4] Klimontovich took advantage of his connection to Balescu and visited him in Brussels in the mid-1970s, when the Belgo-Soviet agreement was in force. It is here that Balescu introduced Irina Veretennikoff to his Russian colleague. She visited IZMIRAN in 1980 and Slava Belyi was in charge of her visit to Moscow. In 1982 it was Slava Belyi's turn to go to Brussels to be hosted by Irina for the next round of the scientific exchange. The three-month visit was filled with memorable moments with his Belgian colleagues. Being a very open-minded person, he established durable friendships. In 1983, Balescu planned a new visit to Moscow—this time he was to accompany Ilya Prigogine, already a globally renowned scientist by then, affiliated with both Belgium and the US. Quite naturally, Migulin asked Slava Belyi to host the Nobel Prize laureate, which conferred a lot of responsibility and personal honor.

Prigogine preferred to combine his academic visit with some personal activities. He brought his son, Pascal, (a curious 14-year-old teenager keen to discover the world) on his visit. He apparently wanted to see some of his distant relatives in Moscow and invited one of them—a woman—to a sight-seeing tour. A mini-bus was arranged for transport, but the consequences of this tour were dire. Belyi was summoned by the Soviet intelligence services and interrogated about inviting Prigogine's relative to the tour when she wasn't registered in the visit protocol. Slava Belyi had to respond with a written explanation of his alleged misbehavior, addressed directly to IZMIRAN's director. Slava had to explain that if he had not allowed the lady into the mini-bus, the important guest might have simply left the Soviet Union to go home. Slava Belyi was certainly right from a human viewpoint, but the authorities thought otherwise because formalities matter more for them. Belyi was banned from leaving the country and being in contact with foreign scientists. However, with the end of the Cold War, the travel bans were subsequently lifted. The Nobel prize laureate deeply appreciated the humanity of Slava's behavior—his choices weren't forgotten and a remarkable friendship commenced.

---

[3] Balescu [3].

[4] Weyssow et al. [4].

3 Doctor of Science and 'a Prize Laureate

Slava Belyi needed to pass the second stage of his Russian academic career: the 'doctor of science', equivalent to a habilitation thesis. This process required a thesis defense in front of a panel of experts in the field. At that time there were two major national research institutes for the theoretical research of plasma: one in Moscow and another in Kiev. An influential scholar specialized in plasma, Nikolai Bogolubov—actually a former thesis supervisor of Yuri Klimontovich—was a Ukrainian, keen to reinforce Kiev's position in the field. He already knew Slava Belyi and was happy to invite him for his thesis defense, which took place in the capital of Ukraine at the Institute of Theoretical Physics (which now holds Bogolubov's name). The venue was meaningful to Slava—Ukraine, the country of the Belyi family's ancestors, had always remained a country close to his heart. At the end, in the fall of 1988, Slava Belyi's thesis defense took place on the subject of 'Carryover effect and kinetic fluctuations in non-equilibrium systems'.

Obtaining the title of 'doctor of science' was an important step toward gaining higher visibility in the national scientific community. Belyi's academic achievements, coupled with his collaboration with Irina Veretennikoff, attracted the attention from Russia's academic leaders. Of note, by the end of the Cold War in the late 1980s, many in Moscow positively viewed Russian scholars who collaborated with Westerners. At the right time, then, Slava and Irina produced new articles, including one published by Cambridge University Press,[5] and attracted the attention from many contemporaries. At some point, they received a recommendation from the Academy of Science for a national scientific Prize which was awarded by the end of 1991 to Belyi, Klimontovich, and Veretennikoff for their work in plasma physics.

4 New Sets of Collaboration

In the early 1990s, in the aftermath of the collapse of the Soviet Union, the work atmosphere at IZMIRAN changed. Migulin retired already in 1989 and his successor did not match his influence over scientific policies, nor his international reputation. In the aftermath of the collapse of the Soviet Union, economic transition to a market-based economy was accompanied by drastic budget cuts for research institutes, IZMIRAN included. The Institute lacked funds to finance existing laboratories and attract new researchers. Researchers oftentimes joked that a state simulates the salaries while researchers simulate their work. As an illustrative example of the lack of scientific funding, one of Slava Belyi's colleagues used IZMIRAN's office facilities to start his business, which had nothing to do with science at all. Slava Belyi started losing collaborative networks around him and often felt isolated in the institute. IZMIRAN's inability to attract new scientists certainly impacted Belyi's ability to create a proper school of thought, driven by younger scholars. The only remaining collaborator was

[5] Belyi and Veretennikoff [5].

Yuri Kukharenko, not even based at IZMIRAN, who collaborated with Belyi on statistical mechanics.

Slava was, however, able to maintain and strengthen a collaboration with his Belgian colleagues. His main co-author Irina Veretennikoff was promoted to head of a research unit at the Flemish-speaking VUB and she subsequently changed her research priorities. She introduced Slava to Jean Wallenborn, a researcher from the French-speaking ULB, in order to keep him close to Balescu's group of researchers. Belyi created a virtual bridge between Jean Wallenborn and Yuri Kukharenko. This trilateral academic union was eased by Slava's ability to bridge between two worlds and make the best results out of it. Overall, eight papers were produced by the three scientists. Perhaps their main result was an article on Pair correlation function and non-linear kinetic equation for a spatially uniform polarizable non-ideal plasma published in Physical Review Letters.[6]

5 The Solvay Institutes and Prigogine's Influence

Contacts with Prigogine played a significant role in Slava's professional life. By the time Slava Belyi was introduced to Belgium, the Nobel Prize laureate had directed the International Solvay Institutes for Physics and Chemistry, an independent scientific body founded by Ernest Solvay in 1911 following his communication with Wilhelm Ostwald in the early twentieth century. The Solvay Institutes of Physics and Chemistry that operated in keeping with Ostwald's worldviews to signify the symbiosis between science and industry. The first scientific council took place in 1911, where the most renowned scientists of that time (such as Albert Einstein, Marie Curie, and Max Planck) were all members of the scientific council. In 1970s his great-grandson, Jacques Solvay, managed to agree with the University of Brussels—with both ULB and VUB at once—on an academic affiliation. Prigogine was indeed the very first scientific director of the newly created institution. Some decades later, at the dawn of his scientific career, Slava Belyi became involved in The Solvay Institutes activities as well.

The administrative live at the Solvay Institutes was predominantly shaped by Ioannis Antoniou, a Greek scholar, who assumed a role of the deputy director since 1994. In tandem with the Institutes' director, Antoniou obtained European funds to initiate a so-called Euro-Russian collaboration on complexity. Belyi wasn't initially included into work packages of the project as he was not specifically focused on complexity studies. However, his ability to understand the two worlds—the East and the West—were indeed noticed by the Institutes management. Both Prigogine and Antoniou appreciated Slava's ability to build intercultural bridges and engaged him as an expert in non-linear physics and Euro-Russian relations.

[6] E.g. Belyi et al. [6].

Even though their collaboration never resulted in co-authorship, Belyi was very much influenced by the Nobel Prize laureate's approach to uncertainty and non-equilibrium. These contacts gave him a strong sense of criticism of scientific measurement systems and models. The Nobel Prize laureate had argued that the world is not outside the scientist's view, but an integral part of it! He argued that one of the basic objectives of scientific study is to explain the general limitations introduced by any measurement processes. Partly, therefore, Belyi often deplored that scientific methods, like probability models, tend to outstrip theories and conceptions. After Prigogine's death in 2003 Belyi's contacts with the Solvay Institutes diminished, but the memory of Prigogine remained deep in his heart. For Prigogine's 100th anniversary, in 2017, Slava wrote an article dedicated to the Nobel Prize laureate with a very warm preface: 'Professor Ilya Prigogine's legacy in science is enduring and indisputable. He was an outstanding philosopher, mathematician, chemist, physicist, and biologist, and he had an in-depth understanding of history, archaeology, and the arts. Ilya Prigogine had a great heart, was open to everyone, and his altruism has vividly contributed to many in our international scientific community. He believed in the human mission to achieve solidarity and progress. His scientific discoveries are actually inherent to these ideas which he defended throughout his life. Undoubtedly, the name of Prigogine remains among the greatest of this world'.[7] Through Slava Belyi's efforts in perpetuating Prigogine's memory, many viewed him as a genuine disciple of the Nobel prize laureate.

6 Major Results and Recognition

Slava Belyi was already a renowned scholar who had seen the evolution of his discipline through two renowned schools of statistical physics, one of Bogolubov and Klimontovich and the other of Balescu. One of his major single-authored outputs was an article in the Physical Reviews Letter,[8] on the fluctuation–dissipation theorem, a long-debated theoretical subject since Albert Einstein's discoveries. In his article, Belyi produced an illustrative chart demonstrating different dynamics of electrons and of electrostatic fields during a fluctuation process produced by an external disturbance. He showed that results of fluctuation differ in non-homogeneous plasma, hence a non-ideal state of plasma needs to be considered differently. The article in Physical Reviews Letters almost coincided with Klimontovich's death. So, Belyi decided to write a new text devoted to one of the most significant achievements of his supervisor—the Klimontovich-Langevin approach to the fluctuation–dissipation processes.[9] In the newer text, Belyi emphasized that fluctuations determine the sensibility of a plasma, hence measuring fluctuations become a necessary tool in diagnostics of plasma processes.

[7] Belyi [7].

[8] Belyi [8].

[9] Belyi [9].

Even towards retirement age, Belyi closely followed scientific debates and carefully read scholarly literature on the topic of plasma diagnostics. In 2016 he noticed an article published in *Nature* addressing Thomson scattering of laser light for general observations in plasma physics.[10] After having assessed the text in depth, Belyi thought that authors attempted to take a specific experimental observation to a more general trend in plasma physics, and didn't sufficiently analyzed the earlier works on fluctuations, including the Klimontovich-Langevin approach. In response he published a refutation article where he argued that this specific interpretation should not violate the fluctuation–dissipation theorem.[11] Together with his own publication, the authors of the article published their response to Slava Belyi's text. In a very respectful terms, the authors highlighted his achievements in the field and agreed that his approach is certainly more general than their observation. After having highlighted some more specific objectives of their article, the authors concluded that the debate he launched would fuel further scholarly attention.[12]

Belyi was happy of launching an important conceptual debate on the subject. He wanted to produce this result for quite a while, he said. His field of study—plasma physics—was gaining renewed interest since the world's policy communities had expressed hope in finding alternative fuel sources to hydrocarbons. Among others, thermonuclear energy extracted from ionized gases—plasma—has been tested at the international center in ITER in France.[13] Once he even told me then that in the past he had worked on some conceptual models for energy extraction from plasma. He probably should have taken the unpublished manuscript on energy sources out of storage. Slava Belyi passed away on 20 May 2020 in the city of Troitsk, near Moscow, where he spent most of his life.

References

1. Y. Klimontovich, The Statistical Theory of Non-Equilibrium Processes in a Plasma (Pergamon, 1967)
2. M. Klimenko, V. Klimenko, D. Kotova, A. Karpachev, V. Ovudenko, K. Ratovsky, Y. Yasyukevich, G. Zbankov, Ionosphere as a Medium of Radio Wave Propagation in Different Applied Tasks, Russian Open Conference on Radiowave Propagation (October 2019)
3. R. Balescu, In memoriam: Yuri Lvovich Klimontovich (2004)
4. B. Weyssow, A. Vulpiani, F. Mainardi, R. Sánchez, D. del Castillo-Negrete (2008). In Memoriam: Radu Balescu. Anomalous Transport: Foundations and Applications, 1–12
5. V. Belyi, I. Veretennikoff, Electrostatic-field fluctuations and form factors in multi-component non-equilibrium plasmas. J. Plasma Phys. **43**(1), 1–21 (1990)
6. V.V. Belyi, Yu. A. Kukharenko, J. Wallenborn, Taking into account the first non-Markovian correction to the Alescu–Lenard equation. Phys. Rev. Lett. **76**, 3554 (1996)
7. V. Belyi, Fluctuations out of equilibrium, in *Philosophical Transactions of The Royal Society A Mathematical Physical and Engineering Sciences*. **376**, 20170833 (2018). Published in the

[10] Kozlowski et al. [10].

[11] Belyi [11].

[12] Kozlowski et al. [12].

[13] ITER webpage. https://www.iter.org/.

theme issue by M. Tlidi, M.G. Clerc, and K. Panajotov, Philo. Trans. Roy. Soc. A, 376, 20180276 (2018).
8. V.V. Belyi, Fluctuation-dissipation relation for a nonlocal plasma. (2017) Phys. Rev. Lett. **88**(25), 255001–4 (2002)
9. V. Belyi, Klimontovich-Langevin approach to the fluctuation-dissipation theorem for a nonlocal plasma. J. Phys: Conf. Ser. **11**, 61–69 (2005)
10. P.M. Kozlowski, B.J.B. Crowley, D.O. Gericke, S.P. Regan, G. Gregori, Theory of Thomson scattering in inhomogeneous media. Sci. Rep. **6**, 24283 (2016)
11. V. Belyi, Thomson scattering in inhomogeneous plasma: the role of the fluctuation-dissipation theorem. Sci. Rep. **8**, 7946 (2018)
12. P.M. Kozlowski, D.O. Gericke, S.P. Regan, G. Gregori, Reply to 'Thomson scattering in inhomogeneous plasmas: the role of the fluctuation-dissipation theorem'. Sci. Rep. **8**, 1 (2018)

General Problems of Non-equilibrium Thermodynamics and Fluctuation Kinetics. Fluctuation Dissipation Theorem

The Many Faces of Fluctuation-Dissipation Relations Out of Equilibrium

Marco Baldovin, Lorenzo Caprini, Andrea Puglisi, Alessandro Sarracino, and Angelo Vulpiani

Abstract In this paper, we offer to the reader an essential review of the theory of Fluctuation-Dissipation Relations (FDR), from the first formulations due to Einstein and Onsager, to the recent developments in the framework of stochastic thermodynamics of non-equilibrium system. We focus on two general approaches, somehow complementary, where out-of-equilibrium contributions to the FDR are expressed in terms of different quantities, related either to the stationary distribution or to the transition rates of the system. In particular, we discuss applications of the FDR in the general field of causation and inference, and in the contexts of non-equilibrium systems, such as spin models, granular media and active matter.

M. Baldovin · A. Vulpiani
Dipartimento di Fisica, Università di Roma Sapienza, P.le Aldo Moro 2, 00185 Rome, Italy
e-mail: angelo.vulpiani@roma1.infn.it

L. Caprini
Scuola di Scienze e Tecnologie, Università di Camerino, Via Madonna delle Carceri, I-62032 Camerino, Italy
e-mail: lorenzo.caprini@gssi.it

Institut für Theoretische Physik II: Weiche Materie, Heinrich-Heine-Universität Düsseldorf, 40225 Düsseldorf, Germany

A. Puglisi (✉)
Istituto dei Sistemi Complessi - CNR and Dipartimento di Fisica, Università di Roma Sapienza, P.le Aldo Moro 2, 00185 Rome, Italy
e-mail: andrea.puglisi@cnr.it

A. Sarracino
Dipartimento di Ingegneria, Università della Campania "L. Vanvitelli", via Roma 29, 81031 Aversa (Caserta), Italy
e-mail: alessandro.sarracino@unicampania.it

L. Brenig et al. (eds.), *Nonequilibrium Thermodynamics and Fluctuation Kinetics*, Fundamental Theories of Physics 208,
https://doi.org/10.1007/978-3-031-04458-8_3

1 Introduction

The Fluctuation-Dissipation Relation (FDR) is among the few pillars of non-equilibrium statistical mechanics. The reason for its great relevance is rather transparent: it allows to compute the statistical response of a system to small external perturbations in terms of correlations of the unperturbed dynamics. In other words, one can understand how the system reacts to an external disturbance just looking at the statistical features in the absence of any perturbation: in such a way it is possible to determine perturbed properties (response) in terms of unperturbed features (correlations).

The FDR has been widely investigated in the context of turbulence (and more generally statistical fluid mechanics): for instance, it plays a key role for the closure problem in the Kraichnan's approach [98]. Moreover, there is a wide interest in the scientific community active in geophysical systems, in particular, for climate dynamics, where it is very important to understand the features of the system under perturbations (such as a volcanic eruption, or a change of CO_2 concentration) in terms of the knowledge based on time series. Another very relevant field where the FDR has been used and investigated is the general theory of stochastic thermodynamics, with particular focus on models for colloidal systems, granular media and active matter. Finally, FDRs play a central role in the study of the non-equilibrium dynamics of slow relaxing systems, such as Ising models or spin glasses.

Since response and dissipation are intimately related (this intuitive fact is made more formal later in this section), in this paper, we use "Fluctuation-Response" and "Fluctuation-Dissipation" in an interchangeable way. Historically, one of the first examples of Fluctuation-Response relation is given by the formula expressing the fluctuations of energy in an equilibrium system at temperature T with a (constant volume) heat capacity C_v, that reads

$$\langle E^2 \rangle - \langle E \rangle^2 = k_B T^2 C_v. \tag{1}$$

On the left-hand side of the formula, one has the fluctuations in an unperturbed system, while on the right-hand side, there is a quantity representing a response (the heat capacity), and the factor of proportionality between the two is the temperature (k_B is the Boltzmann constant). Einstein derived an analogous formula connecting the diffusivity D to the mobility μ for a Brownian particle dispersed in a solvent fluid at thermodynamic equilibrium:

$$D = k_B T \mu, \tag{2}$$

where again the unperturbed fluctuations (diffusivity) are proportional to response (mobility) through a factor of proportionality represented by the bath temperature T.

The two previous examples are instances of the larger class of so-called "static" equilibrium FDR, as they do not involve time-dependent quantities. In the first half of the twentieth century, a series of experimental and theoretical works made longer and longer the list of such kind of relations, always tying in the same way spontaneous fluctuations, response and temperature [101, 120]. A noticeable example from this list is the expression given by Nyquist in 1938, relating the fluctuations of voltage in a conducting wire where no potential differences or currents are externally applied (the so-called Johnson noise) to the resistance of the conductor and the temperature. The resistance is the analogous of the mobility and of the heat capacity in the previous equations, i.e. it represents a response. In this case, it is also particularly simple to appreciate the equivalence between response and dissipation.

A first step towards the generalization to a time-dependent—or dynamic—relation is represented by the regression hypothesis made by Onsager in 1931 [130, 131], which states that—for small perturbations from equilibrium—the system returns to equilibrium at the same rate as a fluctuation does at equilibrium. This fact is already contained in the Einstein relation above. By recalling the general connection between diffusivity and the velocity autocorrelation, i.e. that

$$D = \int_0^\infty dt \langle v(t)v(0)\rangle, \tag{3}$$

we can transform Eq. (2) into

$$\langle v(t)v(0)\rangle = k_B T R_{vF}(t), \tag{4}$$

with the identification

$$\mu = \int_0^\infty dt R_{vF}(t). \tag{5}$$

In the r.h.s. of Eq. (5), we define the so-called response function, $R_{vF}(t)$, which connects the mean variation of the particle's velocity at time t with a perturbation of the external force applied at time 0.

In order to discuss in full generality the FDR, we need to give a general definition of response function, which is the central object of linear response theory. We restrict the discussion to the linear perturbation of stationary states, i.e. states that are invariant under translations of time, so that time-dependent correlation functions and response functions only depend on differences of times. Generalizations to non-steady states are mentioned in Sect. 2.

The response function $R_{\mathcal{O}\mathcal{F}}(t)$ of the observable $\mathcal{O}(t)$ to a time-dependent perturbation of a parameter or degree of freedom $\delta\mathcal{F}(t)$ is implicitly defined in the following relation

$$\overline{\Delta\mathcal{O}(t)} = \int_{-\infty}^t dt' R_{\mathcal{O}\mathcal{F}}(t-t')\delta\mathcal{F}(t'), \tag{6}$$

where $\overline{\Delta\mathcal{O}(t)} = \overline{\mathcal{O}(t)} - \langle\mathcal{O}(t)\rangle_0$ represents the average deviation, at time t, of the observable $\mathcal{O}$ with respect to its average value in the unperturbed stationary system. Here $\overline{f(t)}$ denotes an average of the observable f at time t over many realizations of the same perturbation, while $\langle f\rangle_0$ denotes the average of f in the stationary unperturbed state, which is not time-dependent. It is clear that, taking an impulsive shape for the external perturbation, i.e. $\delta\mathcal{F}(t) = \Delta\mathcal{F}\delta(t)$ (with $\delta(t)$ the Dirac delta distribution), one has

$$\frac{\overline{\Delta\mathcal{O}(t)|_{imp}}}{\Delta\mathcal{F}} = R_{\mathcal{O}\mathcal{F}}(t), \tag{7}$$

which is also an operational definition of the response function. Here we stress that $\Delta\mathcal{F}$ has the dimensions of a time-integral of $\mathcal{F}(t)$. When (for instance) the perturbation has the shape of a Heaviside unit step function, i.e. $\delta\mathcal{F}(t) = \delta\mathcal{F}_0\Theta(t)$, then

$$\frac{\overline{\Delta\mathcal{O}(t)|_{step}}}{\delta\mathcal{F}_0} = \int_0^t dt' R_{\mathcal{O}\mathcal{F}}(t'). \tag{8}$$

If $\mathcal{O}(t)$ is the tracer's velocity along one axis and $\mathcal{F}(t)$ is the external force applied from time 0 to time ∞ to the tracer (parallel to that axis), the final velocity reached by the tracer is exactly $\delta\mathcal{F}_0 \int_0^\infty dt' R_{vF}(t')$, which explains the connection with the identification made in Eq. (5).

The FDR for systems with Hamiltonian $\mathcal{H}$ at equilibrium with a thermostat at temperature T—historically attributed to Callen and Welton and immediately after generalized by Kubo [100]—reads:

$$R_{\mathcal{O}\mathcal{F}}(t) = \frac{1}{k_B T}\langle \mathcal{O}(t)\dot{A}(0)\rangle_0 = -\frac{1}{k_B T}\langle \dot{\mathcal{O}}(t)A(0)\rangle_0, \tag{9}$$

where A is the observable (or degree of freedom) which is coupled to $\mathcal{F}(t)$ in the Hamiltonian to produce the perturbation, i.e. $\mathcal{H}(t) = \mathcal{H}_0 - \mathcal{F}(t)A$. It is easy to verify that if $\mathcal{O}$ is the tracer's velocity and $\mathcal{F}(t)$ is an external force applied to its x coordinate, Eq. (9) becomes Eq. (4). In conclusion the "dynamical" Einstein relation is a particular case of equilibrium FDR. From Eq. (9) one may get several possible variants, which are useful in different physical situations. A large amount of remarkable results concern, for instance, the time-Fourier transform of Eq. (9), as well as the relation connecting currents/flows and transport coefficients in spatially extended systems (the so-called Green-Kubo relations, see below) [101, 120].

The equilibrium FDR is valid also in the framework of stochastic processes, when they describe the dynamics of system fluctuating around thermal equilibrium. The main difference with the case considered by Kubo is that a stochastic process typically describes small systems, far from the thermodynamic limit, but the system size is in fact irrelevant for the purpose of the validity of the FDR. In the case of large systems (without long-range correlations), however, the averages are easily taken by means of one or few experiments, while in a stochastic process where noise is large, one needs to average over many realizations. An illustrative example is the so-called Klein-Kramers model which describes the dynamics of simple particle systems at thermal equilibrium [81]. In one dimension, its stochastic differential equations read:

$$\frac{dx(t)}{dt} = v(t) \tag{10a}$$

$$m\frac{dv(t)}{dt} = -\frac{dU(x)}{dx} - \gamma v(t) + \sqrt{2\gamma k_B T}\xi(t), \tag{10b}$$

where $\xi(t)$ is a white Gaussian noise with $\langle\xi(t)\rangle = 0$ and $\langle\xi(t)\xi(t')\rangle = \delta(t - t')$, γ is the viscous damping, $U(x)$ is an external potential. The model can be easily generalized to $N > 1$ interacting particles in any dimensions. In the absence of the external potential, Eq. (10) coincides with the original Langevin equation proposed a few years after the theories of Einstein [73] and Smoluchovski [161] to explain diffusion in Brownian motion [103]. Its steady probability distribution (achieved with the condition $\gamma > 0$ and confining potential) is given by $P(x, v) \propto e^{-\mathcal{H}(x,v)/(k_B T)}$ with $\mathcal{H}(x, v) = mv^2/2 + U(x)$. Linear response theory, when applied to the Klein-Kramers model in its stationary state, gives exactly the same result as Eq. (9) [120, 147]. The Klein-Kramers process is Markovian with respect to the variables (x, v), a property which is a rough approximation for the dynamics of a tracer which interacts with other particles in a fluid. Typically, it has to be generalized to take into account retarded (hydrodynamic) effects, by the introduction of linear memory terms, e.g. by writing a Generalized Langevin Equation (GLE) [101]:

$$m\frac{dv(t)}{dt} = -\int_{-\infty}^{t} dt'\Gamma(t - t')v(t') + \eta(t), \tag{11}$$

where $\Gamma(t)$ is a memory kernel representing retarded damping, and $\eta(t)$ is a stationary stochastic process with zero average $\langle\eta(t)\rangle = 0$. The noise time-correlation—to comply with the requirement of thermodynamic equilibrium (i.e. steady Gibbs distribution and detailed balance)—must satisfy the so—called FDR of the second kind:

$$\Gamma(t) = \frac{1}{k_B T}\langle\eta(t)\eta(0)\rangle. \tag{12}$$

It is clear that Eq. (12) has the same structure of Eq. (9), and this motivates the name of the relation. The Markovian case (damping with zero memory) is obtained when $\Gamma(t) = 2\gamma\delta(t)$ (recalling that $\int_0^t dt' 2\gamma\delta(t')v(t') = \gamma v(t)$). For a more detailed discussion of the significance of this condition and its connection to detailed balance, we invite to read Sect. 4.1 of [144].

This brief review paper is organized as follows. In Sect. 2, we introduce two different possible approaches to the FDR, which are based either on the knowledge of the stationary distribution or on the knowledge of the dynamical rules of the model. Then, in Sect. 3, we discuss several applications of the FDR, in particular in the field of non-equilibrium systems, such as granular media and active matter. Finally, in Sect. 4, some conclusions are drawn.

2 Two Approaches to Non-equilibrium FDR

The first examples of FDR date back to Einstein's work on Brownian motion (1905), and to Onsager's regression hypothesis (1930s). Since initially the FDR was obtained for Hamiltonian systems in thermodynamic equilibrium, somehow there is a certain confusion on its real validity. Here, we summarize two different generalizations of FDR which both hold for a broad class of systems, including the non-equilibrium cases [120].

2.1 *An Approach Based upon the Knowledge of the Stationary Distribution*

Let us consider a system whose stationary probability density $P_{st}(\mathbf{x})$ is non-vanishing everywhere, and wonder about the time behavior of the mean response of the variable $x_n(t)$ at time t under a small impulsive perturbation $\delta\mathbf{x}(0)$. We can write

$$\overline{\delta x_n(t)} = \left\langle x_n(t)\right\rangle_p - \left\langle x_n(t)\right\rangle$$

where $\langle\,\rangle_p$ and $\langle\,\rangle$ denote the average for the perturbed and the unperturbed systems respectively. For a Markov system, we can write

$$\left\langle x_n(t)\right\rangle_p = \int x_n P_p(\mathbf{y}) W(\mathbf{y}\to\mathbf{x}, t)\, d\mathbf{x}d\mathbf{y} \;, \quad \left\langle x_n(t)\right\rangle = \int x_n P_0(\mathbf{y}) W(\mathbf{y}\to\mathbf{x}, t)\, d\mathbf{x}d\mathbf{y} \;,$$

where $W(\mathbf{y}\to\mathbf{x}, t)$ is the probability of a transition from $\mathbf{y}$ at time 0 to $\mathbf{x}$ at time t, $P_0(\mathbf{y}) = P_{st}(\mathbf{y})$ and $P_p(\mathbf{y})$ is the initial distribution of the perturbed system.

In the case of an impulsive perturbation, the perturbed probability satisfies $P_p(\mathbf{y}) = P_{st}(\mathbf{y} - \delta\mathbf{x}(0))$, which allow us to derive a compact expression for $\overline{\delta x_n}$ when the perturbation is small:

$$\overline{\delta x_n(t)} = -\sum_j \left\langle x_n(t) \frac{\partial \ln P_{st}[\mathbf{x}(0)]}{\partial x_j(0)} \delta x_j(0) \right\rangle, \tag{13}$$

where the average is performed in the unperturbed system. Let us note that the assumption of small perturbation is necessary only in the last step of the derivation of Eq. (13): therefore, such a result can be generalized to the case of non-infinitesimal $\delta\mathbf{x}(0)$ [31]. As by-product we have that it is possible to avoid the criticism of van Kampen according to which it is not possible to rely on an expansion for small perturbations, because chaos makes them grow exponentially [169]. On the contrary, in the derivation of the above result [76], there are only assumptions about $\delta\mathbf{x}(t=0)$ and therefore chaos has no relevance.

We can say that formula (13) summarizes the main results of the linear theory, e.g. in Hamiltonian systems and stochastic processes: in addition one understands the existence of a link between response and a suitable correlation function even in non-equilibrium systems [120]. For instance in inviscid hydrodynamics with an ultraviolet cutoff, in spite of the non-trivial dynamics, since the presence of quadratic invariant, and a Liouville theorem, one has a Gaussian statistics and therefore a FDR holds for each of the variables, i.e. self-response functions to infinitesimal perturbations coincide with the corresponding self-correlation functions. Let us note that although $P_{st}(\mathbf{x})$ is Gaussian the dynamics is non-linear and it is not easy to compute the correlation functions.

Beyond the many conceptual advantages of Eq. (13), there is an obvious practical limit: the difficulty to determine $P_{st}(\mathbf{x})$, which is known only in some specific cases. In the next subsection, we will discuss an approach that does not need the knowledge of $P_{st}(\mathbf{x})$.

Let us open a brief parenthesis on chaotic deterministic dissipative systems: because of the phase space contraction one has that the invariant measure is singular, typically with a multifractal structure, and therefore, Eq. (13) cannot be applied. A quite natural temptation is to add a small amount of noise, so that a smoothing of the invariant probability density allows for the use of the FDR. At a first glance such an approach can appear unfair. On the contrary, the idea of the beneficial role of the noise, which seems to date back to Kolmogorov, has strong bases: a small noisy term in the evolution equations has the role of selecting the natural measure: one can say that in the numerical experiments the round-off errors of the computer play a positive role. It is quite natural to expect that the behaviors of the purely deterministic chaotic system are very close to those obtained by adding a small amount of noise; such a conjecture is widely confirmed by numerical computations [70].

A similar approach was extended by Seifert and Speck, who established interesting connections of the FDR with observables in the framework of stochastic thermodynamics, such as entropy production and housekeeping heat [156, 157, 164] (see also the next section).

2.2 An Approach Based upon the Knowledge of the Dynamical Model

When the dynamics of the system under study is defined in terms of transition rates or Langevin equations, but the stationary probability density function is not known, a complementary approach with respect to the one discussed in the previous subsection can provide a FDR valid also out of equilibrium. These kinds of FDRs have been derived in several different contexts, following different mathematical schemes (see discussion below).

The general approach dates back to the 60s of the twentieth century, when Furutsu and Novikov independently derived, under general conditions, a FDR [80, 125] which expresses the response function of a Gaussian process in terms of the equilibrium time-correlation between the observed variable and the Gaussian noise itself. Nowadays, a method based on similar ideas—sometimes termed Malliavin weight sampling [175]—has been extended to include field theories through the Martin-Siggia-Rose-Jansen-de Dominicis approach [1, 3, 60] and employed in the context of particle-based glassy systems to numerically calculate effective temperature and susceptibility [55, 57, 59]. This allows one to express the response function in terms of suitable correlation functions of the state variables. We mention here examples for non-equilibrium Langevin dynamics driven by a time-dependent force both in overdamped [7–9, 157, 178] and underdamped regimes [6], or even in the presence of a non-linear Stokes force [149]. The non-equilibrium terms appearing in the generalized FDR have been interpreted in several ways: some authors focused on the different roles of entropic and frenetic contributions (for a recent review, see [111]), outlining their different nature with respect to the symmetry under time-reversal transformation; other approaches have focused on the connection with entropy production and heat [93, 106, 164].

The class of generalized FDR so far mentioned is expressed in terms of the correlation between the observable $\mathcal{O}$ and a function of both the state variables and their time-derivatives. Without loss of generality, the starting point for these relations is of the form:

$$R_{\mathcal{O}x_j}(t,s) = \langle \mathcal{O}(t)\mathcal{M}_j[\mathbf{x}(s), \dot{\mathbf{x}}(s)]\rangle \,, \tag{14}$$

where, as usual, the average in the r.h.s. of Eq. (14) is performed through the unperturbed measure. $\mathcal{M}_j$ is a function uniquely determined by the dynamics of the system under consideration that depends both on $\mathbf{x}$ and $\dot{\mathbf{x}}$. Its functional form can be expressed in terms of known observables: for instance, in the case of continuous first order dynamics of the kind

$$\dot{x}_j = f_j(\mathbf{x}) + \sqrt{2D_j}\eta_j, \tag{15}$$

where η_j is a white noise with zero average and unit variance, one has

$$\mathcal{M}_j = \frac{1}{2D_j}(\dot{x}_j - f_j). \tag{16}$$

The above result is general, holding in stationary or transient non-equilibrium regimes. In some cases, i.e. when the quantity $\mathcal{M}_j$ can be measured, Eq. (14) may represent an advantage with respect to Eq. (13) (which depends upon the knowledge of the steady-state probability).

The explicit dependence on the time-derivative of the state variables, $\dot{\mathbf{x}}$, in Eq. (14) may still represent a source of complications. Restricting to the calculation of the response matrix, $R_{x_i x_j}(t)$, i.e. such that $\mathcal{O} = x_i$, from Eq. (14) one can derive [42] a simpler expression for processes with additive Gaussian noises in the stationary state (the result can be easily generalized to the case of non-diagonal diffusion, not reported for conciseness):

$$R_{x_i x_j}(t) = -\frac{1}{2D_j}\left[\langle x_i(t) f_j(0)\rangle + \langle f_i(t) x_j(0)\rangle\right] . \tag{17}$$

Each element of the response matrix is given by the sum of two correlations: (i) the temporal correlation between the observed variable and the force ruling the dynamics of the perturbed variable and (ii) the temporal correlation between the force of the observed variable and the perturbed variable (that for the diagonal elements, $R_{x_i x_i}(t)$, is the same correlation of (i) with swapped times). The two terms are equal only at equilibrium. On the contrary they differ when detailed balance does not hold. Note that the generalized FDR (17), differently from the forms (13) or (14), is not determined by the time-correlation between the observed variable evaluated at t and another observable at $s < t$. Moreover, path-integral FDRs require the explicit knowledge of the microscopic dynamics, at variance with the approach (13) which only requires a model of the stationary probability in phase space: it must be noticed that in experimental situations it can be simpler to formulate a model for the steady state probability rather than for the full dynamics. In both cases, however, one needs to individuate the relevant variables, an often underestimated aspect [86].

The generalized FDR (17) is particularly fascinating because the diagonal elements of the response matrix (r.h.s of Eq. (17)) are expressed in terms of the time-symmetric part of the anticipated/retarded equipartition relations while the non-diagonal elements represent the time-symmetric part of the anticipated/retarded Virial equation [42]. Indeed, because of the causality condition, we have $R_{x_i x_j}(t = 0) = \delta_{ij}$, so that the initial time elements of the response matrix contain the same information as the generalized equipartition and Virial equations holding out of equilibrium, namely:

$$D_i = \langle x_i f_i\rangle , \quad \langle x_i f_j\rangle = -\langle x_j f_i\rangle . \tag{18}$$

This physical interpretation has been discussed in detail in [42] and exploited in well-known examples, such as passive and active colloids both in underdamped and overdamped regimes, see also Sect. 3.4.

Let us also comment on the interesting case of discrete variables, relevant for instance for the Ising model or spin glasses, which requires some care. In particular, for spins $\sigma_i = \pm 1$, with $i = 1, \ldots, N$, evolving according to a Master Equation with unperturbed transition rates from the configuration σ to the configuration σ', $w(\sigma \to \sigma')$, in contact with a reservoir at temperature T, the response of an observable $\mathcal{O}(\sigma)$ at a magnetic field $\mathcal{F}$ switched on at time s on site j takes the following form [109]

$$R_{\mathcal{O}\mathcal{F}}(t, s) = \frac{1}{2T}\left\{\frac{\partial}{\partial s}\langle \mathcal{O}(t)\sigma_j(s)\rangle - \langle \mathcal{O}(t) B_j(s)\rangle\right\} , \tag{19}$$

where the quantity $B_j[\sigma]$ is defined by

$$B_j[\sigma(s)] = \sum_{\sigma'}[\sigma'_j - \sigma_j(s)]w[\sigma(s) \to \sigma']. \tag{20}$$

The equilibrium FDT (9) is obtained exploiting the property

$$\left\langle \mathcal{O}(t) \sum_{\sigma'} [\sigma'_j - \sigma_j(s)] w[\sigma(s) \to \sigma'] \right\rangle_{eq} = -\frac{\partial}{\partial s} \langle \mathcal{O}(t) \sigma_j(s) \rangle_{eq}, \tag{21}$$

valid when the average is taken at equilibrium [107, 109].

3 Applications

In this section, we discuss recent applications of the generalized formulae discussed above to different problems. We start with two more theoretical cases, namely the broad class of spin and disordered systems and the search for causality measurements, and we conclude with applications to paradigmatic macroscopic physical systems, that are granular and active systems, where the dynamics of each particle is intrinsically out of equilibrium.

3.1 The Interesting Case of Causation Through Response

Among the many practical applications of the generalized FDR (13), its use in the field of causal inference has a particular conceptual interest. It is well known that, in order to understand the cause-effect relations holding between different elements of a system, measuring the degree of correlation of the variables may be, in general, of little help: two elements can be highly correlated even in the absence of a causal link, as summarized by the notorious adage "correlation does not imply causation". The right way to characterize causal relations is indeed to *probe* the system under study, i.e. to perturb it in some way and to observe the effects of this external action, comparing them to the usual behavior of the system in the absence of perturbation [5, 16]; this is, for instance, the fundamental idea at the basis of Pearls' formalism of counterfactual inference [134]. When dealing with physical systems, as discussed in the Introduction, the effect of an external perturbation is quantified by response functions, which are therefore natural indicators of causal relations [5, 15]. In this respect, a surprising consequence of Eq. (13) is that these observables can be estimated by measuring proper correlation functions in an *unperturbed* dynamics: in other words, the generalized FDR allows to infer causal relations without operating any external action on the system, i.e. without actually probing it.

To show the above point, let us consider the example of a linear stochastic dynamics for the three-dimensional vector (x_t, y_t, z_t) in discrete time, ruled by the following Markov process:

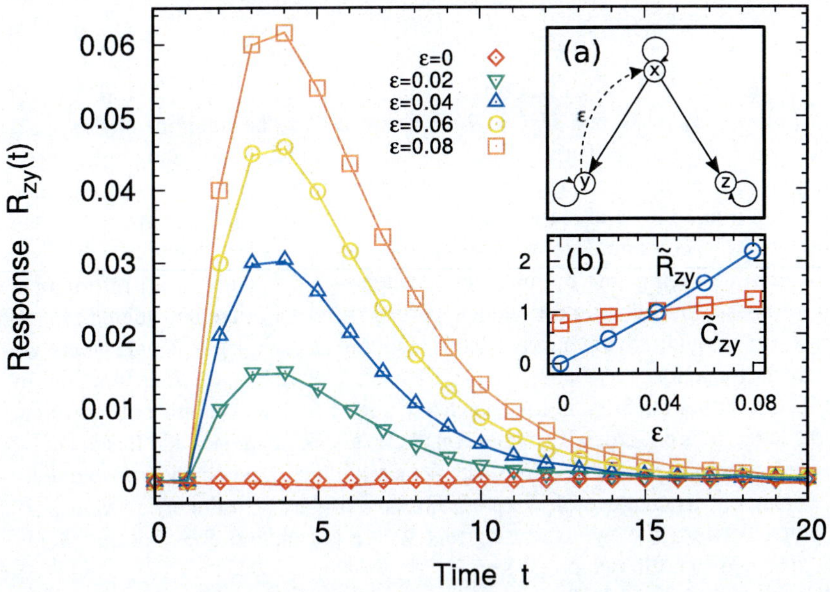

Fig. 1 Relation between causation and response. Main plot: response matrix element $R_{zy}(t)$ of model (22), as a function of time, for several values of the feedback parameter ε; numerical simulations in which the system is actually perturbed (points) are compared to the predictions of the generalized FDR (13) (lines). Inset (a): scheme of the interactions occurring in model (22). Inset (b): correlations (red squares) and responses (blue circles) integrated over time, as functions of ε; both quantities are rescaled with their values at $\varepsilon = 0.04$ for graphical convenience. Parameters: $a = 0.5$, $b = 1$. Perturbation for the computation of response: $\delta y_0 = 0.01$. $M = 10^6$ trajectories have been considered for the averages

$$x_{t+1} = ax_t + \varepsilon y_t + b\eta_t^{(x)} \tag{22a}$$

$$y_{t+1} = ax_t + ay_t + b\eta_t^{(y)} \tag{22b}$$

$$z_{t+1} = ax_t + az_t + b\eta_t^{(z)} \tag{22c}$$

where a, b, and ε are suitable constants and $\eta_t^{(x)}$, $\eta_t^{(y)}$, $\eta_t^{(z)}$ are independent, delta-correlated Gaussian variables with zero mean and unitary variance. In this model the dynamics of y_t and z_t is influenced by x_t, which feels in turn the effect of y_t because of the feedback term εy_t in the r.h.s. of Eq. (22a). A sketch of the interaction scheme is shown in the inset (a) of Fig. 1.

The main plot in Fig. 1 shows the time dependence of the response function between y_t and z_t. As it is clear from the structure of the dynamics, after one time step there is no causal influence (an external perturbation of y_t does not reflect on z_{t+1}). At subsequent times the dynamics of z_t is altered by the perturbation, and the value of the response function crucially depends on the feedback parameter ε, as expected. Due to the linearity of the model, Eq. (13) can be simplified into [120]:

$$R_t = C_t C_0^{-1} \tag{23}$$

where C_t represents the correlation matrix at time t, i.e. $C_t^{ij} = \langle x_i(t) x_j(0) \rangle$ (with $x_1 = x, x_2 = y, x_3 = z$), and C_0^{-1} is the inverse of C_0. The linearity of Eqs. (22) implies that P_{st} is a multi-variate Gaussian and this immediately leads to Eq. (23). Exploiting this version of the generalized FDR, as shown in Fig. 1, one can estimate $R_{zy}(t)$ from a suitable combination of correlation functions: the agreement with the actual responses, computed from numerical simulations, is excellent.

It is worth noticing that the mere knowledge of $C(t)$ is not at all informative about the causal links among the elements of the system. For the considered model, this fact can be qualitatively appreciated by looking at inset (b) of Fig. 1, where we compare the behavior of $\tilde{R}_{zy} = \int_0^\infty R_{zy}(t)\, dt$ and $\tilde{C}_{zy} = \int_0^\infty C_{zy}(t)\, dt$ as functions of ε: while the former quantity, in the considered $\varepsilon \ll 1$ regime, is almost proportional to ε, the latter does not crucially depend on the feedback parameter and it is different from zero also for $\varepsilon = 0$. This difference is clearly due, in the considered example, to the common dependence of y_t and z_t on the variable x_t, inducing a "spurious" correlation between the two (meaning that such a correlation does not unveil any causal link between the two processes).

Using the generalized FDR is not the only way to get some insight into the causal structure of a physical system without perturbing its dynamics. A widely employed method is due to Granger [91] and relies on the computation of the forecasting uncertainty for a given variable of a system, using linear regression models; if it is possible to improve the prediction's precision by including in the model a second, different variable of the system, one may assume a cause-effect relation between the two. A different approach (which has been shown to be equivalent to Granger's method in the case of linear dynamics [16, 22]) is based on the analysis of information transfer between the variables, a process quantified by the so-called *transfer entropy* and by other related observables [32, 148, 155, 166]. Despite the useful information provided by these approaches, response functions appear to be more accurate in characterizing causal relations, at least from a physical point of view; indeed they quantify the (average) consequence of an actual intervention on the system, at variance with Granger's method and transfer entropy analysis, which face the problem from the point of view of predictability and uncertainty [15, 17, 21, 116]. In this respect, generalized FDRs as Eq. (13) are, to the best of our knowledge, the only way to deduce the causal structure of a system, in a proper physical sense, by only observing its spontaneous evolution.

3.2 *Spin and Disordered Systems*

Here we focus on some applications of the FDR in the contexts of spin models and disordered systems. As already underlined, the main aim of an FDR is to give a tool to calculate a response without applying the perturbation. The direct calculation of a linear response function, for instance in numerical simulations (but the same can be true for experiments), is a very time-demanding task: indeed, the signal fluctuations generally increase when the applied field is small, a condition required for the linear regime to hold. Therefore, the application of FDRs in numerical computations is an effective shortcut to get information on the response function from the measure of the correlations in the unperturbed state. This shortcut has been frequently used to develop field-free algorithms in the context of spin systems [50, 53, 56, 109, 145], and glasses [27] or active matter [167]. Let us note that, at variance with previous attempts, specifically designed for a numerical implementations [27, 50, 145], the FDR reported in Eq. (19) involves the quantity B defined in (20), which is an observable quantity because it only depends on the state of the system at a given time and therefore can be in principle measured in real experiments.

3.2.1 Non-linear FDRs

The FDRs in the form (19) can be also derived at non-linear orders in the perturbation, involving multi-point correlation functions. Non-linear response functions play a central role in the context of glassy systems [28, 30, 65, 163], where usually two-point correlators remain always short-ranged due to the presence of disorder. In particular, in a spinglass, the linear susceptibility does not diverge at the critical temperature, whereas non-linear susceptibilities show a divergence when the low temperature phase is approached, signaling a growing amorphous order in the system. Therefore, the relation between non-linear responses and multi-point correlation functions can be an important tool in the context, as initially proposed in [33]. A general derivation of non-linear FDRs valid for arbitrary order was presented in [107, 108]. We report here the form of the second-order response for spin variables perturbed by two fields $\mathcal{F}_1$ and $\mathcal{F}_2$ at sites j_1 and j_2 at times t_1 and t_2 [107]

$$
\begin{aligned}
R^{(2)}_{\mathcal{O}\mathcal{F}}(t, t_1, t_2) &\equiv \left.\frac{\delta\langle \mathcal{O}(t)\rangle_{\mathcal{F}}}{\delta\mathcal{F}_1(t_1)\delta\mathcal{F}_2(t_2)}\right|_{h=0} \\
&= \frac{1}{4T^2}\left\{\frac{\partial}{\partial t_1}\frac{\partial}{\partial t_2}\langle \mathcal{O}(t)\sigma_{j_1}(t_1)\sigma_{j_2}(t_2)\rangle - \frac{\partial}{\partial t_1}\langle \mathcal{O}(t)\sigma_{j_1}(t_1)B_{j_2}(t_2)\rangle \right. \\
&\left. - \frac{\partial}{\partial t_2}\langle \mathcal{O}(t)B_{j_1}(t_1)\sigma_{j_2}(t_2)\rangle + \langle \mathcal{O}(t)B_{j_1}(t_1)B_{j_2}(t_2)\rangle\right\}. \qquad (24)
\end{aligned}
$$

Let us note that at equilibrium, exploiting the property (21), Eq. (24) simplifies to

$$
R^{(2)}_{\mathcal{O}\mathcal{F}}(t, t_1, t_2) = \frac{1}{2T^2}\left\{\frac{\partial}{\partial t_1}\frac{\partial}{\partial t_2}\langle \mathcal{O}(t)\sigma_{j_1}(t_1)\sigma_{j_2}(t_2)\rangle - \frac{\partial}{\partial t_2}\langle \mathcal{O}(t)B_{j_1}(t_1)\sigma_{j_2}(t_2)\rangle\right\}, \qquad (25)
$$

with $t > t_1 > t_2$. Therefore, the presence of the model-dependent quantity B is not canceled, making the higher order FDRs somehow less general than the linear one. As suggested in [23] and [94], this observation can provide information on the dynamical rules governing the system from the study of the equilibrium non-linear responses.

Other interesting applications of non-linear FDRs are related to the study of the thermal response of the system (namely, a perturbation applied to the noise intensity) as discussed in [75], or in the wide field of non-linear optics and quantum spectroscopy [67, 99].

3.2.2 Effective Temperature

One of the main theoretical applications of the FDRs is the possibility to introduce an effective temperature, from the ratio between response and correlation. Review articles on this interesting subjects are [54, 57, 104, 144, 179]. Here we illustrate such a concept for a spin system, where the linear susceptibility, using the FDR (19), can be written as

$$\chi(t, t_w) \equiv \int_{t_w}^{t} ds R_{\sigma\mathcal{F}}(t, s) = \frac{\beta}{2}\int_{t_w}^{t} ds \left[\frac{\partial}{\partial s}C(t, s) - \langle \sigma_i(t) B_i(s)\rangle\right], \tag{26}$$

where $C(t, s) = \langle \sigma_i(t)\sigma_i(s)\rangle$ and t_w is a reference waiting time. Observing that the quantity

$$\psi(t, t_w) = \int_{t_w}^{t} ds \frac{\partial}{\partial s} C(t, s) = 1 - C(t, t_w), \tag{27}$$

for fixed t_w, is a monotonously increasing function of time, one can reparametrize t in terms of ψ and write $\chi(\psi, t_w)$.

In equilibrium, there is no dependence on the waiting time t_w and one obtains a linear parametric representation

$$\chi(\psi) = \beta\psi, \tag{28}$$

yielding

$$\beta = \frac{d\chi(\psi)}{d\psi}. \tag{29}$$

Out of equilibrium, a non-linear dependence can arise and an effective temperature can be introduced generalizing Eq. (29)

$$\beta_{eff}(\psi, t_w) = \frac{\partial\chi(\psi, t_w)}{\partial\psi}, \tag{30}$$

with $\beta_{eff} = 1/T_{eff}$. Then one can define a Fluctuation-Dissipation ratio with respect to the temperature T of the dynamics (after the quench)

$$X(\psi, t_w) = \frac{T}{T_{eff}(\psi, t_w)}, \tag{31}$$

which represents a measure of the deviation from equilibrium. In the limit of large waiting time, the functional dependence of X on the correlation function can show different behaviors, shedding light on the relevance of different characteristic time scales in the system. A detailed discussion of this quantity in the context of aging and glassy systems can be found in [56]. More recent applications of the FDR to equilibrium and non-equilibrium properties of spin glasses have been reported in [10].

The concept of effective temperature has been also applied to systems in the stationary state, such as driven granular media or active particles (see for instance [110, 158]). In this case, the problem is to understand the meaning and the role played by the effective temperature. In some situations, usually when the system is gently driven and the entropy production flux is small, the relevant features of the system behavior can be successfully interpreted in terms of this parameter, leading to an equilibrium-like description. In other cases, the effective temperature can represent an evocative or appealing concept but does not significantly help in the understanding of the underlying physical mechanisms, see next section.

3.3 Granular Materials

Granular materials appear in our everyday life and in several industrial applications, posing deep questions to statistical physics and technology [2, 83, 95, 138]. A granular medium is an ensemble of macroscopic "grains", which interact (among each other and with the surroundings) through non-conservative forces. Several orders of magnitude separate the average energy of internal thermal fluctuations at room temperature—$k_B T \sim 5 \cdot 10^{-21} J$—and the macroscopic energy of a grain (e.g. that related to the position and motion of center of mass): for instance $mgr \sim 10^{-5} J$ for a steel sphere with $r = 2mm$, g being the gravity acceleration. Granular media can display "phase" behaviors: when diluted and under strong shaking a granular "gas" is realized, but when allowed volume and/or the intensity of shaking are reduced, the granular system behaves as a dense "liquid" or a slowly deforming "solid" [96]. The slow-dense phase, close to the so-called *jamming* transition, is difficult to be analyzed: we refer the reader to different theoretical approaches [18, 24, 52, 71, 72, 113, 123, 126, 129, 146]. We briefly summarize the more clear situation established for granular gases and liquids.

A granular gas is realized when the packing fraction is small, typically of the order of 1% or less, such that one can assume instantaneous inelastic binary collisions with restitution coefficient $\alpha \le 1$ (the value 1 is for elastic collisions). In experiments, usually done under gravity, it is necessary to shake the container with accelerations much larger than gravity in order to keep the packing fraction small everywhere [38, 41, 83, 138]. The three main categories of gas regimes are: (1) cooling granular gases, non-steady states which are initially prepared as at equilibrium, and leaving the total energy dissipate under repeated inelastic collisions [37, 92, 171]; (2) boundary-driven gases, where at least one wall injects energy into the gas (e.g. vibration in experiments, thermostats in theory), reaching a non-homogeneous steady state [74, 97, 136]; (3) bulk driven granular gases, where each particle is in repeated contact with some source of energy, for instance bouncing above a vibrating rough plate [90, 128, 141, 142, 172], reaching a homogeneous steady state.

In granular gases, it is customary to define a kinetic "granular temperature" [11, 87, 102, 127]

$$k_B T_g = \frac{m \langle |\mathbf{v}|^2 \rangle}{d}, \tag{32}$$

with $\mathbf{v}$ the velocity of each particle, d the dimensionality of space and k_B is usually replaced with 1. Such a temperature is not expected to have a wide thermodynamic meaning, and also in statistical mechanics it has not a role equivalent to that played for molecular gases, for instance deviations from a Maxwellian are inevitable in the presence of inelastic collisions, a kurtosis excess (or second Sonine coefficient) is observed—larger or smaller—in many regimes [168, 171]. In all gas and liquid regimes, moreover, there is no equipartition of energy among different degrees of freedom (e.g. in a mixture or under non-isotropic external forces), unless they have identical properties [20, 78, 85, 118, 119, 122, 133, 176].

Linear response relations have been frequently studied for granular gases and liquids, particularly in steady states [19, 35, 68, 82, 86, 139, 140, 160, 173, 174], while a few studies also considered cooling regimes [35, 68, 69]. In dilute homogeneously driven granular gases, the equilibrium FDR is empirically observed, provided that the canonical temperature is replaced with the tracer granular temperature T_0 which—in general—can be different from T_g [19, 82, 139, 173, 174]. For instance, a granular tracer under the action of a weak perturbing force in a dilute driven granular system satisfies the dynamical Einstein relation, Eq. (4) with $T = T_0$. Such a result is surprising as, on the basis of the FDR discussed above, Eq. (13) and of the non-Gaussian distribution of velocities, one would expect a correction to it. Nevertheless, in many different dilute cases, such corrections are not observed or—in certain solvable models—can even be proven to vanish [173]. A possible explanation to such a general result comes through the *molecular chaos* which is likely to be valid in dilute cases and which implies that a particle 1 meets particle 2 only once: any collision rule, if restricted to a single particle (that is, disregarding the fate of particle 2) is equivalent to an *elastic* collision with effective masses [144]. For a massive intruder (mass much larger than the other particles), the validity of the Einstein relation is recovered in the context of the derivation of an effective Langevin equation model [151]

The liquid (non-dilute) case is perhaps more interesting. The first experiment focusing on a Brownian-like description of a large intruder in a granular liquid is discussed in Ref. [64]. Most recent studies, both theoretical [140, 143, 152, 173, 174] and experimental [86], have shown that when the granular is a liquid and not a gas, deviations from the equilibrium Fluctuation-Dissipation relation are observed. In granular liquids, as a matter of fact, granular temperature is much less useful than

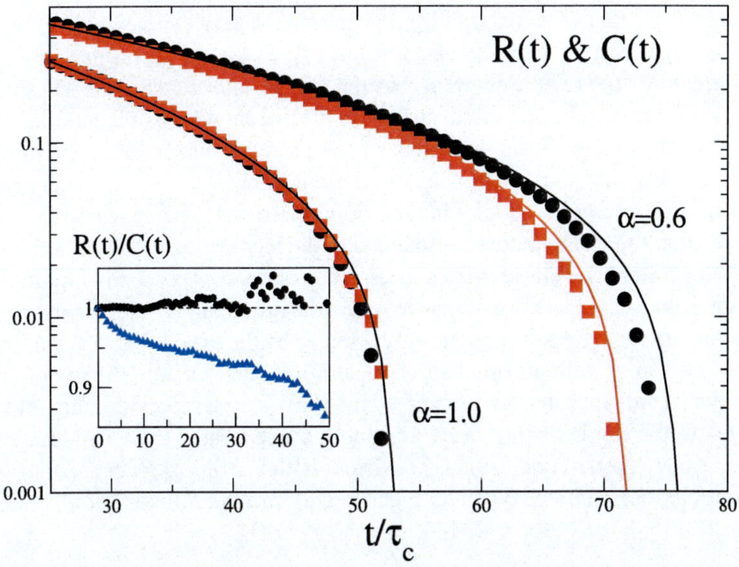

Fig. 2 Response function of the tracer's velocity V under a perturbing force F, $R_{VF}(t)$ and autocorrelation function $C(t) = \langle V(t)V(0)\rangle/\langle V(0)V(0)\rangle$ as a function of time, measured in molecular dynamics simulations of a system composed of a massive intruder interacting with a driven granular fluid [152]. In the main plot an elastic case with restitution coefficient $\alpha = 1$ (where the two functions superimpose as in equilibrium FDR) and an inelastic case $\alpha < 1$ (where equilibrium FDR is violated) are shown. In the inset, the ratio between the two curves is shown for the two cases (black is elastic, blue is inelastic)

in gases, and cannot be replaced by some other temperature for the purpose of an effective description (Fig. 2).

An interesting example, in theory and in experiments is provided, again, by a massive intruder $M \gg m$ [143, 152]. For the purpose of describing, in numerical simulations, the autocorrelation of the velocity V of the tracer and its linear response, the following model provides a fair description for packing fractions smaller than 40%:

$$M\dot{V}(t) = -\Gamma[V(t) - U(t)] + \sqrt{2\Gamma T_{tr}}\mathcal{E}_v(t) \tag{33a}$$

$$M'\dot{U}(t) = -\Gamma' U(t) - \Gamma V(t) + \sqrt{2\Gamma' T_b}\mathcal{E}_U(t), \tag{33b}$$

where $U(t)$ is an auxiliary variable representing the memory effect due to the average velocity field of the particles surrounding the tracer, Γ and T_{tr} are the effective drag coefficient and tracer temperature (both can be derived by kinetic theory in the dilute limit), Γ' and M' are parameters to be determined, for instance from the measured autocorrelation function, and T_b is the value of T_g in the elastic limit (for instance the external bath temperature [142]). Equations (33) can be mapped into a generalized Langevin equation, Eq. (11), with exponential memory kernel. In the dilute limit (parameters such that U is negligible), the massive tracer evolves according to a simple Langevin equation. In the elastic limit ($T_{tr} = T_b = T_g$), on the other side, the coupling with U is still important, but the equilibrium Fluctuation-Dissipation relation is recovered. The numerical simulations have shown that the auxiliary field $U(t)$ is a local average of the velocities of the particles surrounding the intruder. When the density increases numerical simulations suggest $T_{tr} \to T_g$, likely due to a reduction of effective inelasticity in recolliding particles. The appearance of T_b is also interesting: the "temperature" associated to the local velocity field U is equal to the bath temperature and this seems a consequence of the conservation of momentum in collisions, implying that the average velocity of a group of particles is not changed by collisions among themselves and is only affected by the external bath and a (small) number of collisions with outside particles. Summarizing, model (33) suggests that in a granular liquid—at some level of approximation—*two* temperatures are relevant, one related to the single particle scale and another one related to many particles, or collective, scale. Such a conclusion is consistent with a series of recent results about spatial velocity correlations, typically measured as structure factors of the velocity field [12, 13, 36, 84, 89, 90, 132, 137, 141, 170, 172].

3.4 Application to Biological Systems and Active Particles

The results of the FDR have been also applied to several biological systems, for instance in an evolution experiment in bacteria [154] or in the prediction of heart rate response [51]. Another recent application has been proposed in the context of brain activity. Indeed, one can wonder whether, at some scale, the evoked activity in the brain by an external stimulus can be somehow predicted from the observation of the spontaneous, rest activity. In order to quantitatively address this issue, one needs an effective model to describe the brain dynamics at the considered scale. In the work [150], the authors considered the stochastic version of the Wilson-Cowan model [26], describing at a coarse-grained level the dynamics of populations of exitatory and inhibitory neurons. In the linearized version, this model consists in two coupled linear Langevin equations for the two populations. The prediction of the FDR for this model was compared to experimental Magnetoencephalography (MEG) data for rest and evoked activity in healthy subjects. Whereas the behavior of the temporal autocorrelation function of the total rest activity (excitatory plus inhibitory neurons) showed a double exponential decay characterized by two typical times, the decay of the response function was described by a single exponential decay, in qualitative agreement with the prediction of the FDR. These results suggest that some information of the brain response to external stimuli can be obtained from the observation of its spontaneous activity.

A different field that is in large part contained in biology and biophysics is that of self-propelled particles, where non-equilibrium stochastic dynamics has been employed as a main modeling tool [88, 117]. These systems, known as "active", are usually out of equilibrium and store energy from the environment, for instance, taking advantage of chemical reactions or mechanical agents (such as bacterial cilia and flagella), to produce directed motion [25]. The intrinsic non-equilibrium nature of the class of models proposed to describe active systems makes them the ideal platforms to test any version of the generalized FDR [39, 153]. Since their steady-state properties are quite rich, involving unexpected spatial correlations in density, velocity and polarization fields, the use of Eq. (13) can be challenging. For this reason, this method has been applied only in the limit of small activity [46] when the steady probability distribution is known perturbatively or using effective equilibrium-like approaches. This allows one to derive a near-equilibrium expression for the susceptibility [79] and approximated predictions for the transport coefficients of active particles, such as their mobility [63]. In addition, the Malliavin weight sampling has been recently generalized to the more common models used to describe the active particle dynamics [167]. This technique was particularly useful to explore numerically far from equilibrium regimes, calculating (i) the mobility of an interacting active system at low density [63] (ii) the response function due to a shear flow [4] and, finally, (iii) the active effective temperature [58, 105, 124, 135].

In this section, going beyond the approximated approaches explained so far, we apply the technique reported in Sect. 2.2 to obtain exact expressions for the generalized FDR valid in active matter systems [42, 48]. Specifically, we focus on particle systems in the framework of dry active matter without momentum conservation. In this context, the evolution of an active particle of mass m is described by a set of stochastic equations for its position, $\mathbf{x}$, and its velocity, $\mathbf{v}$, given by [47, 115]:

$$\dot{\mathbf{x}} = \mathbf{v} \tag{34a}$$

$$m\dot{\mathbf{v}} = -\gamma \mathbf{v} - \nabla U + \mathbf{f}^a + \sqrt{2T\gamma}\,\boldsymbol{\eta}\,, \tag{34b}$$

while, in the more common overdamped version, such that $m/\gamma \ll 1$, reads:

$$\gamma\dot{\mathbf{x}} = \mathbf{F} + \mathbf{f}^a + \sqrt{2T\gamma}\,\boldsymbol{\eta}\,. \tag{35}$$

In both the dynamics, $\mathbf{f}^a$ is a non-gradient force, called "active force" for simplicity, that models at a coarse-grained level the system-dependent mechanism responsible for the active dynamics so that its complex physical or biological origin is not explicitly considered. This term is chosen as a time-dependent force that provides a certain degree of persistence to the particle trajectory in agreement with the experimental observations of active colloids, bacteria, and other biological microswimmers. The most popular models to account for this persistence in the framework of continuous stochastic processes are the Active Brownian Particles (ABP) [34, 40, 44, 66, 77, 162, 165] and the Active Ornstein-Uhlenbeck particles (AOUP) [29, 43, 61, 112, 121, 177]. In both cases, the active force is expressed as:

$$\mathbf{f}^a = \gamma v_0 \mathbf{n}\,, \tag{36}$$

where v_0 is the swim velocity induced by the active force and $\mathbf{n}$ is a vector representing the particle orientation that evolves stochastically. In the ABP model, $\mathbf{n}$ is a unit vector that evolves as

$$\dot{\mathbf{n}} = \sqrt{2D_r}\mathbf{n} \times \boldsymbol{\xi} \,, \tag{37}$$

while in the AOUP model, $\mathbf{n}$ follows an Ornstein-Uhlenbeck process with unitary variance:

$$\tau\dot{\mathbf{n}} = -\mathbf{n} + \sqrt{2\tau}\boldsymbol{\xi} \,. \tag{38}$$

In both equations, $\boldsymbol{\xi}$ is a vector of δ-correlated white noises with zero average. The coefficient D_r is the rotational diffusion coefficient while τ is simply named persistence time since it coincides with the autocorrelation time of the active force. The models reproduce consistent results by choosing $(d-1)D_r = 1/\tau$ where $d > 1$ is the dimension of the system [77].

In general, the active force pushes the system out of equilibrium, producing entropy with a rate that grows with τ [45, 49, 62, 114, 159]. Applying Eq. (17) to the dynamics (34), the elements of the response matrix after perturbing the x component of the velocity, read [42]:

$$\mathcal{R}_{v,v}(t) = \frac{m}{T}\langle v(t)v(0)\rangle + \frac{m}{2T\gamma}\left(\langle v(t)\nabla_x U(0)\rangle + \langle \nabla_x U(t) v(0)\rangle - \langle v(t)\mathrm{f}^a(0)\rangle - \langle \mathrm{f}^a(t)v(0)\rangle\right) \tag{39a}$$

$$\mathcal{R}_{x,v}(t) = \frac{m}{2T}\langle x(t)v(0)\rangle + \frac{m}{2T\gamma}\langle x(t)\nabla_x U(0)\rangle - \frac{m}{2T\gamma}\langle x(t)\mathrm{f}^a(0)\rangle - \frac{m^2}{2T\gamma}\langle v(t)v(0)\rangle \,, \tag{39b}$$

where we have suppressed the spatial indices for simplicity. Equation (39a) is determined by the generalized retarded kinetic energy and the time-symmetric retarded power injected by the gradient force and the active force. In Eq. (39b), we can identify the retarded mechanical pressure (second term), the so-called retarded swim/active pressure (third term) and, finally, the retarded/anticipated kinetic energy (fourth term). Applying Eq. (17) to the dynamics (35), the response after perturbing the coordinate x of the particle position reads [42]:

$$\mathcal{R}_{x,x}(t) = \frac{1}{2T}\left(\langle x(t)\nabla_x U(0)\rangle + \langle \nabla_x U(t)x(0)\rangle\right) - \frac{1}{2T}\left(\langle x(t)\mathrm{f}^a(0)\rangle + \langle \mathrm{f}^a(t)x(0)\rangle\right) . \tag{40}$$

In the overdamped case, the response is determined by the sum of the time-symmetric part of the retarded/anticipated mechanical and swim pressures. In overdamped systems with $T = 0$, the above formulation of the FDR cannot be directly applied, because the dynamics is not of the Langevin form. In this athermal case, another version of the generalized FDR can be derived using a modified path-integral method developed in [48] in the case of AOUP (FDR for athermal ABP are still unknown), obtaining:

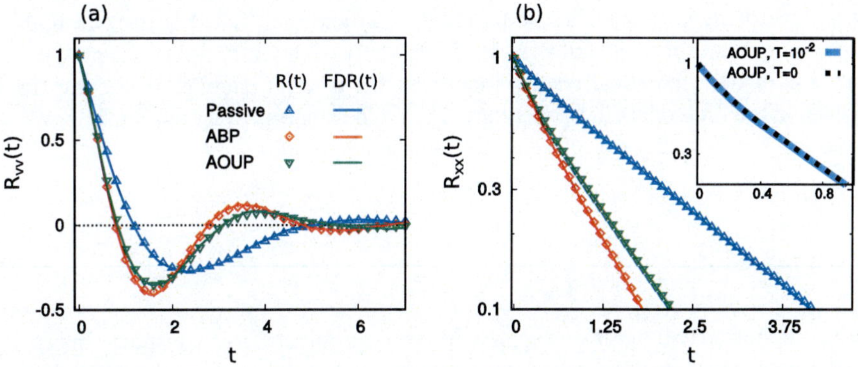

Fig. 3 Comparison between response and FDR for a two-dimensional particle confined in a quartic potential, $U(\mathbf{x}) = k|\mathbf{x}|^4$. Panel (a): $R_{vv}(t)$ (colored points) calculated perturbing the velocity of the underdamped dynamics, Eq. (34). Panel (b): $R_{xx}(t)$ (colored points) calculated perturbing the position of the overdamped dynamics, Eq. (35). The responses are shown for passive, ABP and AOUP as explained in the legend which is shared by both panels. Solid color lines plot the FDR, obtained using Eq. (39a) and (40), for panels (a) and (b), respectively. The inset of panel (b) shows a comparison between Eq. (40) (calculated at $T = 10^{-2}$) and Eq. (41) (holding for $T = 0$). The other parameters of the simulations are $k = 3$, $\gamma = 1$, $T = 10^{-1}$, $v_0 = 1$, and $\tau = 1$

$$
\begin{aligned}
D_a \gamma \mathcal{R}_{x,x}(t) =& \frac{1}{2}\left[\langle x(t)\nabla_x U(0)\rangle + \langle \nabla_x U(t) x(0)\rangle\right] \\
&+ \frac{\tau^2}{2}\sum_{\alpha}\left[\langle v_\alpha(t)\nabla_\alpha\nabla_x U(t) v_x(0)\rangle + \langle v_x(t)\nabla_x\nabla_\alpha U(0) v_\alpha(0)\rangle\right],
\end{aligned}
\tag{41}
$$

where we have introduced the particle velocity $v_\alpha = \dot{\alpha}$, with $\alpha = x, y$. According to our notation, repeated indices are summed, $U(s) = U(\mathbf{x}(s))$. The first line of Eq. (41) coincides with the equilibrium FDR holding for passive particles where the detailed balance holds. The second line contains two additional terms, involving the particle velocity and the second derivative of the potential, that disappear in the equilibrium limit $\tau \to 0$. At variance with the equilibrium scenario, in thermal active systems, the generalized FDR is not only determined by a time correlation involving the position but is affected by the correlations between the other variables, such as the velocity.

To validate the generalized FDR in the case of active particles, we consider both AOUP and ABP dynamics confining the particle through a non-linear force due to an external potential. To go beyond the harmonic case that can be solved analytically [48], we chose a quartic potential, $U(\mathbf{x}) = k|\mathbf{x}|^4/4$, where the constant k determines the strength of U. In Fig. 3, we show the diagonal elements of the response matrix numerically obtained by their definitions (i.e. perturbing the dynamics) and the FDR numerically calculated from the unperturbed system. In particular, in panel (a), we show the results in the underdamped case, reporting the profile of $R_{vv}(t)$ and the FDR calculated from Eq. (39a), while, in panel (b), the analog study is reported for the overdamped dynamics, comparing $R_{x,x}(t)$ and the FDR, Eq. (40). In both cases, the FDRs exactly match with the direct study of the response confirming the exactness of our theoretical results. Finally, in the inset of panel (b), we compare Eq. (40) in the limit of small temperature, T, and the athermal relation, Eq. (41). We reveal that the former converges onto the latter for $T \to 0$.

4 Conclusions

We have reviewed two significant approaches to the problem of linear response in general systems, when the constraint of thermodynamic equilibrium for the unperturbed state is removed. We have also sketched some of the interesting recent applications of such approaches. We cannot avoid to stress, again, the evident fact that— given the system, the observable of interest and the applied perturbation—the linear response function is unique and therefore the two approaches lead to the same result, and in fact an analytical connection can be demonstrated [14]. The difference between the two schemes relies on the required information: in one case, formula (13), one needs some knowledge about the probability distribution at initial time (e.g. the steady-state one) for the relevant degrees of freedom; in the other case, formulas (14), (17) and (19), one needs knowledge about the system's dynamical model (e.g. noise distributions, forces involved, transition rates, etc.). It is not always evident when one approach is more useful than the other. In lucky cases, where both the dynamical model and its probability distribution are known, the two formulas can express different information and one can be more useful than the other (for instance correlations with state variables can be more transparent than correlations with noises or time-derivative of state variables).

In experimental situations, where the underlying model is not known, an empirical approach to retrieve the main features of the probability distribution of the relevant degrees of freedom can be simpler than retrieving information about forces and noises in the system, suggesting the first approach as the more useful. If a dynamical model is known for the relevant degrees of freedom, while the generated probability distribution is unknown, then the second approach should be more direct. However, it is clear that, even when a dynamical model is fully available, the first approach may have some advantage: for instance, in a system with many particles and a massive tracer whose response is investigated, the knowledge of the dynamics of all the particles can be too detailed and result, when inserted in the second approach, in quite a complicate formula, or even not very informative and/or transparent ones; an empirical study of the probability distribution of the relevant degrees of freedom (e.g. those of the tracer and some coarse-grained observable for the surrounding fluid) can provide, sometimes, an approximate but more informative route through the first approach (see for instance the example discussed in Sect. 3.3).

We also recall that an FDR does not give an explicit prediction for the response, but only an expression of it in terms of unperturbed correlations. Once an FDR is known, the problem of obtaining (empirically or analytically) the required correlations remains. An FDR, however, can have already a theoretical meaning, even without the explicit knowledge of the time-dependence of the involved unperturbed correlations, i.e. it is already significant to know *which* correlations are involved, as well illustrated by the application described in Sect. 3.1 for the problem of causation and also in the closure problem in the Kraichnan's approach to turbulence [98].

Acknowledgements The authors acknowledge partial financial support from MIUR through the PRIN 2017 grant number 201798CZLJ. AP also acknowledges financial support from Regione Lazio through the Grant "Progetti Gruppi di Ricerca" N. 85-2017-15257.

References

1. A. Andreanov, G. Biroli, A. Lefevre, Dynamical field theory for glass-forming liquids, self-consistent resummations and time-reversal symmetry. J. Stat. Mech. **2006**, P07008 (2006)
2. B. Andreotti, Y. Forterre, O. Pouliquen, *Granular Media* (Cambridge University Press, 2013)
3. C. Aron, G. Biroli, L.F. Cugliandolo, Symmetries of generating functionals of Langevin processes with colored multiplicative noise. J. Stat. Mech. P11018 (2010)
4. K. Asheichyk, A.P. Solon, C.M. Rohwer, M. Krüger, Response of active brownian particles to shear flow. J. Chem. Phys. **150**, 144111 (2019)
5. E. Aurell, G. Del Ferraro, Causal analysis, correlation-response, and dynamic cavity. J. Phys. Conf. Ser. **699**, 012002 (2016)
6. M. Baiesi, E. Boksenbojm, C. Maes, B. Wynants, Nonequilibrium linear response for markov dynamics, ii: inertial dynamics. J. Stat. Phys. **139**, 492–505 (2010)
7. M. Baiesi, C. Maes, B. Wynants, Fluctuations and response of nonequilibrium states. Phys. Rev. Lett. **103**, 010602 (2009)
8. M. Baiesi, C. Maes, B. Wynants, Nonequilibrium linear response for markov dynamics, i: jump processes and overdamped diffusions. J. Stat. Phys. **137**, 1094 (2009)
9. M. Baiesi, C. Maes, B. Wynants, Nonequilibrium linear response for markov dynamics, II: inertial dynamics. J. Stat. Phys. **139**, 492 (2010)
10. M. Baity-Jesi, E. Calore, A. Cruz, L.A. Fernandez, J.M. Gil-Narvión, A. Gordillo-Guerrero, D. Iñiguez, A. Maiorano, E. Marinari, V. Martin-Mayor, J. Monforte-Garcia, A. Muñoz Sudupe, D. Navarro, G. Parisi, S. Perez-Gaviro, F. Ricci-Tersenghi, J.J. Ruiz-Lorenzo, S.F. Schifano, B. Seoane, A. Tarancón, R. Tripiccione, D. Yllanes, A statics-dynamics equivalence through the fluctuation–dissipation ratio provides a window into the spin-glass phase from nonequilibrium measurements. Proc. Natl. Acad. Sci. U.S.A. **114**, 1838–1843 (2017)
11. A. Baldassarri, A. Barrat, G. D'Anna, V. Loreto, P. Mayor, A. Puglisi, What is the temperature of a granular medium? J. Phys. Condens. Matter **17**, S2405 (2005)
12. A. Baldassarri, U.M.B. Marconi, A. Puglisi, Cooling of a lattice granular fluid as an ordering process. Phys. Rev. E **65**, 051301 (2002)
13. A. Baldassarri, U.M.B. Marconi, A. Puglisi, Influence of correlations on the velocity statistics of scalar granular gases. Europhys. Lett. **58**, 14 (2002)
14. M. Baldovin, L. Caprini, A. Vulpiani, Handy fluctuation-dissipation relation to approach generic noisy systems and chaotic dynamics. Phys. Rev. E **104**, L032101 (2021)
15. M. Baldovin, F. Cecconi, A. Vulpiani, Understanding causation via correlations and linear response theory. Phys. Rev. Res. **2**, 043436 (2020)
16. L. Barnett, A.B. Barrett, A.K. Seth, Granger causality and transfer entropy are equivalent for gaussian variables. Phys. Rev. Lett. **103**, 238701 (2009)
17. L. Barnett, A.B. Barrett, A.K. Seth, Misunderstandings regarding the application of granger causality in neuroscience. **115**, E6676–E6677 (2018)

18. A. Barrat, V. Colizza, V. Loreto, Fluctuation-dissipation ratio for compacting granular media. Phys. Rev. E **66**, 011310 (2002)
19. A. Barrat, V. Loreto, A. Puglisi, Temperature probes in binary granular gases. Physica A **334**, 513 (2004)
20. A. Barrat, E. Trizac, Lack of energy equipartition in homogeneous heated binary granular mixtures. Granular Matter **4**, 57 (2002)
21. A. Barrett, L. Barnett, Granger causality is designed to measure effect, not mechanism. Front. Neuroinform. **7**, 6 (2013)
22. A.B. Barrett, L. Barnett, A.K. Seth, Multivariate granger causality and generalized variance. Phys. Rev. E **81**, 041907 (2010)
23. U. Basu, M. Krüger, A. Lazarescu, C. Maes, Frenetic aspects of second order response. Phys. Chem. Chem. Phys. **17**, 6653 (2015)
24. A. Baule, F. Morone, H.J. Herrmann, H.A. Makse, Edwards statistical mechanics for jammed granular matter. Rev. Mod. Phys. **90**, 015006 (2018)
25. C. Bechinger, R. Di Leonardo, H. Löwen, C. Reichhardt, G. Volpe, G. Volpe, Active particles in complex and crowded environments. Rev. Mod. Phys. **88**, 045006 (2016)
26. M. Benayoun, J.D. Cowan, W. van Drongelen, E. Wallace, Avalanches in a stochastic model of spiking neurons. PLoS Comput. Biol. **6**(7), e1000846 (2010)
27. L. Berthier, Efficient measurement of linear susceptibilities in molecular simulations: Application to aging supercooled liquids. Phys. Rev. Lett. **98**, 220601 (2007)
28. L. Berthier, G. Biroli, J.-P. Bouchaud, L. Cipelletti, D. El Masri, D. L'Hôte, F. Ladieu, M. Pierno, Direct experimental evidence of a growing length scale accompanying the glass transition. Science **310**, 1797 (2005)
29. L. Berthier, E. Flenner, G. Szamel, Glassy dynamics in dense systems of active particles. J. Chem. Phys. **150**, 200901 (2019)
30. G. Biroli, J.-P. Bouchaud, F. Ladieu, Amorphous order & non-linear susceptibilities in glassy materials. J. Phys. Chem. B **125**, 7578–7586 (2021)
31. G. Boffetta, G. Lacorata, S. Musacchio, A. Vulpiani, Relaxation of finite perturbations: Beyond the fluctuation-response relation. Chaos **13**, 806 (2003)
32. T. Bossomaier, L. Barnett, M. Harré, J.T. Lizier, *An Introduction to Transfer Entropy* (Springer, 2016)
33. J.-P. Bouchaud, G. Biroli, Non linear susceptibility in glassy systems: a probe for cooperative dynamical length scales. Phys. Rev. B **72**, 064204 (2005)
34. D. Breoni, M. Schmiedeberg, H. Löwen, Active brownian and inertial particles in disordered environments: short-time expansion of the mean-square displacement. Phys. Rev. E **102**, 062604 (2020)
35. J.J. Brey, M.I.G. de Soria, P. Maynar, Breakdown of the fluctuation-dissipation relations in granular gases. Europhys. Lett. **84**, 24002 (2008)
36. J.J. Brey, P. Maynar, M.I.G. de Soria, Fluctuating hydrodynamics for dilute granular gases. Phys. Rev. E **79**, 051305 (2009)
37. J.J. Brey, M.J. Ruiz-Montero, D. Cubero, Homogeneous cooling state of a low-density granular flow. Phys. Rev. E **54**, 3664 (1996)
38. N.V. Brilliantov, T. Pöschel, *Kinetic Theory of Granular Gases* (Oxford University Press, 2004)
39. E.W. Burkholdera, J.F. Brady, Fluctuation-dissipation in active matter. J. Chem. Phys. **150**, 184901 (2019)
40. I. Buttinoni, J. Bialké, F. Kümmel, H. Löwen, C. Bechinger, T. Speck, Dynamical clustering and phase separation in suspensions of self-propelled colloidal particles. Phys. Rev. Lett. **110**, 238301 (2013)
41. C.S. Campbell, Rapid granular flows. Annu. Rev. Fluid Mech. **22**, 57–92 (1990)
42. L. Caprini, Generalized fluctuation-dissipation relations holding in non-equilibrium dynamics. J. Stat. Mech. 063202 (2021)
43. L. Caprini, U.M.B. Marconi, Active particles under confinement and effective force generation among surfaces. Soft Matter **14**, 9044–9054 (2018)

44. L. Caprini, U.M.B. Marconi, C. Maggi, M. Paoluzzi, A. Puglisi, Hidden velocity ordering in dense suspensions of self-propelled disks. Phys. Rev. Res. **2**, 023321 (2020)
45. L. Caprini, U.M.B. Marconi, A. Puglisi, A. Vulpiani, The entropy production of ornstein-uhlenbeck active particles: a path integral method for correlations. J. Stat. Mech. **2019**, 053203 (2019)
46. L. Caprini, U.M.B. Marconi, A. Vulpiani, Linear response and correlation of a self-propelled particle in the presence of external fields. J. Stat. Mech. **2018**, 033203 (2018)
47. L. Caprini, U.M.B. Marconi, Inertial self-propelled particles. J. Chem. Phys. **154**, 024902 (2021)
48. L. Caprini, A. Puglisi, A. Sarracino, Fluctuation-dissipation relations in active matter systems. Symmetry **13**, 81 (2021)
49. S. Chaki, R. Chakrabarti, Effects of active fluctuations on energetics of a colloidal particle: superdiffusion, dissipation and entropy production. Phys. A Stat. Mech. Appl. **530**, 121574 (2019)
50. C. Chatelain, A far-from-equilibrium fluctuation-dissipation relation for an Ising-Glauber-like model. J. Phys. A **10739**, 36 (2003)
51. M. Chen, L.R. Niestemski, R. Prevost, M. McRae, S. Cholleti, G. Najarro, T.G. Buchman, M.W. Deem, Prediction of heart rate response to conclusion of the spontaneous breathing trial by fluctuation dissipation theory. Phys. Biol. **10**, 016006 (2013)
52. M.P. Ciamarra, A. Coniglio, M. Nicodemi, Thermodynamics and statistical mechanics of dense granular media. Phys. Rev. Lett. **97**, 158001 (2006)
53. F. Corberi, E. Lippiello, A. Sarracino, M. Zannetti, Fluctuation-dissipation relations and field-free algorithms for the computation of response functions. Phys. Rev. E **81**, 011124 (2010)
54. F. Corberi, E. Lippiello, M. Zannetti, Fluctuation dissipation relations far from equilibrium. J. Stat. Mech **2007**, P07002 (2007)
55. A. Crisanti, F. Ritort, Violation of the fluctuation-dissipation theorem in glassy systems: basic notions and the numerical evidence. J. Phys. A Math. Theor. **36**, R181 (2003)
56. A. Crisanti, F. Ritort, Violation of the fluctuation-dissipation theorem in glassy systems: basic notions and the numerical evidence. J. Phys. A **36**, R181 (2003)
57. L.F. Cugliandolo, The effective temperature. J. Phys. A Math. Theor. **44**, 483001 (2011)
58. L.F. Cugliandolo, G. Gonnella, I. Petrelli, Effective temperature in active brownian particles. Fluct. Noise Lett. **18**, 1940008 (2019)
59. L.F. Cugliandolo, J. Kurchan, G. Parisi, Off equilibrium dynamics and aging in unfrustrated systems. J. de Physique I **4**, 1641 (1994)
60. L.F. Cugliandolo, P. Le Doussal, Large time nonequilibrium dynamics of a particle in a random potential. Phys. Rev. E **53**, 1525 (1996)
61. L. Dabelow, S. Bo, R. Eichhorn, Irreversibility in active matter systems: fluctuation theorem and mutual information. Phys. Rev. X **9**, 021009 (2019)
62. L. Dabelow, S. Bo, R. Eichhorn, How irreversible are steady-state trajectories of a trapped active particle? J. Stat. Mech. **2021**, 033216 (2021)
63. S. Dal Cengio, D. Levis, I. Pagonabarraga, Linear response theory and green-kubo relations for active matter. Phys. Rev. Lett. **123**, 238003 (2019)
64. G. D'Anna, P. Mayor, G. Gremaud, A. Barrat, V. Loreto, F. Nori, Observing brownian motion in vibration-fluidized granular matter. Nature **424**, 909 (2003)
65. G. Diezemann, Nonlinear response theory for Markov processes: simple models for glassy relaxation. Phys. Rev. E **85**, 051502 (2012)
66. P. Digregorio, D. Levis, A. Suma, L.F. Cugliandolo, G. Gonnella, I. Pagonabarraga, Full phase diagram of active brownian disks: from melting to motility-induced phase separation. Phys. Rev. Lett. **121**, 098003 (2018)
67. K.E. Dorfman, F. Schlawin, S. Mukamel, Nonlinear optical signals and spectroscopy with quantum light. Rev. Mod. Phys. **88**, 045008 (2016)
68. J.W. Dufty, J.J. Brey, Green-Kubo expressions for a granular gas. J. Stat. Phys. **109**, 433 (2002)

69. J.W. Dufty, V. Garzó, Mobility and diffusion in granular fluids. J. Stat. Phys. **105**, 723–744 (2001)
70. J. Eckmann, D. Ruelle, Rev. Mod. Phys. **57**, 617 (1985)
71. S.F. Edwards, C.C. Mounfield, The statistical mechanics of granular systems composed of spheres and elongated grains. Physica A **210**, 290 (1994)
72. S.F. Edwards, R.B.S. Oakeshott, Theory of powders. Physica A **157**, 1080 (1989)
73. A. Einstein, On the movement of small particles suspended in a stationary liquid demanded by the molecular-kinetic theory of heat. Ann. D. Phys. **17**, 549 (1905)
74. P. Eshuis, K. van der Weele, D. van der Meer, R. Bos, D. Lohse, Phase diagram of vertically shaken granular matter. Phys. Fluids **19**, 123301 (2007)
75. G. Falasco, M. Baiesi, Nonequilibrium temperature response for stochastic overdamped systems. New J. Phys. **18**, 043039 (2016)
76. M. Falcioni, S. Isola, A. Vulpiani, Correlation functions and relaxation properties in chaotic dynamics and statistical mechanics. Phys. Lett. A **144**, 341–346 (1990)
77. T.F. Farage, P. Krinninger, J.M. Brader, Effective interactions in active brownian suspensions. Phys. Rev. E **91**, 042310 (2015)
78. K. Feitosa, N. Menon, Breakdown of energy equipartition in a 2d binary vibrated granular gas. Phys. Rev. Lett. **88**, 198301 (2002)
79. É. Fodor, C. Nardini, M.E. Cates, J. Tailleur, P. Visco, F. van Wijland, How far from equilibrium is active matter? Phys. Rev. Lett. **117**, 038103 (2016)
80. K. Furutsu. *On the Statistical Theory of Electromagnetic Waves in a Fluctuating Medium (II)*, v. 79 (US Government Printing Office, 1964)
81. C. Gardiner, *Handbook of Stochastic Methods for Physics (Chemistry and the Natural Sciences* (Springer, Berlin, 1990)
82. V. Garzó, On the Einstein relation in a heated granular gas. Physica A **343**, 105 (2004)
83. V. Garzó, *Granular Gaseous Flows: A Kinetic Theory Approach to Granular Gaseous Flows* (Springer, 2019)
84. V. Garzó, M.G. Chamorro, F.V. Reyes, Transport properties for driven granular fluids in situations close to homogeneous steady states. Phys. Rev. E **87**, 032201 (2013)
85. V. Garzó, J. Dufty, Homogeneous cooling state for a granular mixture. Phys. Rev. E **60**, 5706 (1999)
86. A. Gnoli, A. Puglisi, A. Sarracino, A. Vulpiani, Nonequilibrium Brownian motion beyond the effective temperature. Plos One **9**, e93720 (2014)
87. I. Goldhirsch, Introduction to granular temperature. Powder Technol. **182**, 130 (2008)
88. G. Gompper, R.G. Winkler, T. Speck, A. Solon, C. Nardini, F. Peruani, H. Löwen, R. Golestanian, U.B. Kaupp, L. Alvarez et al., The 2020 motile active matter roadmap. J. Phys. Condens. Matter **32**, 193001 (2020)
89. G. Gradenigo, A. Sarracino, D. Villamaina, A. Puglisi, Fluctuating hydrodynamics and correlation lengths in a driven granular fluid. J. Stat. Mech. P08017 (2011)
90. G. Gradenigo, A. Sarracino, D. Villamaina, A. Puglisi, Non-equilibrium length in granular fluids: from experiment to fluctuating hydrodynamics. Europhys. Lett. **96**, 14004 (2011)
91. C.W. Granger, Investigating causal relations by econometric models and cross-spectral methods. Econometrica **37**, 424–438 (1969)
92. P.K. Haff, Grain flow as a fluid-mechanical phenomenon. J. Fluid. Mech. **134**, 401 (1983)
93. T. Harada, S.-I. Sasa, Equality connecting energy dissipation with a violation of the fluctuation-response relation. Phys. Rev. Lett. **95**, 130602 (2005)
94. L. Helden, U. Basu, M. Krüger, C. Bechinger, Measurement of second-order response without perturbation. EPL **116**, 60003 (2016)
95. H.M. Jaeger, S.R. Nagel, Physics of the granular state. Science **255**, 1523 (1992)
96. H.M. Jaeger, S.R. Nagel, R.P. Behringer, Granular solids, liquids, and gases. Rev. Mod. Phys. **68**, 1259 (1996)
97. E. Khain, B. Meerson, Onset of thermal convection in a horizontal layer of granular gas. Phys. Rev. E **67**, 021306 (2003)
98. R.H. Kraichnan, Classical fluctuation-relaxation theorem. Phys. Rev. **113**, 118 (1959)

99. M. Kryvohuz, S. Mukamel, Multidimensional measures of response and fluctuations in stochastic dynamical systems. Phys. Rev. A **86**, 043818 (2012)
100. R. Kubo, Statistical-mechanical theory of irreversible processes. I. General theory and simple applications to magnetic and conduction problems. J. Phys. Soc. Jpn. **12**, 570 (1957)
101. R. Kubo, M. Toda, N. Hashitsume, *Statistical Physics II: Nonequilibrium Stastical Mechanics* (Springer, 1991)
102. V. Kumaran, Temperature of a granular material "fluidized" by external vibrations. Phys. Rev. E **57**, 5660–5664 (1998)
103. P. Langevin, Sur la theorie du mouvement brownien. C. R. Acad. Sci. (Paris) **146**, 530 (1908). Translated in Am. J. Phys. **65**, 1079 (1997)
104. L. Leuzzi, A stroll among effective temperatures in aging systems: limits and perspectives. J. Non-Cryst. Solids **355**, 686 (2009)
105. D. Levis, L. Berthier, From single-particle to collective effective temperatures in an active fluid of self-propelled particles. EPL (Europhysics Letters) **111**, 60006 (2015)
106. E. Lippiello, M. Baiesi, A. Sarracino, Nonequilibrium fluctuation-dissipation theorem and heat production. Phys. Rev. Lett. **112**, 140602 (2014)
107. E. Lippiello, F. Corberi, A. Sarracino, M. Zannetti, Nonlinear response and fluctuation-dissipation relations. Phys. Rev. E **78**, 041120 (2008)
108. E. Lippiello, F. Corberi, A. Sarracino, M. Zannetti, Nonlinear susceptibilities and the measurement of a cooperative length. Phys. Rev. B **77**, 212201 (2008)
109. E. Lippiello, F. Corberi, M. Zannetti, Off-equilibrium generalization of the fluctuation dissipation theorem for ising spins and measurement of the linear response function. Phys. Rev. E **71**, 036104 (2005)
110. D. Loi, S. Mossa, L.F. Cugliandolo, Effective temperature of active matter. Phys. Rev. E **77**, 051111 (2008)
111. C. Maes, Response theory: a trajectory-based approach. Front. Phys. **8**, 00229 (2020)
112. C. Maggi, M. Paoluzzi, L. Angelani, R. Di Leonardo, Memory-less response and violation of the fluctuation-dissipation theorem in colloids suspended in an active bath. Sci. Rep. **7**, 1–7 (2017)
113. H. Makse, J. Brujić, S.F. Edwards, Statistical mechanics of jammed matter, in *The Physics of Granular Media* (Wiley-VCH, 2004), pp. 45–85
114. D. Mandal, K. Klymko, M.R. DeWeese, Entropy production and fluctuation theorems for active matter. Phys. Rev. Lett. **119**, 258001 (2017)
115. S. Mandal, B. Liebchen, H. Löwen, Motility-induced temperature difference in coexisting phases. Phys. Rev. Lett. **123**, 228001 (2019)
116. P. Manshour, G. Balasis, G. Consolini, C. Papadimitriou, M. Paluš, Causality and information transfer between the solar wind and the magnetosphere-ionosphere system. Entropy **23**, 390 (2021)
117. M. Marchetti, J. Joanny, S. Ramaswamy, T. Liverpool, J. Prost, M. Rao, R.A. Simha, Hydrodynamics of soft active matter. Rev. Mod. Phys. **85**, 1143–1189 (2013)
118. U.M.B. Marconi, A. Puglisi, Mean-field model of free-cooling inelastic mixtures. Phys. Rev. E **65**, 051305 (2002)
119. U.M.B. Marconi, A. Puglisi, Steady-state properties of a mean-field model of driven inelastic mixtures. Phys. Rev. E **66**, 011301 (2002)
120. U.M.B. Marconi, A. Puglisi, L. Rondoni, A. Vulpiani, Fluctuation-dissipation: response theory in statistical physics. Phys. Rep. **461**, 111 (2008)
121. D. Martin, J. O'Byrne, M.E. Cates, É. Fodor, C. Nardini, J. Tailleur, F. van Wijland, Statistical mechanics of active Ornstein-Uhlenbeck particles. Phys. Rev. E **103**, 032607 (2021)
122. P.A. Martin, J. Piasecki, Thermalization of a particle by dissipative collisions. Europhys. Lett. **46**, 613 (1999)
123. A. Mehta, S.F. Edwards, Statistical mechanics of powder mixtures. Physica A **157**, 1091 (1989)
124. S.K. Nandi, N. Gov, Effective temperature of active fluids and sheared soft glassy materials. Eur. Phys. J. E **41**, 117 (2018)

125. E.A. Novikov, Functionals and the random-force method in turbulence theory. Soviet Physics-JETP **20**, 1290 (1965)
126. E.R. Nowak, J.B. Knight, E. Ben-Naim, H.M. Jaeger, S.R. Nagel, Density fluctuations in vibrated granular materials. Phys. Rev. E **57**, 1971–1982 (1998)
127. S. Ogawa, A. Umemura, N. Oshima, On the equations of fully fluidized granular materials. J. Appl. Math. Phys. **31**, 483 (1980)
128. J.S. Olafsen, J.S. Urbach, Clustering, order, and collapse in a driven granular monolayer. Phys. Rev. Lett. **81**, 4369–4372 (1998)
129. I.K. Ono, C.S. O'Hern, D.J. Durian, S.A. Langer, A.J. Liu, S.R. Nagel, Effective temperatures of a driven system near jamming. Phys. Rev. Lett. **89**, 095703 (2002)
130. L. Onsager, Reciprocal relations in irreversible processes. I. Phys. Rev. **37**, 405–426 (1931)
131. L. Onsager, Reciprocal relations in irreversible processes. II. Phys. Rev. **38**, 2265–2279 (1931)
132. E.T.P. Maynar, M.I.G. de Soria, Fluctuating hydrodynamics for driven granular gases. Eur. Phys. J. Special Topics **179**, 123 (2009)
133. R. Pagnani, U.M.B. Marconi, A. Puglisi, Driven low density granular mixtures. Phys. Rev. E **66**, 051304 (2002)
134. J. Pearl, *Causality* (Cambridge University Press, 2009)
135. I. Petrelli, L.F. Cugliandolo, G. Gonnella, A. Suma, Effective temperatures in inhomogeneous passive and active bidimensional brownian particle systems. Phys. Rev. E **102**, 012609 (2020)
136. G. Pontuale, A. Gnoli, F.V. Reyes, A. Puglisi, Thermal convection in granular gases with dissipative lateral walls. Phys. Rev. Lett. **117**, 098006 (2016)
137. A. Prados, A. Lasanta, P.I. Hurtado, Nonlinear driven diffusive systems with dissipation: fluctuating hydrodynamics. Phys. Rev. E **86**, 031134 (2012)
138. A. Puglisi, *Transport and Fluctuations in Granular Fluids* (Springer, 2015)
139. A. Puglisi, A. Baldassarri, V. Loreto, Fluctuation-dissipation relations in driven granular gases. Phys. Rev. E **66**, 061305 (2002)
140. A. Puglisi, A. Baldassarri, A. Vulpiani, Violations of the Einstein relation in granular fluids: the role of correlations. J. Stat. Mech. P08016 (2007)
141. A. Puglisi, A. Gnoli, G. Gradenigo, A. Sarracino, D. Villamaina, Structure factors in granular experiments with homogeneous fluidization. J. Chem. Phys. **014704**, 136 (2012)
142. A. Puglisi, V. Loreto, U.M.B. Marconi, A. Petri, A. Vulpiani, Clustering and non-Gaussian behavior in granular matter. Phys. Rev. Lett. **81**, 3848 (1998)
143. A. Puglisi, A. Sarracino, G. Gradenigo, D. Villamaina, Dynamics of a massive intruder in a homogeneously driven granular fluid. Granul. Matter **14**, 235 (2012)
144. A. Puglisi, A. Sarracino, A. Vulpiani, Temperature in and out of equilibrium: a review of concepts, tools and attempts. Phys. Rep. **709**, 1–60 (2017)
145. F. Ricci-Tersenghi, Measuring the fluctuation-dissipation ratio in glassy systems with no perturbing field. Phys. Rev. E **065104**(R), 68 (2003)
146. P. Richard, M. Nicodemi, R. Delannay, P. Ribière, D. Bideau, Slow relaxation and compaction of granular systems. Nat. Mater. **4**, 121 (2005)
147. H. Risken, *The Fokker-Planck Equation: Methods of Solution and Applications* (Springer, Berlin, 1989)
148. J. Runge, J. Heitzig, V. Petoukhov, J. Kurths, Escaping the curse of dimensionality in estimating multivariate transfer entropy. Phys. Rev. Lett. **108**, 258701 (2012)
149. A. Sarracino, Time asymmetry of the kramers equation with nonlinear friction: fluctuation-dissipation relation and ratchet effect. Phys. Rev. E **88**, 052124 (2013)
150. A. Sarracino, O. Arviv, O. Shriki, L. de Arcangelis, Predicting brain evoked response to external stimuli from temporal correlations of spontaneous activity. Phys. Rev. Res. **2**, 033355 (2020)
151. A. Sarracino, D. Villamaina, G. Costantini, A. Puglisi, Granular Brownian motion. J. Stat. Mech. P04013 (2010)
152. A. Sarracino, D. Villamaina, G. Gradenigo, A. Puglisi, Irreversible dynamics of a massive intruder in dense granular fluids. Europhys. Lett. **92**, 34001 (2010)

153. A. Sarracino, A. Vulpiani, On the fluctuation-dissipation relation in non-equilibrium and non-hamiltonian systems. Chaos **29**, 083132 (2019)
154. K. Sato, Y. Ito, T. Yomo, K. Kaneko, On the relation between fluctuation and response in biological systems. Proc. Natl. Acad. Sci. U.S.A. **100**, 14086–14090 (2003)
155. T. Schreiber, Measuring information transfer. Phys. Rev. Lett. **85**, 461–464 (2000)
156. U. Seifert, Stochastic thermodynamics, fluctuation theorems and molecular machines. Rep. Prog. Phys. **75**, 126001 (2012)
157. U. Seifert, T. Speck, Fluctuation-dissipation theorem in nonequilibrium steady states. EPL (Europhysics Letters) **89**, 10007 (2010)
158. A. Shakerpoor, E. Flenner, G. Szamel, The einstein effective temperature can predict the tagged active particle density. J. Chem. Phys. **154**, 184901 (2021)
159. S. Shankar, M.C. Marchetti, Hidden entropy production and work fluctuations in an ideal active gas. Phys. Rev. E **98**, 020604 (2018)
160. Y. Shokef, G. Bunin, D. Levine, Fluctuation-dissipation relations in driven dissipative systems. Phys. Rev. E **73**, 046132 (2006)
161. M. Smoluchowski, Zur kinetischen Theorie der Brownschen Molekularbewegung und der Suspensionen. Ann. d. Phys. **21**, 756 (1906)
162. A.P. Solon, Y. Fily, A. Baskaran, M.E. Cates, Y. Kafri, M. Kardar, J. Tailleur, Pressure is not a state function for generic active fluids. Nat. Phys. **11**, 673–678 (2015)
163. T. Speck, Modeling non-linear dielectric susceptibilities of supercooled molecular liquids (2021)
164. T. Speck, U. Seifert, Restoring a fluctuation-dissipation theorem in a nonequilibrium steady state. Europhys. Lett. **74**, 391 (2006)
165. J. Stenhammar, D. Marenduzzo, R.J. Allen, M.E. Cates, Phase behaviour of active brownian particles: the role of dimensionality. Soft Matter **10**, 1489–1499 (2014)
166. J. Sun, D. Taylor, E.M. Bollt, Causal network inference by optimal causation entropy. SIAM J. Appl. Dyn. Syst. **14**, 73–106 (2015)
167. G. Szamel, Evaluating linear response in active systems with no perturbing field. Europhys. Lett. **117**, 50010 (2017)
168. E. Trizac, I. Pagonabarraga, T.P.C. van Noije, M.H. Ernst, Randomly driven granular fluids: collisional statistics and short scale structure. Phys. Rev. E **65**, 011303 (2001)
169. N. van Kampen, A discussion on linear response theory. Phys. Norv. **5**, 279 (1971)
170. T.C.P. van Noije, M.H. Ernst, R. Brito, J.A.G. Orza, Mesoscopic theory of granular fluids. Phys. Rev. Lett. **79**, 411 (1997)
171. T.P.C. van Noije, M.H. Ernst, Velocity distributions in homogeneous granular fluids: the free and the heated case. Granular Matter **1**, 57–64 (1998)
172. T.P.C. van Noije, M.H. Ernst, E. Trizac, I. Pagonabarraga, Randomly driven granular fluids: large-scale structure. Phys. Rev. E **59**, 4326 (1999)
173. D. Villamaina, A. Baldassarri, A. Puglisi, A. Vulpiani, Fluctuation dissipation relation: how does one compare correlation functions and responses? J. Stat. Mech. P07024 (2009)
174. D. Villamaina, A. Puglisi, A. Vulpiani, The fluctuation-dissipation relation in sub-diffusive systems: the case of granular single-file diffusion. J. Stat. Mech. L10001 (2008)
175. P.B. Warren, R.J. Allen, Malliavin weight sampling: a practical guide. Entropy **16**, 221 (2014)
176. R. Wildman, D. Parker, Coexistence of two granular temperatures in binary vibrofluidized beds. Phys. Rev. Lett. **88**, 064301 (2002)
177. R. Wittmann, C. Maggi, A. Sharma, A. Scacchi, J.M. Brader, U.M.B. Marconi, Effective equilibrium states in the colored-noise model for active matter i. Pairwise forces in the fox and unified colored noise approximations. J. Stat. Mech. **2017**, 113207 (2017)
178. C. Yolcu, A. Bérut, G. Falasco, A. Petrosyan, S. Ciliberto, M. Baiesi, A general fluctuation-response relation for noise variations and its application to driven hydrodynamic experiments. J. Stat. Phys. **167**, 29–45 (2017)
179. M. Zannetti, Aging in domain growth, in *Kinetics of Phase Transitions*, ed. by S. Puri, V. Wadhawan, chapter 5 (CRC Press Taylor & Francis Group, 2009), p. 153

Mandelbrot's Fractal Structure in Decaying Process of a Matter-field Interacting System

Tomio Petrosky, Daisuke Kotaka, and Satoshi Tanaka

Abstract A decaying process of an unstable matter-field interacting system is analyzed by the complex spectral analysis of the Hamiltonian, which is a microscopic analysis of irreversibility based on the basic laws of physics. By using the iterative method to solve a nonlinear eigenvalue problem of an effective Hamiltonian in the complex energy plane, we found a fractal structure described by a Mandelbrot set associated to the broken time-symmetry for the unstable dynamical system.

1 Introduction

It is our pleasure to contribute to the special issue of a Springer Nature book in honor of Professor Slava Belyi on Non-equilibrium Thermodynamics and Fluctuation Kinetics. One of the authors (T.P.) especially expresses his thanks to have several occasions to discuss with Slava on the broken time-symmetry that is the main subject in non-equilibrium statistical physics. We had a wonderful moment to discuss this problem whenever Slava visited the International Institute for Physics and Chemistry in Brussels, Belgium. We have also spent a wonderful moment to drink an excellent vodka together. Slava has many interesting contributions to Non-Equilibrium physics.

Due to a remarkable development in the complex spectral analysis of the Liouville-von Neumann operator (Liouvillian, in short), as well as Hamiltonian, we now understand that irreversible processes are exact dynamical processes occurring outside the Hilbert space. This is not coming from some approximation, such as a coarse graining

T. Petrosky (✉)
Center for Complex Quantum Systems, The University of Texas at Austin, Texas, USA
e-mail: petrosky@physucs.utexas.edu

Institute of Industrial Science, The University of Tokyo, Kashiwa, Japan

D. Kotaka
IDC Service Division, Panasonic Information Systems Co., Ltd., Kadoma, Osaka, Japan

S. Tanaka
Department of Physical Science, Osaka Prefecture University, Sakai, Japan

L. Brenig et al. (eds.), *Nonequilibrium Thermodynamics and Fluctuation Kinetics*, Fundamental Theories of Physics 208,
https://doi.org/10.1007/978-3-031-04458-8_4

procedure of fundamental equations of motions due to human's limitation to control informations of complex nature. Indeed, in spite of the fact that Liouvillian and Hamiltonian in the Hilbert space are Hermitian operators, they may have complex eigenvalues in an extended Hilbert space outside the Hilbert space, due to the resonance singularity appearing in the solution of the equation of motion. Moreover, we have shown a set of eigenstates of the Liouvillian or Hamiltonian with the complex eigenvalue may span bi-complete and bi-orthonormal sets in an extended Hilbert space outside the Hilbert space [1–3]. We have shown that the imaginary part of a complex eigenvalue gives a transport coefficient for irreversible processes, such as a diffusion coefficient, as well as viscosity in a kinetic equation [4–11], or a decay rate of spontaneous decay process of an excited atom in quantum system [3, 13–18], and radiation dumping in classical system [19–21].

In this paper, we will consider the Friedrichs model that has been introduced to analyze a quantum decaying process in a matter-field interacting system where a discrete spectrum is coupled with a continuous spectrum through the resonance interaction [3, 12, 22]. The Hamiltonian H of the Friedrichs model consists of an unperturbed part H_0 of the matter and the field without interaction and the interaction part gV,

$$H = H_0 + gV, \tag{1}$$

where gV is a Hermitian operator, and a real number g is a dimensionless coupling constant. The explicit form of the Hamiltonian is given in (2) and (3).

In this model, if the unperturbed discrete state does not overlap with the unperturbed continuous states, one can show that the complete set of the eigenstates of the full Hamiltonian H consists of a discrete stable state associated to the matter as well as continuous states associated to the field. In this situation, there is no resonance between the discrete spectrum and the unperturbed continuous states. Then, the system is stable.

On the other hand, if the resonance occurs, i.e., if the unperturbed discrete state overlaps with the unperturbed continuous states, the system becomes unstable. In this case, the discrete state with a real eigenvalue disappears in the complete set of the eigenstates of the full Hamiltonian. Nevertheless, Friedrichs has shown that one can solve exactly the Schrödinger equation and shown that the decaying process that breaks time-symmetry is an exact solution of the Schrödinger equation which is symmetric in time-inversion. As mentioned above, we have shown for this unstable case that a set of eigenstates of the full Hamiltonian H with the complex eigenvalue spans bi-complete and bi-orthonormal set in an extended Hilbert space [3]. Moreover, we have shown the transition from stable case to the unstable case (i.e., the transition from reversible process to the irreversible process) can be understood as a dynamical phase transition [15].

As we will see, to find the complex eigenvalue of the full Hamiltonian reduces to a new nonlinear eigenvalue problem of the effective Hamiltonian which is closely related to the self-energy part in the dispersion equation associated to the unstable state. Since the self-energy part is generally a nonlinear function of the eigenvalue of the full Hamiltonian, it is often difficult to solve the nonlinear eigenvalue problem of the effective Hamiltonian. For this case, an iterative method in terms of a nonlinear map has been often used to obtain an approximate value of the complex eigenvalue. However, a remarkable feature of the Friedrichs model which we consider in this paper is that we can algebraically solve this nonlinear equation. Hence, we will first show the explicit form of the exact complex eigenvalue.

After that, we will analyze the iterative method with a comparison to the exact solution. We shall see that the nonlinear map for the iterative method shows a strong nonlinear behavior near the dynamical phase transition from the time-symmetric regime to the broken time-symmetric regime. More amazingly, we will show that the inverse map of the iterative method used in this model exactly reduces to the nonlinear map that generates Mandelbrot sets [23]. This indicates that a well-analyzed process of the spontaneous decay process of an excited unstable atom still shows an interesting new aspect of the fractal in nonlinear dynamics.

2 Complex Spectral Analysis of Friedrichs' Hamiltonian

We will use Dirac's braket notation for the states. Let us consider a simple electron system with a single discrete state $|a\rangle$ with energy ϵ_a coupled with a three-dimensional free-electron band-state $|\mathbf{k}\rangle$ with a continuous wave vector $\mathbf{k}$ where the Hamiltonian of the system is given by (1) with the unperturbed Hamiltonian,

$$H_0 = \epsilon_a |a\rangle\langle a| + \int d\mathbf{k}\, \epsilon_k\, |\mathbf{k}\rangle\langle\mathbf{k}|, \tag{2}$$

and the interaction,

$$gV = \int d\mathbf{k}\, gV_k\big(|a\rangle\langle\mathbf{k}| + |\mathbf{k}\rangle\langle a|\big), \tag{3}$$

where the real number g is a dimensionless coupling constant. We assume that $\langle a|a\rangle = 1$, $\langle a|\mathbf{k}\rangle = \langle\mathbf{k}|a\rangle = 0$, and $\langle\mathbf{k}|\mathbf{k}'\rangle = \delta(\mathbf{k} - \mathbf{k}')$ with the three-dimensional delta function, and they satisfy the completeness relation

$$|a\rangle\langle a| + \int d\mathbf{k}\, |\mathbf{k}\rangle\langle\mathbf{k}| = 1. \tag{4}$$

Moreover, we assume that the dispersion relation of the free electron is given by

$$\epsilon_k = \frac{\hbar^2 k^2}{2m}, \tag{5}$$

with $k = |\mathbf{k}|$, and the interaction is given by

$$V_k = \frac{1}{2\sqrt{2}k_c^2}\theta(k_c - k), \tag{6}$$

where we have introduced the cutoff wavenumber k_c to obtain a finite value of the self-energy by renormalizing the ultraviolet divergence in the self-energy. Indeed, this choice of interaction gives the cutoff-independent contribution from the upper bound of the integration over the wavenumber k in the self-energy part. The numerical factor front of the interaction is for our convenience in the numerical calculation. Hereafter, we will use the units $\hbar = 2m = 1$.

In the case ϵ_a is a real number, the Hamiltonian H is a Hermitian operator in the Hilbert space. For this case, one can show that if ϵ_a is negative, or greater than ϵ_{k_c}, the eigenvalues of H are real numbers. Then, the system is stable. However, when the condition $\epsilon_a < \epsilon_{k_c}$ with $\epsilon_a > 0$ is satisfied, the system becomes unstable due to the resonance singularity expressed by $1/(\epsilon_k - \epsilon_a)$ in the usual perturbation analysis. For this unstable case, the discrete eigenstate disappears from the spectrum of H. In this case, the solution of the Schrödinger equation breaks time-symmetry. Nevertheless, the Hamiltonian may have complex eigenvalues in an extended Hilbert space that is spanned by a dual base of a set of eigenstates (see (7) and (8)) [3].

In the complex spectral analysis, we construct the solution to the eigenvalue problem of the generator of motion (such as Hamiltonian, or Liouvillian, which is a Hermitian operator in the Hilbert space) outside the Hilbert space. Since the function space is extended outside the Hilbert space, the symmetric operator may have complex eigenvalues. For this solution, the left-eigenstate belonging to the same complex eigenvalue is not a Hermitian conjugate to the left-eigenstate. Hence, we have to solve a pair of these dual eigenvalue problems,

$$H|\Psi_\alpha\rangle = Z_\alpha|\Psi_\alpha\rangle, \quad \langle\tilde{\Psi}_\alpha|H = Z_\alpha\langle\tilde{\Psi}_\alpha|, \tag{7}$$

where the argument $\alpha = a$, or $\mathbf{k}$. In general, Z_α is a complex number. (For the Friedrichs model that we consider, one can show that Z_a is a complex number, but the continuous spectrum $Z_\mathbf{k} = \epsilon_k$ is a real number). Hence, from (7), we see that $\langle\tilde{\Psi}_\alpha| \neq \langle\Psi_\alpha|$ because of the fact that H is a Hermitian operator $H^\dagger = H$.

These eigenstates span a bi-orthogonal and bi-complete set of an extended Hilbert space, i.e., $\langle\tilde{\Psi}_a|\Psi_a\rangle = 1$, $\langle\tilde{\Psi}_a|\Psi_\mathbf{k}\rangle = \langle\tilde{\Psi}_\mathbf{k}|\Psi_a\rangle = 0$, and $\langle\tilde{\Psi}_\mathbf{k}|\Psi_{\mathbf{k}'}\rangle = \delta(\mathbf{k}-\mathbf{k}')$, and

$$|\Psi_a\rangle\langle\tilde{\Psi}_a| + \int_C d\mathbf{k}\, |\Psi_\mathbf{k}\rangle\langle\tilde{\Psi}_\mathbf{k}| = 1, \tag{8}$$

where the integration over the wave vector is chosen with a suitable contour C when it is performed by a contour deformation in the complex plane. (In this paper, we will not show the explicit form of eigenstate with continuous spectrum, since our discussion on the Mandelbrot set is only associated to the complex discrete state. One can find the explicit form and the contour C in [3]).

In order to solve the eigenvalue problem (7), we extend the Brillouin-Wigner type of the projection-operator method to the extended Hilbert space and introduce the following projection operator P,

$$P \equiv |a\rangle\langle a|, \quad Q = 1 - P. \tag{9}$$

They satisfy the relations,

$$P^2 = P, \ Q^2 = Q, \ PQ = QP = 0. \tag{10}$$

Then, the right-eigenvalue equation in (7) leads to a set of equations for P and Q components of the eigenstates,

$$(PHP + PHQ)|\Psi_a\rangle = Z_a P|\Psi_a\rangle, \tag{11}$$

$$(QHP + QHQ)|\Psi_a\rangle = Z_a Q|\Psi_a\rangle. \tag{12}$$

This set of equation leads to a new eigenvalue equation for the $P|\Psi_a\rangle$ component,

$$H_{\text{eff}}(Z_a)P|\Psi_a\rangle = Z_a P|\Psi_a\rangle. \tag{13}$$

Here, $H_{\text{eff}}(z)$ is the effective Hamiltonian defined by

$$H_{\text{eff}}(z) \equiv PHP + PHQ\frac{1}{z - QHQ}QHP, \tag{14}$$

with a suitable analytic continuation of the denominator $1/(z - QHQ)$ which is consistent to the decaying process oriented to our future.

Note that this new eigenvalue equation for the effective Hamiltonian is a nonlinear equation with respect to the eigenvalue Z_a. We also note that the second part leads to the well-known self-energy part as

$$\Sigma_a(z) = \langle a|PHQ\frac{1}{z - QHQ}QHP|a\rangle. \tag{15}$$

After solving the nonlinear eigenvalue problem (13), one can obtain a right-eigenstate of the full Hamiltonian H with the use of the set of equation (12) as

$$|\Psi_a\rangle = N_a\Big[P + Q\frac{1}{z - QHQ}QHP\Big]P|\Psi_a\rangle, \tag{16}$$

where N_a is a normalization constant determined by $\langle\tilde{\Psi}_a|\Psi_a\rangle = 1$ with a similar expression of the left-eigenstate $\langle\tilde{\Psi}_a|$ to (16). The explicit forms of the complex eigenvalues and these dual eigenstates for a general Friedlichs model are presented in [3].

Before going to the main subject on the Mandelbrot set in this paper, let us mention a remarkable feature of expression on the Liouvillian parallel to the above expression for the eigenvalue problem of the Hamiltonian. As is well known, the Liouvillian L_H associated to a given Hamiltonian H in the Liouville-von Neumann equation for the distribution function for a classical system, or the density matrix for a quantum system is also a linear operator, just as a Hamiltonian in a suitable vector space. Indeed, the Liouvillian is defined as the Poisson bracket with the Hamiltonian for the classical case, and the commutation relation with the Hamiltonian for the quantum case. Hence, one can construct a similar type of complex eigenstate of the Liouvillian parallel to the Hamiltonian just constructed as above. Then, the complex eigenvalue problem of the right-eigenstate of the Liouvillian for example given by

$$L_H|F_\beta\rangle = Z_\beta|F_\alpha\rangle \tag{17}$$

is reduced to the nonlinear eigenvalue problem of the effective Liouvillian as

$$\psi(Z_\beta)\hat{P}|F_\beta\rangle = Z_\beta\hat{P}|F_\beta\rangle. \tag{18}$$

Here, $\psi(z)$ is defined by

$$\psi(z) \equiv \hat{P}L_H\hat{P} + \hat{P}L_H\hat{Q}\frac{1}{z - \hat{Q}L_H\hat{Q}}\hat{Q}L_H\hat{P}, \tag{19}$$

with a suitable projection operator $\hat{P}$ and $\hat{Q}$ associated to a set of the eigenstate for the unperturbed Liouvillian. Then, the second team of (19) is the self-frequency part of the Liouvillian. If the reader is familiar to the non-equilibrium statistical mechanics, the reader may notice that this self-frequency part of the Liouvillian is just the collision operator which is the central object in non-equilibrium statistical physics. Indeed for a weakly coupled gas or the low density limit of a gas, the effective Liouvillian reduces to the Boltzmann collision operator in the molecular dynamics. In this sense, our complex spectral analysis of the generator of motion, such as the Hamiltonian, or the Liouvillian, gives a quantitative analysis of the irreversible process based on the fundamental laws of physics [1, 2].

Let us now come back to our main problem. By substituting the dispersion relation (5) and the explicit form of the interaction (6), one can easily perform the integration over the wave vector **k** in the self-energy part (15),

$$\Sigma_a(z) = \int d\mathbf{k}\,\frac{g^2V_k^2}{z - \epsilon_\mathbf{k}} = -\frac{\pi g^2}{4k_c^4}(2k_c + i\pi\sqrt{z}). \tag{20}$$

Because we could obtain the explicit form of the self-energy part as in (20), we can find the complex eigenvalue of the nonlinear eigenvalue equation by solving the following equation (see (13)),

$$z = \epsilon_a - \frac{\pi g^2}{4k_c^4}(2k_c + i\pi\sqrt{z}). \tag{21}$$

We can easily solve this equation, and we have

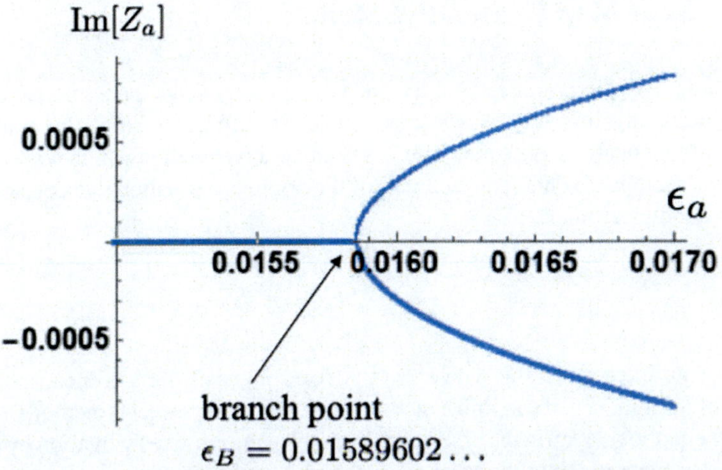

Fig. 1 Imaginary part of the eigenvalue as a function of the unperturbed energy ϵ_a for $g = 0.1$, $k_c = 1$. The branch point is a transition point from reversible process to irreversible process. Since the Hamiltonian H is a symmetric operator, we have always complex-conjugate pair of the complex eigenvalues when their imaginary parts do not vanish. The value of ϵ_a at the branch point is indicated as ϵ_B in the figure

$$z = \epsilon_a - \frac{\pi g^2}{2k_c^3} - \frac{1}{2}\left(\frac{\pi^2 g^2}{4k_c^4}\right)^2 \pm \frac{\pi^2 g^2}{4k_c^4}\sqrt{-\epsilon_a + \frac{\pi g^2}{2k_c^3} + \frac{1}{4}\left(\frac{\pi^2 g^2}{4k_c^4}\right)^2}. \tag{22}$$

The decay rate of the unstable state $|a\rangle$ is given by the imaginary part of this solution of the eigenvalue. In Fig. 1, we show the imaginary part of the eigenvalue as a function of the unperturbed energy ϵ_a for the parameters $g = 0.1$, $k_c = 1$. Hereafter, we will always use these values for g and k_c. One can see that there is a branch point at $\epsilon_a = \epsilon_B$ with the value,

$$\epsilon_B = 0.0158602\ldots \tag{23}$$

for the special choice of the parameters. Below this value, the imaginary part of the eigenvalue of the full Hamiltonian is zero, and the perturbed eigenstate is stable. But above this value, the imaginary part of the eigenvalue of the full Hamiltonian is not zero, and there appears two branches of the complex eigenvalue, one is the complex conjugate to the other. The negative branch of the imaginary part is associated with the decaying process oriented to our future, while the positive branch of the imaginary part is associated to the decaying process oriented to our past.

3 Nonlinear Map for Iterative Method

In our example of the Friedrichs model, we could exactly solve the nonlinear eigenvalue equation algebraically. However, it is usually difficult to solve the nonlinear equation algebraically. Hence, one often uses an iterative method which we will discuss in this section. To obtain the solution of the nonlinear equation, we consider the following nonlinear map,

$$z_{r+1} = \epsilon_a - \frac{\pi g^2}{4k_c^4}(2k_c + i\pi\sqrt{z_r}), \tag{24}$$

where r is an integer starting with $r = 0$. A fixed point of the nonlinear map is a solution of (21). Indeed, the iterative method is often a quite powerful method to solve nonlinear equations, even in the case one cannot solve the nonlinear equation by using an algebraic method. Note, however, our nonlinear map contains $\sqrt{z_r}$. Therefore, we should be careful to perform this iterative method to specify the branch of the Riemann sheet of the analytic continuation of $\sqrt{z_r}$.

In Fig. 2, we show two examples of the result of the iterative procedure for (a) the unstable case where ϵ_a is chosen far from the branch point where the dynamical phase transitions from reversible process to the irreversible process, while (b) is chosen near the branch point for $g = 0.1$, $k_c = 1$. The value of ϵ_a is indicated in the figure. In both figures (a) and (b), the large circle is the location of the exact eigenvalue in the complex Z_a-plane obtained by (22). We have chosen the initial value z_0 of the iteration near some point of the exact location. For case (a) far from the branch point, we see that the convergence to the exact point is rapid. On the other hand, for case (b) near the phase transition, the convergence to the exact point is slow, and there are lots of spiral rotations to approach the exact fixed point.

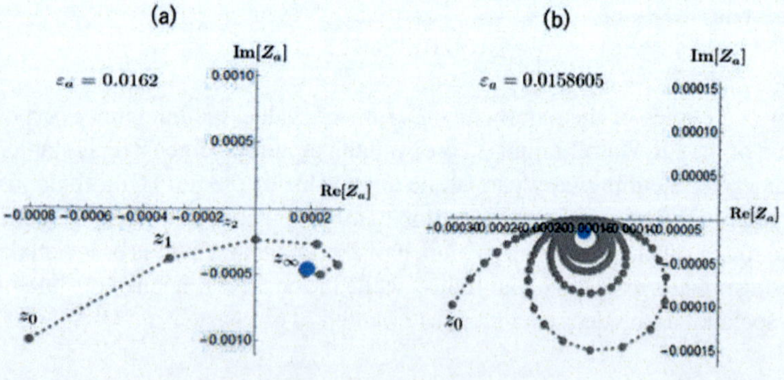

Fig. 2 Trajectories of the iterative procedure for the nonlinear map (24) to obtain the complex eigenvalue in the complex Z_a-plane. The large circles are locations of the eigenvalues. (a) is the unstable case where ϵ_a is chosen far from the branch point, while (b) is chosen near the branch point for $g = 0.1$, $k_c = 1$. The values ϵ_a for these examples are shown in these figures

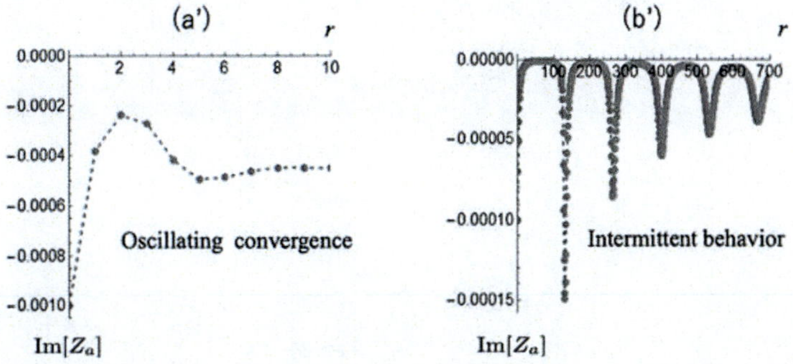

Fig. 3 The imaginary part of Z_a obtained by the iterative procedure of the nonlinear map (24) as a function of the iterative number r. (a') corresponds to (a), and (b') corresponds to (b) in Fig. 2, respectively, with same choice $g = 0.1$ and $k_c = 1$. Approach to the fixed point of the map is more intermittent in (b') near the branch point than (a') far from the branch point

In Fig. 3, we show results of the iterative procedure for the imaginary part of the map for the unstable case, where (a') corresponds to (a), and (b') corresponds to (b) in Fig. 2, respectively. The horizontal axis indicates the iteration number r. The figure (a') shows a smooth oscillating convergence to the fixed point for the case far from the branch point. On the other hand, the figure (b') shows a strong nonlinear intermittent behavior to the approach to the fixed point.

In Fig. 4, we show examples of the result of the iterative procedure for an interesting stable case with a real eigenvalue. In these examples, we have chosen as $\epsilon_a = 0.01572 < \epsilon_B$, where ϵ_B is the value of the branch point indicated at (23) for our special choice of the parameters as $g = 0.1$, $k_c = 1$. For this case, the unperturbed energy of the discrete spectrum ϵ_a is positive. Hence, the unperturbed energy of the continuous spectrum $\epsilon_{\mathbf{k}}$ can overlap with the unperturbed energy of the discrete spectrum. Nevertheless, since we have chosen the value of ϵ_a being moderately close to the branch point, there appears a stable solution on the eigenvalue problem of the full Hamiltonian H due to our choice of a moderately a strong interaction with $g = 0.1$. In these figures, (c) corresponds to (a) and (b) in Fig. 2, and (d) corresponds to (a') and (b') in Fig. 3. These figures show a monotonic convergence to the fixed point for the stable case with a real eigenvalue. Hence, the nonlinearity is weak as compared with the cases (a) and (a'). This monotonic convergence is observed for the negative choice of ϵ_a where the resonance singularity does not play a role.

4 Mandelbrot Set

Let us now consider the inverse map of (24). We will show that the inverse map exactly reduces to the nonlinear map that generate the Mandelbrot set. One can obtain its inverse map by replacing z_{r+1} by ξ_r, and z_r by ξ_{r+1}. Then, we have the

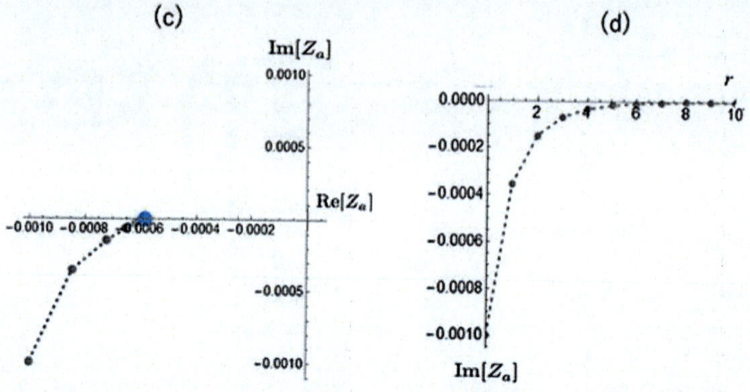

Fig. 4 The imaginary part of Z_a obtained by the iterative procedure of the nonlinear map (24) for a stable case as a function of the iterative number r with same choice $g = 0.1$ and $k_c = 1$ as the above figures. For this case, the imaginary part of the fixed point of the map is 0. The figure shows the approach to the fixed point is monotonic

inverse map,

$$\xi_{r+1} = -\left(\frac{4k_c^4}{\pi^2 g^2}\right)^2 \left(\xi_r - \epsilon_a + \frac{\pi g^2}{2k_c^3}\right)^2. \tag{25}$$

Moreover, we change the variable as

$$p_r \equiv -\left(\frac{4k_c^4}{\pi^2 g^2}\right)^2 \left(\xi_r - \epsilon_a + \frac{\pi g^2}{2k_c^3}\right). \tag{26}$$

Then, we obtain a simple nonlinear map,

$$p_{r+1} = p_r^2 + A(\epsilon_a, g), \tag{27}$$

with

$$A(\epsilon_a, g) \equiv \left(\frac{4k_c^4}{\pi^2 g^2}\right)^2 \epsilon_a - \frac{8k_c^5}{\pi^3 g^2}. \tag{28}$$

This is just the nonlinear map for the Mandelbrot set [23].

It is well known that starting with $p_0 = 0$ there are two domains of complex values $A(\epsilon_a, g)$, one of which gives the convergent value of $|p_r|$ in the limit $r \to \infty$, while the other of which gives the diverging value of $|p_r|$ in the same limit. In Fig. 5, we show these two domains in the parameter space for complex values of ϵ_a for $g = 0.1$, $k_c = 1$. The domain indicated by white is the domain which gives the convergent value of $|p_r|$ in the limit $r \to \infty$. On the other hand, the domain indicated by black is the domain which gives the diverging value of $|p_r|$ in the same limit. In the figure, we indicate the location of the branch point where the broken time-symmetry starts in our Friedrichs model.

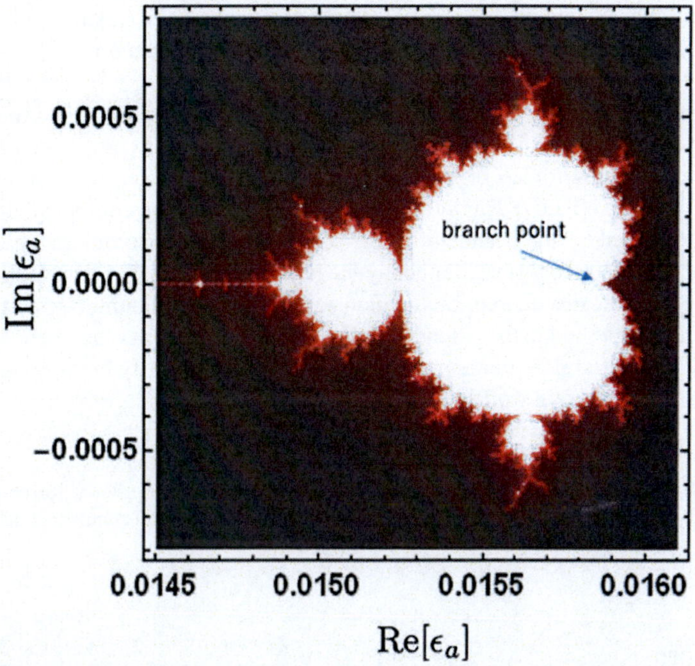

Fig. 5 Mandelbrot set obtained by the inverse map (27) of the nonlinear map (24) with the same choice $g = 0.1$ and $k_c = 1$ as the above figures. This figure is obtained by extending the value ϵ_a into the complex plane. Mandelbrot set is a border line of the convergent and divergent value of $|p_r|$ in the limit $r \to \infty$. The domain indicated by white is the converging domain, while the domain indicated by black is the diverging domain. The location of the branch point where the broken time-symmetry starts in our Friedrichs model is indicated in the figure

5 Concluding Remarks

We have investigated a decaying process of the Friedrichs model that is a simple model of unstable matter-field interacting system, by using the complex spectral analysis of the Hamiltonian. The complex spectral analysis of the generator of motion for conservative dynamical systems offers a microscopic analysis of irreversibility based on the basic laws of physics. An advantage of the Friedrichs model is that one can exactly solve the eigenvalue problem of the Hamiltonian.

In spite of the fact that the Hamiltonian of the Friedrichs mode is a Hermitian operator in the Hilbert space, this operator may have a complex eigenvalue in the dual space of the extended Hilbert space which lies outside the Hilbert space. Transport coefficients, such as the decay rate of the unstable state, in the irreversible process can be evaluated as an imaginary component of the complex eigenvalue of the generator of motion. This reveals the fact that irreversibility is not an approximate concept in dynamics, but it is an exact dynamical property of dynamics coming from the fundamental microscopic equation of motions, such as the Schrödinger equation or the Liouville-von Neumann equation. The authors recommend to read the references [1–3] for more details on this essential aspect of the complex spectral analysis of the generator of motion in the fundamental laws of physics.

As has been explained, the eigenvalue problem of the generator of motion, such as the Hamiltonian or the Liouvillian, reduces to the nonlinear eigenvalue problem of the effective Hamiltonian or the effective Liouvillian. We have shown that the nonlinearity of the effective Hamiltonian leads to a surprisingly complicated fractal structure described by the Mandelbrot set when we solve the nonlinear eigenvalue problem of the effectively Hamiltonian by the iterative method. The fractal structure we found here indicates the renormalization scaling in the Brillouin-Wigner type of the perturbation method in the extended Hilbert space. This shows that there are still lots of uninvestigated rich problems associated to the nonlinearity in the problem of broken time-symmetry in well-investigated decaying process in unstable dynamical systems.

Acknowledgements The authors would like to express their thanks to Professor Kazuki Kanki, Professor Savannah Garmon, and Professor Tetsuro Konishi for their helpful comments and discussions.

References

1. T. Petrosky, I. Prigogine, Chaos, Solitons and Fractals **7**(4), 441 (1996)
2. I. Prigogine, S. Rice (eds.), *Advances in Chemical Physics*, vol. 99 (Wiley and Sons, New York, 1997), p. 1
3. T. Petrosky, I. Prigogine, S. Tasaki, Physica **173A**, 175 (1991)
4. I. Prigogine, S. Rice (eds.), *Advances in Chemical Physics*, vol. 112 (Wiley and Sons, New York, 2003), p. 129
5. K. Hashimoto, K. Kanki, S. Tanaka, T. Petrosky, Phys. Rev. E **93**, 02213 (2016)
6. K. Hashimoto, K. Kanki, S. Tanaka, T. Petrosky, Chaos **27**(10), 104616 (2017)
7. G. Ordonez, T. Petrosky, I. Prigogine, Phys. Rev. A **63**, 05210052106 (2001)
8. G. Ordonez, T. Petrosky, I. Prigogine, Phys. Rev. A **64**, 062101 (2001)
9. I. Prigogine, T. Petrosky, G. Ordonez, *Advances in Chemical Physics*, (Special Volume in Memory of Ilya Prigogine), vol. 135 (Wiley and Sons, New York, 2007), p. 135
10. S. Nakade, K. Kanki, S. Tanaka, T. Petrosky, Phys. Rev. E **102**, 032137 (2020)
11. S. Nakade, K. Kanki, S. Tanaka, T. Petrosky, Symmetry **13**(3), 506 (2021)
12. E. Karpov, I. Prigogine, T. Petrosky, G. Pronko, J. Math. Phys. **41**, 118 (2000)
13. S. Tanaka, S. Garmon, G. Ordonez, T. Petrosky, Phys. Rev. B **76**, 153308 (2007)
14. K. Noba, Y. Uesaka, N. Yamada, S. Tanaka, T. Petrosky, J. Phys. A **85**, 38530 (2014)
15. S. Tanaka, S. Garmon, K. Kanki, T. Petrosky, Phys. Rev. A **94**, 022105 (2016)
16. H. Yamane, S. Tanaka, Symmetry **2018**(10), 313 (2018)
17. S. Tanaka, K. Kanki, Prog. Theor. Exp. Phys. **2020**, 12A107 (2020)
18. Y. Dunham, K. Kanki, S. Garmon, S. Tanaka, G. Ordonez, Phys. Rev. A **103**, 043513 (2021)
19. G. Ordonez, T. Petrosky, I. Prigogine, Phys. Rev. A **68**, 022107 (2003)
20. T. Petrosky, Int. J. Quant. Chem. **98**, 103 (2004)
21. T. Petrosky, J. Adv. Simul. Sci. Eng. **6**(1), 56 (2019)
22. K.O. Friedrichs, Commun. Pure Appl. Math. **1**, 361 (1948)
23. B.B. Mandelbrot, *The Fractal Geometry of Nature* (W. H. Freeman, San Francisco, 1982)

Thermodynamic Flux-Force Closure Relations for Systems out of the Onsager Region

Giorgio Sonnino

Abstract The first attempts to develop non-equilibrium thermodynamics theory occurred after the first observations of some coupled phenomena of thermal diffusion and thermoelectric phenomena. However, the big obstacle to overcome is that the number of unknowns is greater than the number of equations expressing the conservation laws. So, it is crucial to determine the *closure relations* to make the problem solvable. The objective of this work is to determine the nonlinear flux-force relations for systems out of Onsager's region that respect the existing thermodynamic theorems for systems far from equilibrium. To this aim, a thermodynamic theory for irreversible processes [referred to as the *Thermodynamical Field Theory* (TFT)] has been developed. The TFT rests upon the concept of equivalence between thermodynamic systems. The equivalent character of two alternative descriptions of a thermodynamic system is ensured if, and only if, the two sets of thermodynamic forces are linked with each other by the so-called *Thermodynamic Covariant Transformations* (TCT). The TCT are the most general thermodynamic force transformations which leave unaltered both the entropy production and the Glansdorff-Prigogine dissipative quantity. In this work, we describe the Lie group and the group representations associated with the TCT. The TCT leads to the so-called *Thermodynamic Covariance Principle* (TCP): *The nonlinear closure equations, i.e., the flux-force relations, must be covariant under* TCT. In this chapter, we provide the explicit form of the nonlinear PDEs, subjected to the appropriate boundary conditions, which have to be satisfied by transport coefficients when the skew-symmetric piece is absent. The solution of these equations allows determining the flux-force closure relations for systems out of the Onsager region. Since the proposed PDEs are obtained without neglecting any term present in the balance equations (i.e., the mass, momentum, and energy balance equations), we propose them as a good candidate for describing

G. Sonnino (✉)
Department of Physics, Université Libre de Bruxelles (ULB), Bvd du Triomphe, Campus de la Pliane, C.P. 231 - Build. NO - 1050, Brussels, Belgium
e-mail: giorgio.sonnino@ulb.be

L. Brenig et al. (eds.), *Nonequilibrium Thermodynamics and Fluctuation Kinetics*, Fundamental Theories of Physics 208,
https://doi.org/10.1007/978-3-031-04458-8_5

transport in thermodynamic systems also in turbulent regimes. As a special case, we derive the nonlinear PDEs for transport coefficients when the thermodynamic system is subjected to two thermodynamic forces. The obtained PDE is, in Thermodynamical Field Theory (TFT), analogous to Liouville's equation in Riemannian (or pseudo-Riemannian) geometry. A preliminary test is carried out by analysing a concrete example where Onsager's relations manifestly disagree with experience: losses in magnetically confined Tokamak-plasmas. More specifically, we compute the mass and energy losses in FTU (*Frascati Tokamak Upgrade*)-plasmas subjected to two thermodynamic forces. We show a good agreement between the theoretical (TFT) predictions and the experimental data. The aim is to apply our approach to the *Divertor Tokamak Test facility* (DTT), to be built in Italy, and to ITER. Other applications of the TFT to the thermoelectric effects or to out-of-equilibrium chemical reactions can be found in the references cited at the end of the chapter.

1 Introduction

When there are more unknowns than equations expressing conservation laws, additional *closure laws* are needed to make the problem solvable. Generally, these additional closure relations are not derivable from one of the physical equations being solved. Several approaches to getting the closure relations are currently applied. Among them, we cite the so-called *truncation schemes* and the *Asymptotic schemes*. In *truncation schemes*, higher order moments are arbitrarily assumed to vanish, or simply negligible with respect to the terms of lower moments. Truncation schemes can often provide quick insight into fluid systems, but always involve uncontrolled approximation. This method is often used in transport processes in Tokamak-plasma (see, for instance, the book [1]). The *asymptotic schemes* are based on the rigorous exploitation of some small parameters. They have the advantage of being systematic, and providing some estimate of the error involved in the closure. However, as the title itself suggests, these methods are effective only when small parameters enter, by playing a crucial role, in the dynamic equations. These schemes are often used for solving numerically kinetic equations (refer, for instance, to the book [2]). Another possibility is to obtain the closure relations by formulating a specific theory or *ad hoc* models. The most important closure equations are the so-called *transport equations* (or the *flux-force relations*), relating the thermodynamic forces with the conjugate dissipative fluxes that produce them. The thermodynamic forces are related to the spatial inhomogeneity and (in general) they are expressed as gradients of the thermodynamic quantities. The study of these relations is the object of *non-equilibrium thermodynamics*. Morita and Hiroike eased this task for a closure relation by providing the formally exact closure formula [3]

$$J_\nu(X) = \varpi_{\mu\nu}(X)X^\mu \tag{1}$$

Here, X^μ and J_μ denote the thermodynamic forces and thermodynamic fluxes, respectively. Coefficients $\varpi_{\mu\nu}(X)$ are the *transport coefficients*, where it is clearly highlighted that the transport coefficients may depend on the thermodynamic forces. We suppose that all quantities appearing in Eq. (1) are dimensionless. Note that in this equation, as well as in the sequel, the Einstein summation convention on the repeated indices is understood. Matrix $\varpi_{\mu\nu}(X)$ can be decomposed into a sum of two matrices, one symmetric and the other skew symmetric, which we denote by $g_{\mu\nu}(X)$ and $f_{\mu\nu}(X)$, respectively. The second law of thermodynamics requires that $g_{\mu\nu}(X)$ is a positive-definite matrix. Note that, in general, the dimensionless entropy production, denoted by σ, with $\sigma = \varpi_{\mu\nu}(X)X^\mu X^\nu = g_{\mu\nu}(X)X^\mu X^\nu$, may not be a simply bilinear expression of the thermodynamic forces (since the transport coefficients may depend on the thermodynamic forces). For conciseness, in the sequel we drop the symbol X in $g_{\mu\nu}$ as well as in the skew-symmetric piece of the transport coefficients $f_{\mu\nu}$ being implicitly understood that these matrices may depend on the thermodynamic forces.

In previous works, a macroscopic *Thermodynamic Field Theory* (TFT) for deriving the closure relations valid for thermodynamic systems out of Onsager's region has been proposed. More specifically, the aim of the TFT in [4–11] is to determine the nonlinear flux-force relations which are valid for thermodynamic systems out of the thermodynamic linear region (commonly referred to as the *Onsager region*) [12, 13]. This task is accomplished by means of three hypotheses: two constraints 1. and 2., and one assumption 3. In order to establish the vocabulary and notations that shall be used in the sequel of this work, we briefly recall these hypotheses.

1. *The thermodynamic laws and the theorems demonstrated for systems far from equilibrium must be satisfied.*
2. *The validity of the Thermodynamic Covariance Principle (TCP) must be ensured.* The TCP stems from the concept of equivalent systems from the thermodynamic point of view. Thermodynamic equivalence was originally introduced by Th. De Donder and I. Prigogine [14–16]. However, the De Donder-Prigogine definition of thermodynamic equivalence, based only on the invariance of the entropy production, is not sufficient to guarantee the equivalence character between two sets of thermodynamic forces and conjugate thermodynamic fluxes. In addition, it is known that there exists a large class of flux-force transformations such that, even though they leave unaltered the expression of the entropy production, they may lead to certain paradoxes [17, 18]. The equivalent character of two alternative descriptions of a thermodynamic system is ensured if, and only if, the two sets of thermodynamic forces are linked with each other by the so-called *Thermodynamic Covariant Transformations* (TCT). The TCT are the most general thermodynamic force transformations which leave unaltered both the entropy production σ and the Glansdorff-Prigogine dissipative quantity P [for the definition of P, see the forthcoming Eq. (12)]. In this work, we also describe the Lie group and the group representations associated with the TCT. The TCT leads in a natural way to postulate the validity of the so-called *Thermodynamic Covariance Principle* (TCP):

The nonlinear closure equations, i.e., the flux-force relations must be covariant under the Thermodynamic Covariant Transformations (TCT).

3. *Close to the steady states, the nonlinear closure equations can be derived by the principle of least action.*

This theory, based on 1., 2., and 3. is referred to as the *Thermodynamical Field Theory* (TFT). The three hypotheses 1., 2., and 3. allow determining the nonlinear TFT-Partial Differential Equations (PDEs) for transport coefficients $\varpi_{\mu\nu}$. In this chapter, we shall limit ourselves to the case in which the transport coefficients possess only the symmetric piece (i.e., $f_{\mu\nu} = 0$). We shall show the explicit form of the TFT-PDEs for $g_{\mu\nu}$. Later, inspired by the theory of Jackiw and Teitelboim [19–24], we shall also derive the explicit form of the TFT-PDEs for $g_{\mu\nu}$ for the two-dimensional case, i.e., when the system is subjected to two independent thermodynamic forces.

The final part of the chapter is devoted to the application of the theory to some relevant examples of systems out of equilibrium. More precisely, we shall apply the derived TFT-PDEs to Tokamak-plasmas in a collisional regime. This is a very interesting example of application since, in this case, the Onsager relations strongly disagree with experimental data. One of the main issues in Fusion Science is the computation of energy and mass losses in Tokamak-plasmas. It is well-known that there is a strong disagreement, of several orders of magnitude, especially for electron mass and energy losses, between the theoretical predictions of the Onsager theory (at the basis of the so-called *neoclassical theory*) and experiments. This discrepancy is even more pronounced in the case of magnetically confined plasmas in turbulent regimes. The aim is to compute the electron heat loss in Tokamak-plasmas by considering the contribution of the nonlinear terms in the flux-force relations derived by the TFT. In order to test the validity of our results, we have computed the electron heat loss for Frascati Tokamak Upgrade (FTU)-plasmas in a fully collisional transport regime. We have compared the theoretical profile obtained by the nonlinear theory satisfying the TCP with the experimental data for the FTU-plasmas (provided by the ENEA C.N.R.—EuroFusion in Frascati) and with the theoretical predictions of the linear theory (the Onsager theory). We found that there is a fairly good agreement between the TFT and experiments (in contrast with Onsager's theory). However, disagreements appear in the region where the dimensionless entropy $\tilde{\sigma}_{(L)}$ is of order 1. In particular, we found that the disagreement appears in the region of the tokamak where the plasma is in a turbulent regime. Incidentally, this corresponds also to the region where $\tilde{\sigma}_{(L)} \sim 1$. Preliminary calculations and theoretical results in the region $\tilde{\sigma}_{(L)} \sim 1$ have also been performed. We are currently comparing the theoretical predictions of the TFT with the experimental data for FTU-plasma in a turbulent regime. Other examples of application of the TFT to unimolecular triangular chemical reactions (i.e., three isomerisations take place) and to materials subjected to temperature and electric potential gradients, and to chemical reactions out of equilibrium and to Hall effect can be found in Refs. [4, 8] and in [25–29], respectively.

The chapter is organised as follows. In Sect. 2, we recall the basic concepts of the Thermodynamical Field Theory (TFT). To this aim, we quickly introduce the definition of the *space of the thermodynamic forces*, and we describe the *Thermodynamic*

Covariance Principle (TCP) and the TCT-*symmetry group*. Successively, in Sect. 3 we derive the explicit form of the nonlinear TFT-PDEs for transport coefficients in the absence of the skew-symmetric part (i.e., when $f_{\mu\nu} = 0$). In Sect. 4, we derive the nonlinear PDEs for transport coefficients $g_{\mu\nu}$ when the system is subjected to two independent thermodynamic forces (i.e., when $n = 2$). Section 5 provides the linearised TFT transport equations. The physical meaning of the gauge invariance in TFT is reported in Sect. 6. The analytic solution of the two-dimensional linearised homogeneous TFT-PDE is obtained in Sect. 7. The solution of the TFT-PDE for collisional FTU-plasmas, subjected to two independent thermodynamic forces, can be found in Sect. 8. In this section, we shall show the good agreement between the theoretical predictions and the experimental data. Finally, the main results are concluded in Sect. 9. Here, we also specify the boundary conditions for turbulent Tokamak-plasmas. The determination of the boundary conditions, which have to be satisfied by the TFT-PDEs for a general system out of thermodynamic equilibrium, is obtained in Appendix. In Appendix, we can also find the analytic solutions of the linearised, two-dimensional, inhomogeneous TFT-PDE.

2 The Thermodynamical Field Theory (TFT)

2.1 The Space of the Thermodynamic Forces

The first task is to define the space where we may perform calculations. To this aim, it is not enough to specify the nature of the axes, we must also determine two quantities: the *metric tensor* and the *affine connection* (denoted by symbol $\Gamma^{\lambda}_{\mu\nu}$).

- The metric tensor is a central object in the theory; it describes the local geometry of space. The metric tensor is a symmetric tensor used to raise and lower the indicative tensors and generate the connections used to construct the PDEs and the curvature tensor of the space.
- The curvature of a space can be identified by taking a vector at some point and transporting it parallel along a curve in space-time. An affine connection is a rule that describes how to legitimately move a vector along a curve on the variety without changing its direction.

The metric tensor and the affine connection are determined by physics, i.e., by ensuring the validity of the thermodynamic theorems valid for systems out of equilibrium (in accordance with the above-mentioned assumption 1). More precisely, we must take into account the validity of the *second law of thermodynamics* and the *General Evolution Criterion* (GEC) [30, 31]. We adopt the following definitions [4, 11]:

- *The space of the thermodynamic forces* (or, simply, *the thermodynamic space*) *is the space spanned by the thermodynamic forces.*

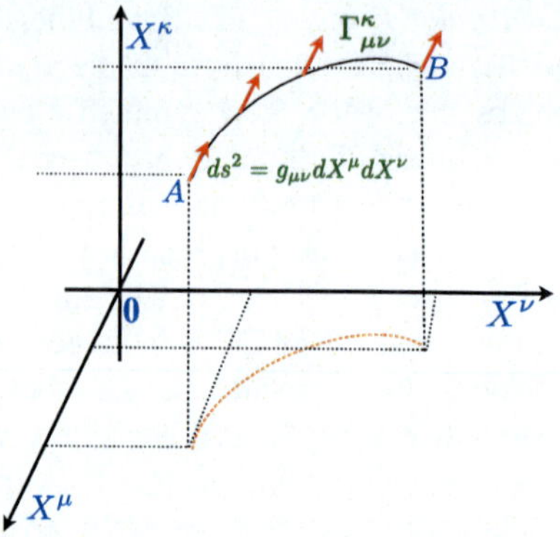

Fig. 1 The thermodynamic space. The space is spanned by the thermodynamic forces, the metric tensor is identified with the symmetric piece of the transport coefficients, and the expression of the affine connection is determined by the *General Evolution Criterion*

- *The metric tensor is identified with the symmetric piece $g_{\mu\nu}$ of the transport coefficients.*
  - Note that this definition takes into account the second law of thermodynamics as, for the second law of thermodynamics, the square (infinitesimal) distance $ds^2 = d\mathbf{s} \cdot d\mathbf{s}$ is always a non-negative quantity—see Fig. 1.
- *The expression of the thermodynamic affine connection $\widetilde{\Gamma}^\kappa_{\mu\nu}$ is determined by requiring the validity of the Glansdorff-Prigogine General Evolution Criterion.* In Ref. [11] it is shown that, when $f_{\mu\nu} = 0$, we get

$$\widetilde{\Gamma}^\mu_{\alpha\beta} \equiv \left\{ {\mu \atop \alpha\beta} \right\} + \frac{1}{2\sigma} X^\mu X^\eta g_{\alpha\beta,\eta} - \frac{1}{2(n+1)\sigma}\left(\delta^\mu_\alpha X^\nu X^\eta g_{\beta\nu,\eta} + \delta^\mu_\beta X^\nu X^\eta g_{\alpha\nu,\eta}\right) \tag{2}$$

where commas stand for partial derivatives with respect to the thermodynamic forces. δ^j_i denotes the Kronecker delta and

$$\left\{ {\mu \atop \alpha\beta} \right\} = \frac{1}{2} g^{\mu\lambda}\left(\frac{\partial g_{\lambda\alpha}}{\partial X^\beta} + \frac{\partial g_{\lambda\beta}}{\partial X^\alpha} - \frac{\partial g_{\alpha\beta}}{\partial X^\lambda}\right) \tag{3}$$

the *Levi-Civita affine connection.*

2.2 The De Donder-Prigogine Thermodynamic Invariance

Onsager's theory is based on three assumptions: (i) The probability distribution function for the fluctuations of thermodynamic quantities (temperature, pressure, degree of advancement of a chemical reaction, etc.) is a Maxwellian, (ii) Fluctuation decay according to a linear law, and (iii) The detailed balance principle (or the microscopic reversibility) is satisfied. Onsager showed the equivalence of the assumptions (i), (ii), and (iii) with the equations [12, 13]

$$J_\nu = L_{\mu\nu} X^\mu \quad \text{with} \quad \frac{\partial L_{\mu\nu}}{\partial X^\lambda} = 0 \quad \text{and} \quad L_{\mu\nu} = L_{\nu\mu} \tag{4}$$

where the coefficients of matrix $L_{\mu\nu}$ are the Onsager transport coefficients. $L_{\mu\nu}$ is a symmetric matrix and the elements are independent of the thermodynamic forces. The limit of validity of Eq. (4) establishes the limit of validity of the *Onsager region*. Assumption **(iii)** allows deriving the *reciprocity relations* $L_{\mu\nu} = L_{\nu\mu}$. The Onsager theory of fluctuations starts from the Einstein formula linking the probability of a fluctuation, W, with the entropy change, ΔS, associated with the fluctuations from the state of equilibrium

$$W = W_0 \exp[\Delta S / K_B] \tag{5}$$

In Eq. (5), K_B is the Boltzmann constant and W_0 is a normalisation constant which ensures that the sum of all probabilities equals one [12, 13]. Prigogine generalised Eq. (5), which applies to adiabatic or isothermal transformations, by introducing the entropy production due to fluctuations. Denoting by ξ_i $(i = 1 \cdots m)$, the m deviations of the thermodynamic quantities from their equilibrium value, Prigogine proposed that the probability distribution of finding a state in which the values ξ_i lie between ξ_i and $\xi_i + d\xi_i$ is given by [15]

$$W = W_0 \exp[\Delta_I S / K_B] \quad \Delta_I S = \int_E^F d_I s \quad ; \quad \frac{d_I s}{dt} \equiv \int_\Omega \sigma d\text{v} \tag{6}$$

Here, $d\text{v}$ is the spatial volume element of the system, and the integration is over the entire space Ω occupied by the system. E and F indicate the equilibrium state and the state to which a fluctuation has driven the system, respectively. Note that this *probability distribution remains unaltered for flux-force transformations leaving invariant the entropy production.* Concrete examples of chemical reactions, equivalent from the thermodynamic point of view, have also been analysed in the literature. As an example, among these, we choose the simplest of all. Let us consider, for example, the following chemical system in which two isomerisations

(a): $A \to B$ and $B \to C$.

take place [15]. Of course, from the macroscopic point of view, the chemical changes in **(a)** are equivalent to the two isomerisations

(b): $A \to C$ and $B \to C$.

It can be checked that, under a linear transformation of the thermodynamic forces (which in this case corresponds to a linear transformation of the chemical affinities), the entropy productions for the two chemical reactions **(a)** and **(b)** are equal. Indeed, the corresponding affinities of the reactions **(a)** read: $A^1 = \mu_A - \mu_B$ and $A^2 = \mu_B - \mu_C$, with A^i and μ_i $(i = A, B, C)$ denoting the *chemical affinities* and the *chemical potentials*, respectively. The change per unit time of the mole numbers is given by

$$\frac{dn_A}{dt} = -v_1 \quad ; \quad \frac{dn_B}{dt} = v_1 - v_2 \quad ; \quad \frac{dn_C}{dt} = v_2 \tag{7}$$

with v_i ($i = 1, 2$) denoting the chemical reaction rates. In this case, the thermodynamic forces and the thermodynamic fluxes are the chemical affinities (over temperature) and the chemical reaction rates, respectively, i.e., $X^\mu = A^\mu/T$ and $J_\mu = v_\mu$. Hence, the corresponding entropy production reads $d_I S/dt = (A^1/T)v_1 + (A^2/T)v_2 \geq 0$. The affinities corresponding to reactions **(b)** are related to the previous ones by

$$A'^1 = \mu_A - \mu_C = A^1 + A^2 \quad ; \quad A'^2 = \mu_B - \mu_C = A^2 \tag{8}$$

By taking into account that

$$\frac{dn_A}{dt} = -v'_1 \quad ; \quad \frac{dn_B}{dt} = -v'_2 \quad ; \quad \frac{dn_C}{dt} = v'_1 + v'_2 \tag{9}$$

we get

$$v_1 = v'_1 \quad ; \quad v_2 = v'_1 + v'_2 \tag{10}$$

where the invariance of the entropy production is manifestly shown. Indeed,

$$\frac{d_I S}{dt} = (A^1/T)v_1 + (A^2/T)v_2 = (A'^1/T)v'_1 + (A'^2/T)v'_2 = \frac{d_I S'}{dt} \tag{11}$$

or $J_\mu X^\mu = J'_\mu X'^\mu$ (where, as usual, the Einstein summation convention on the repeated indices is adopted). On the basis of the above observations, Th. De Donder and I. Prigogine formulated, for the first time, the concept of equivalent systems from the thermodynamical point of view. For Th. De Donder and I. Prigogine, thermodynamic systems are thermodynamically equivalent if, under flux-force transformation, the bilinear form of the entropy production remains unaltered, i.e., $\sigma = \sigma'$ [14–16].

2.3 Remarks on De Donder-Prigogine's Thermodynamic Invariance Formulation

The Thermodynamic Invariance Principle formulated by De Donder-Prigogine, based only on the invariance of the entropy production, is not sufficient to guarantee the equivalence character of the two descriptions (J_μ, X^μ) and (J'_μ, X'^μ). Indeed, we can easily convince ourselves that there exists a large class of transformations such that, even though they leave unaltered the expression of the entropy production, they may lead to certain paradoxes to which J. E. Verschaffelt and R. O. Davies have called attention [17, 18]. This obstacle can be removed if one takes into account one of the most fundamental and general theorems valid in the thermodynamics of irreversible processes: the General Evolution Criterion. Glansdorff and Prigogine have shown that: *For time-independent boundary conditions, a thermodynamic system, even in strong non-equilibrium conditions, relaxes to a stable stationary state in such a way that the following General Evolution Criterion is satisfied*

$$P = \int_{\Omega} J_{\mu} \frac{\partial X^{\mu}}{\partial t} dv \leq 0 \tag{12}$$

In addition

$$P = \int_{\Omega} J_{\mu} \frac{\partial X^{\mu}}{\partial t} dv = 0 \quad \text{at the steady state} \tag{13}$$

Quantity P may be referred to as the *Glansdorff-Prigogine dissipative quantity*. Let us check the validity of this theorem by considering two, very simple, examples. Let us consider, for instance, a closed system containing m components ($i = 1 \cdots m$) among which chemical reactions are possible. The temperature, T, and the pressure, p, are supposed to be constant in time. The chance in the number of moles n_i, of component i, is

$$\frac{dn_i}{dt} = \nu_i^j v_j \tag{14}$$

with ν_i^j denoting the *stoichiometric coefficients*. By multiplying both members of Eq. (14) by the time derivative of the chemical potential of component i, we get

$$\frac{d\mu^i}{dt}\frac{dn_i}{dt} = \left(\frac{\partial \mu^i}{\partial n_\iota}\right)_{(pT)} \frac{dn_i}{dt}\frac{dn_\iota}{dt} = \left(\frac{\partial \mu^i}{\partial n_\iota}\right)_{(pT)} \nu_i^j \nu_\iota^\kappa v_j v_\kappa \geq 0 \tag{15}$$

where the positive sign of the term on the right-hand side is due to the second law of thermodynamics. By taking into account the De Donder law between the affinities A^j and the chemical potentials, i.e., $A^j = -\nu_i^j \mu^i$, and that the chemical thermodynamic force (X^j) and its chemical conjugate flux (J_j) read $X^i = A^j/T$ and $J_i = v_j$ respectively, we finally get

$$P = \int_{\Omega} J_{\mu} \frac{\partial X^{\mu}}{\partial t} dv = \Omega J_{\mu} \frac{dX^{\mu}}{dt} = \Omega v_j \frac{d}{dt}\left(\frac{A^j}{T}\right) = -\frac{\Omega}{T}\left(\frac{\partial \mu^i}{\partial n_\iota}\right)_{(pT)} \nu_i^j \nu_\iota^\kappa v_j v_\kappa \leq 0 \tag{16}$$

Hence, the Glansdorff-Prigogine dissipative quantity P is always negative and it vanishes at the stationary state. As a second example, we analyse the case of heat conduction in non-expanding solid. In this case, the thermodynamic forces and the conjugate fluxes are the (three) components of the gradient of the inverse of the temperature, $X^{\mu} = \nabla(1/T)$, and the (three) components of the heat flux, $J_{\mu} = \boldsymbol{J}_{(q)}$ (with $\mu = 1, 2, 3$), respectively. Hence,

$$P = \int_{\Omega} J_{\mu} \frac{\partial X^{\mu}}{\partial t} dv = \int_{\Omega} \boldsymbol{J}_{(q)} \cdot \nabla(1/T) dv \tag{17}$$

The heat flux, $\boldsymbol{J}_{(q)}$, is linked to the (partial) time derivative of temperature by the Fourier law (expressing the energy balance equation)

$$\rho c_v \frac{\partial T}{\partial t} = -\nabla \cdot \boldsymbol{J}_q \tag{18}$$

with ρ and c_v denoting the mass density and the specific heat at volume constant of the fluid, respectively. By performing the integration by parts, and by assuming that the heat flux vanishes at the boundary, we easily get

$$P = -\int_{\Omega} \frac{\rho c_v}{T^2} \left(\frac{\partial T}{\partial t} \right)^2 dv \leq 0 \tag{19}$$

with $P = 0$ at the steady state. By summarising, *for all thermodynamic systems, without using the Onsager reciprocal relations, and even if the transport coefficients are dependent on the thermodynamic forces, the dissipative quantity P is always a negative quantity*. This quantity vanishes at the steady state. In the two above-mentioned examples, the thermodynamic forces are the chemical affinities (over temperature) and the gradient of the inverse of temperature, respectively. However, we could have adopted a different choice of thermodynamic forces. If we analyse, for instance, the case of heat conduction in non-expanding solid, where chemical reactions take place simultaneously, we can choose as thermodynamic forces a combination of the (dimensionless) chemical affinities (over temperature) and the (dimensionless) gradient of the inverse of temperature. Clearly, this representation is thermodynamically equivalent to the previous one (where the thermodynamic forces are simply the chemical affinities over temperature and the gradient of the inverse of temperature) only if the negative sign of the dissipative quantity P is preserved. In other words, the equations providing the stationary states (i.e., Eq. (13)) must admit exactly the same solutions.

2.4 The Thermodynamic Covariant Transformations (TCT) and the Thermodynamic Covariance Principle (TCP)

One of the central aspects of the TFT is the concept of invariance of physics' laws. This invariance can be described in many ways, for example, in terms of local covariance or covariance of diffeomorphism. A more explicit description can be given through the use of tensors. The characteristic of the tensors that proves to be crucial is the fact that, once the metric is given, the operation of contracting a tensor of rank r on all indices r provides a number—an *invariant*—which is independent of the coordinates used to perform the contraction. Physically, this means that the invariant calculated by choosing a specific coordinate system (i.e., in a specific set of the thermodynamic forces) will have the same value if calculated in another—thermodynamically equivalent—coordinate system (i.e., in another equivalent set of thermodynamic forces). *According to the Thermodynamical Field Theory* (TFT), *two set of thermodynamic forces are equivalent if the following two conditions are satisfied*[1] [11]:

(i) The entropy production σ must be invariant under transformation of the thermodynamic forces $\{X^{\mu}\} \to \{X'^{\mu}\}$.
(ii) The Glansdorff-Prigogine dissipative quantity P must also be invariant under the force transformations $\{X^{\mu}\} \to \{X'^{\mu}\}$.

Condition (ii) stems from the fact that a stable steady state must be transformed into the same stable steady-state, with the *same degree of stability*. In mathematical terms, this implies

$$\sigma = J_{\mu}X^{\mu} = J'_{\mu}X'^{\mu} = \sigma' \; ; \; P = P' \to J_{\mu}\delta X^{\mu} = J'_{\mu}\delta X'^{\mu} \quad \text{and} \quad t = t' \tag{20}$$

Equations (20) are satisfied iff the transformed thermodynamic forces and conjugate fluxes read as [11, 32, 33]

$$X'^{\mu} = \frac{\partial X'^{\mu}}{\partial X^{\nu}} X^{\nu} \quad , \quad J'_{\mu} = \frac{\partial X^{\nu}}{\partial X'^{\mu}} J_{\nu} \tag{21}$$

Transformations (21) are referred to as the *Thermodynamic Covariant Transformations* (TCT) [11]. The thermodynamic equivalence principle leads, naturally, to the following *Thermodynamic Covariance Principle* (TCP) [32, 33]:

The nonlinear closure equations, i.e., the flux-force relations, must be covariant under TCT.

The essence of the TCP is the following. The equivalent character between two representations is warranted iff the fundamental thermodynamic equations (e.g., the transport equations) are covariant under the Thermodynamic Covariant Transformations (TCT).

2.5 *The TCT-Symmetry Group*

2.5.1 Topological Structure the TCT-Group

The invariance of a system under TCP is intimately related to the existence of a group, which we refer to as the *TCT-group* [34, 35]. The TCT-group with its properties can be identified by analysing the solution of Eq. (21). The solution of Eq. (21) reads

$$X'^{\mu} = X^{1} F^{\mu}\left(\frac{X^2}{X^1}, \frac{X^3}{X^2}, \cdots \frac{X^n}{X^{n-1}}\right) \tag{22}$$

[1] According to the TFT conditions, **(i)** and **(ii)** establish the *equivalent character* between two different representations (i.e., between two different set of thermodynamic forces.)

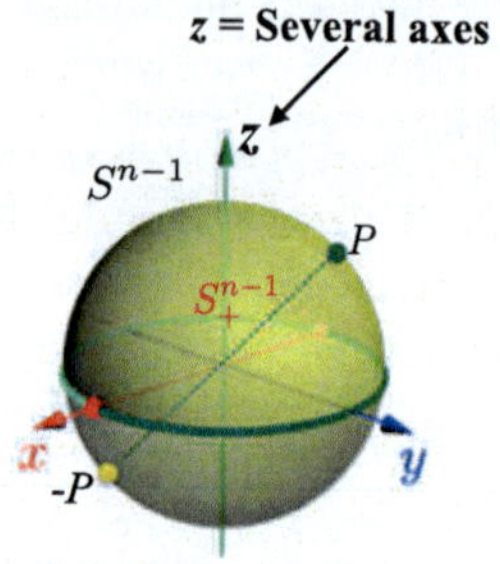

Fig. 2 The projective space. The projective space $\mathbb{RP}^{n-1}$ is diffeomorphic to S_+^{n-1} made by the upper hemisphere + half equator (without the red and yellow points) + the red point

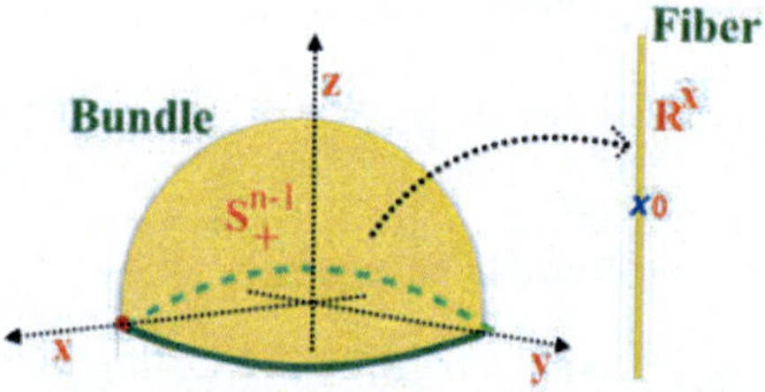

Fig. 3 The TCT-group. The TCT-group G^n is the application F^μ, from the *bundle* (the projective space $\mathbb{RP}^{n-1}$) to the *fibre* ($\mathbb{R}^x$)

with F^μ denoting arbitrary functions. Hence, the ratio $\{X^\mu/X^{\mu-1}\}$ are the coordinates for a different space: the *Real Projective Space* RP^{n-1}, which is defined to be the quotient of R^n minus the origin by the scaling map $X^\mu \to \alpha X^\mu$ with α denoting any nonzero real number—see Fig. 2. The TCT-group is then *the product of* diff$(\mathbb{RP}^{n-1})$ *with the multiplicative group of the map from* $\mathbb{RP}^{n-1} \to \mathbb{R}^\times$—see Fig. 3 [34, 35].

2.5.2 Algebraic Structure of the TCT-Group

In the previous subsection, we have seen that the TCT group, denoted by G^n, is a specific subgroup of the homogeneous diffeomorphisms from diff$(\mathbb{R}\backslash(\{0\})$. In algebraic terms, the result of the previous Subsection may be expressed as follows: *The* G^n *results from the application*:
diff$(\mathbb{R}\backslash(\{0\}) \ni X \vdash Y_g(X) \in$ diff$(\mathbb{R}\backslash(\{0\})$, *with* $Y_g \in G^n$ iff $Y_g(\lambda X) = \lambda Y_g(X)$ with $\lambda \in \mathbb{R}$.

It is possible to demonstrate that the TCT-group, G^n, may be split in a semidirect product of two subgroups where the first one is a *normal*, Abelian, subgroup, denoted by N^n, and the second one is the *reflection* subgroup. The demonstration of this theorem can be found in Ref. [34]. More specifically, let us introduce two subgroups N^n and H^n defined as follows. Let N^n denote the *normal subgroup* of G^n defined as

$$N^n : \; Y_g(X) = Xr_g(X) \text{ with } r_g(\lambda X) = r_g(X) > 0 \tag{23}$$

with $\lambda > 0$. Here, $r_g(X)$ is a positive $\mathbb{C}^\infty(\mathbb{R}\backslash(\{0\}))$ homogeneous function.

Let H^n denote the *reflection* subgroup of G^n defined as

$$H^n : \| Y_h(\mathbf{X}) \| = \| X \| \; ; \; Y_h(-X) = -Y_h(X) \quad \text{with } h \in H^n \tag{24}$$

In Ref. [34], it is proved that

$$G^n = N^n \rtimes H^n \tag{25}$$

The irreducible representations of the group G are then related to the irreducible representations of the subgroups N^n and H^n.

2.6 *The Thermodynamic Action Principle*

Constraint 2. and assumption 3., reported in Introduction, lead to the following *thermodynamic action principle* [11]:

- *There exists a thermodynamic action I, scalar under Thermodynamic Covariant Transformations (TCT), which is stationary with respect to arbitrary variations in the transport coefficients and the affine connection.*

This action, scalar under TCT, must be constructed only by the transport coefficients, the affine connection, and their first derivatives. In addition, it must be linear in the second derivatives of the transport coefficients, and it cannot contain second or higher derivatives of the affine connection. We also require that the action is stationary when the affine connection takes the following expression [11]:

$$\Gamma^\lambda_{\mu\nu} = \widetilde{\Gamma}^\lambda_{\mu\nu} \tag{26}$$

Hence, our Lagrangian density $\mathcal{L}$ depends on three sets of dynamical variables: $\mathcal{L} = \mathcal{L}(g_{\mu\nu}, f_{\mu\nu}, \Gamma^\lambda_{\mu\nu})$. The simplest action satisfying these requirements is

$$I = \int \mathcal{L} \sqrt{g} d^n X = \int \left[B - (\Gamma^\lambda_{\mu\nu} - \widetilde{\Gamma}^\lambda_{\alpha\beta}) S^{\mu\nu}_\lambda + \widetilde{\mathcal{L}}(g_{\mu\nu}, f_{\mu\nu}) \right] \sqrt{g} d^n X \tag{27}$$

with B denoting the *scalar curvature of the thermodynamic space*[2]:

$$B = B_{\mu\nu} g^{\mu\nu} \quad ; \quad B_{\mu\nu} = \frac{\partial \Gamma^\lambda_{\mu\lambda}}{\partial X^\nu} - \frac{\partial \Gamma^\lambda_{\mu\nu}}{\partial X^\lambda} + \Gamma^\eta_{\mu\lambda}\Gamma^\lambda_{\eta\nu} - \Gamma^\eta_{\mu\nu}\Gamma^\lambda_{\eta\lambda} \tag{28}$$

and the expressions of $S^{\alpha\beta}_\lambda$ is [11]

[2] To avoid misunderstanding with the Riemannian (or Pseudo-Riemannian) geometry, we adopt the Eisenhart notations [37].

$$S_{\lambda}^{\mu\nu} = \Psi_{\lambda\alpha}^{\mu} g^{\nu\alpha} + \Psi_{\lambda\alpha}^{\nu} g^{\mu\alpha} - \frac{1}{2}\Psi_{\alpha\beta}^{\mu} g^{\alpha\beta}\delta_{\lambda}^{\nu} - \frac{1}{2}\Psi_{\alpha\beta}^{\nu} g^{\alpha\beta}\delta_{\lambda}^{\mu} \tag{29}$$

with

$$\Psi_{\lambda\kappa}^{\mu} = \frac{1}{2}\tau^{\mu\eta} g_{\lambda\kappa,\eta} - \frac{1}{2(n+1)}\tau^{\eta\alpha} g_{\lambda\eta,\alpha}\delta_{\kappa}^{\mu} - \frac{1}{2(n+1)}\tau^{\eta\alpha} g_{\kappa\eta,\alpha}\delta_{\lambda}^{\mu}$$
$$\tau^{\mu\nu} = \frac{X^{\mu}X^{\nu}}{\sigma} \quad ; \quad \sigma = g_{\mu\nu}X^{\mu}X^{\nu} \tag{30}$$

$\widetilde{\mathcal{L}}(g_{\mu\nu}, f_{\mu\nu})$ is a Lagrangian density that may depend on the transport coefficients but not on the affine connection. Note that $\tau^{\mu\nu}$ is a second-rank thermodynamic tensor. The physical meaning of the Lagrangian density stems from its (strict) connection with the curvature of the thermodynamic space. So, we require that the Lagrangian density must coincide with the scalar curvature B when the affine connection takes the expression $\widetilde{\Gamma}_{\mu\nu}^{\lambda}$. This is because the scalar B is the simplest curvature scalar, and the only one that is linear in the curvature of the space. This implies that $\widetilde{\mathcal{L}} \equiv 0$ and the final expression of the thermodynamic action reads

$$I = \int \left[B - (\Gamma_{\mu\nu}^{\lambda} - \widetilde{\Gamma}_{\mu\nu}^{\lambda})S_{\lambda}^{\mu\nu}\right]\sqrt{g}d^{n}X \tag{31}$$

2.7 The Privileged Thermodynamic Coordinate System

By definition, a *thermodynamic coordinate system is a complete set of independent thermodynamic forces*. Once a particular set of thermodynamic coordinates is chosen, the other sets of coordinates are linked to the first one through a *Thermodynamic Coordinates Transformation* (TCT). The simplest way to determine a particular set of coordinates is to quote the entropy balance equation

$$\frac{\partial \rho s}{\partial t} + \nabla \cdot \mathbf{J}_s = \sigma \tag{32}$$

Here, ρs is the local total entropy per unit volume (ρ is the mass density) and $\mathbf{J}_s$ is the entropy flux, respectively. Let us consider, as an example, a thermodynamic system confined in a rectangular box where chemical reactions, diffusion of matter, macroscopic motion of the volume element (convection), and heat current take place simultaneously. The entropy flux and the entropy production read [16, 38, 39]

$$\mathbf{J}_s = \frac{1}{T}(\mathbf{J}_q - \sum_i \mathbf{J}_i \mu_i) + \sum_i \rho_i v_i s_i$$
$$\sigma = \mathbf{J}_q \cdot \nabla \frac{1}{T} - \frac{1}{T} \sum_i \mathbf{J}_i \cdot \Big[T\nabla\Big(\frac{\mu_i}{T}\Big) - \mathbf{F}_i\Big] + \sum_i \frac{w_i A_i}{T} - \frac{1}{T}\sum_{ij} \Pi_{ij} \partial_{\mathbf{r}_i} v_j \geq 0 \tag{33}$$

where μ_i, $\rho_i s_i$, and A_i are the chemical potential, the local entropy, and the affinity of species "i", respectively. $\mathbf{J}_q$ is the heat flux; $\mathbf{J}_i$ and w_i are the diffusion flux and the chemical reaction rate of species i, respectively. Moreover, Π_{ij} are the components of the dissipative part of the pressure tensor $\mathcal{M}_{ij}$ ($\mathcal{M}_{ij} = p\delta_{ij} + \Pi_{ij}$; p is the hydrostatic pressure), $\mathbf{F}_i$ is the external force per unit mass acting on i, and v_j denotes the component of the hydrodynamic velocity [39]. The set of the thermodynamic coordinates reads

$$X^\mu = \left\{\nabla\Big(\frac{1}{T}\Big);\ -\frac{1}{T}\Big(T\nabla\Big(\frac{\mu_i}{T}\Big) - \mathbf{F}_i\Big);\ \frac{A_i}{T};\ -\frac{1}{T}\partial_{\mathbf{r}_i} v_j\right\} \tag{34}$$

For this particular example, this set may be referred to as the *privileged thermodynamic coordinates system*. Other examples of privileged thermodynamic coordinates systems, related to magnetically confined plasmas, can be found in Refs. [1, 40, 41].

3 Transport Equations

Action (31) is stationary with respect to small, and arbitrary, variations of the *dynamical variable* $g_{\mu\nu}$ and $\Gamma^\lambda_{\mu\nu}$ (we set $f_{\mu\nu} = 0$). We recall that action (31) has been constructed in such a way that it is stationary for $\Gamma^\lambda_{\mu\nu} = \widetilde{\Gamma}^\lambda_{\mu\nu}$. Indeed, by variational calculations, we get that the action is stationary with respect to small, independent, variations of $g_{\mu\nu}$ and $\Gamma^\lambda_{\mu\nu}$ if [11]

$$\begin{aligned} B_{\mu\nu} - \frac{1}{n-2}\, g_{\mu\nu} B &= -S^{\lambda\kappa}_\eta \frac{\delta \widetilde{\Gamma}^\eta_{\lambda\kappa}}{\delta g^{\mu\nu}} \equiv T_{\mu\nu} \\ \Gamma^\lambda_{\mu\nu} &= \widetilde{\Gamma}^\lambda_{\mu\nu} \end{aligned} \tag{35}$$

Equation (35) is valid for $n \neq 2$. Much less easy is to compute the explicit expression of $T_{\mu\nu}$. After (quite long) calculations, we get

$$\begin{aligned} T_{\mu\nu} &\equiv -S^{\lambda\kappa}_\eta \frac{\delta \widetilde{\Gamma}^\eta_{\lambda\kappa}}{\delta g^{\mu\nu}} = \\ &\frac{1}{2} g^{-1/2} (S^{\lambda\kappa}_\alpha g^{\alpha\beta} g^{1/2})_{,\beta}\, g_{\lambda\mu} g_{\kappa\nu} - \frac{1}{2} g^{-1/2}\Big(g^{1/2}(S^{\kappa\beta}_\mu g_{\kappa\nu} + S^{\kappa\beta}_\nu g_{\kappa\mu})\Big)_{,\beta} \\ &- \frac{1}{4} S^{\eta h}_\mu (g_{\nu\eta,h} + g_{\nu h,\eta} - g_{\eta h,\nu}) - \frac{1}{4} S^{\eta h}_\nu (g_{\mu\eta,h} + g_{\mu h,\eta} - g_{\eta h,\mu}) \end{aligned} \tag{36}$$

$$
-\frac{1}{2}g^{-1/2}\left(g^{1/2}\tau^{\eta\beta}S_{\eta}^{\lambda\kappa}-\frac{g^{1/2}}{n+1}\left(\tau^{\lambda\beta}S_{\eta}^{\eta\kappa}+\tau^{\kappa\beta}S_{\eta}^{\eta\lambda}\right)\right)_{,\beta}g_{\lambda\mu}g_{\kappa\nu}
$$
$$
-\frac{1}{2}\tau^{\alpha\beta}\tau^{\lambda\kappa}S_{\alpha}^{\eta\iota}g_{\eta\iota,\beta}g_{\lambda\mu}g_{\kappa\nu}+\frac{1}{n+1}\tau^{\alpha\beta}\tau^{\lambda\kappa}S_{\eta}^{\eta\iota}g_{\alpha\iota,\beta}g_{\lambda\mu}g_{\kappa\nu} \tag{37}
$$

After a little algebra, we find

$$
-\frac{1}{2}g^{-1/2}\left(g^{1/2}\tau^{\eta\beta}S_{\eta}^{\lambda\kappa}-\frac{g^{1/2}}{n+1}\left(\tau^{\lambda\beta}S_{\eta}^{\eta\kappa}+\tau^{\kappa\beta}S_{\eta}^{\eta\lambda}\right)\right)_{,\beta}g_{\lambda\mu}g_{\kappa\nu}=
$$
$$
-\frac{1}{2}\left(\Psi_{\eta\alpha}^{\lambda}g_{,\beta}^{\kappa\alpha}+\Psi_{\eta\alpha}^{\kappa}g_{,\beta}^{\lambda\alpha}\right)\tau^{\eta\beta}g_{\lambda\mu}g_{\kappa\nu}
$$
$$
+\frac{1}{2(n+1)}g^{-1/2}\left(g^{1/2}\left(\Psi_{\eta\alpha}^{\lambda}g^{\eta\alpha}\tau^{\kappa\beta}+\Psi_{\eta\alpha}^{\kappa}g^{\eta\alpha}\tau^{\lambda\beta}\right)\right)_{,\beta}g_{\lambda\mu}g_{\kappa\nu}
$$
$$
-\frac{1}{2}g^{-1/2}g_{\kappa\mu}\left(g^{1/2}\Psi_{\eta\nu}^{\kappa}\tau^{\eta\beta}\right)_{,\beta}-\frac{1}{2}g^{-1/2}g_{\kappa\nu}\left(g^{1/2}\Psi_{\eta\mu}^{\kappa}\tau^{\eta\beta}\right)_{,\beta} \tag{38}
$$

where we note that the trace of the last expression of Eq. (38) vanishes. We also have

$$
-\frac{1}{2}\tau^{\alpha\beta}\tau^{\lambda\kappa}S_{\alpha}^{\eta\iota}g_{\eta\iota,\beta}g_{\lambda\mu}g_{\kappa\nu}+\frac{1}{n+1}\tau^{\alpha\beta}\tau^{\lambda\kappa}S_{\eta}^{\eta\iota}g_{\alpha\iota,\beta}g_{\lambda\mu}g_{\kappa\nu}=
$$
$$
-\Psi_{\alpha\gamma}^{\eta}g^{\iota\gamma}g_{\eta\iota,\beta}\tau^{\alpha\beta}\tau^{\lambda\kappa}g_{\lambda\mu}g_{\kappa\nu}+\frac{1}{n+1}\Psi_{\eta\gamma}^{\iota}g^{\eta\gamma}g_{\alpha\iota,\beta}\tau^{\alpha\beta}\tau^{\lambda\kappa}g_{\lambda\mu}g_{\kappa\nu} \tag{39}
$$

Hence, tensor $T_{\mu\nu}$ can be brought into the form

$$
T_{\mu\nu}=\frac{1}{2}g^{-1/2}(S_{\alpha}^{\lambda\kappa}g^{\alpha\beta}g^{1/2})_{,\beta}g_{\lambda\mu}g_{\kappa\nu}-\frac{1}{2}g^{-1/2}\Big(g^{1/2}(S_{\mu}^{\kappa\beta}g_{\kappa\nu}+S_{\nu}^{\kappa\beta}g_{\kappa\mu})\Big)_{,\beta}
$$
$$
-\frac{1}{4}S_{\mu}^{\eta h}(g_{\nu\eta,h}+g_{\nu h,\eta}-g_{\eta h,\nu})-\frac{1}{4}S_{\nu}^{\eta h}(g_{\mu\eta,h}+g_{\mu h,\eta}-g_{\eta h,\mu})
$$
$$
-\frac{1}{2}g^{-1/2}\left(g^{1/2}\tau^{\eta\beta}S_{\eta}^{\lambda\kappa}-\frac{g^{1/2}}{n+1}\left(\tau^{\lambda\beta}S_{\eta}^{\eta\kappa}+\tau^{\kappa\beta}S_{\eta}^{\eta\lambda}\right)\right)_{,\beta}g_{\lambda\mu}g_{\kappa\nu}
$$
$$
-\Psi_{\alpha\gamma}^{\eta}g^{\iota\gamma}g_{\eta\iota,\beta}\tau^{\alpha\beta}\tau^{\lambda\kappa}g_{\lambda\mu}g_{\kappa\nu}+\frac{1}{n+1}\Psi_{\eta\gamma}^{\iota}g^{\eta\gamma}g_{\alpha\iota,\beta}\tau^{\alpha\beta}\tau^{\lambda\kappa}g_{\lambda\mu}g_{\kappa\nu} \tag{40}
$$

and the trace of tensor $T_{\mu\nu}$ reads

$$
T\equiv T_{\mu\nu}g^{\mu\nu}=\frac{n-2}{2}g^{-1/2}(g^{1/2}\Psi_{\lambda\kappa}^{\beta}g^{\lambda\kappa})_{,\beta}+\frac{1}{n+1}g_{\eta\alpha}g^{-1/2}(g^{1/2}\Psi_{\lambda\kappa}^{\alpha}g^{\lambda\kappa}\tau^{\eta\beta})_{,\beta}
$$
$$
-\Psi_{\alpha\gamma}^{\eta}\tau^{\alpha\beta}\tau^{\lambda\kappa}g^{\iota\gamma}g_{\lambda\kappa}g_{\eta\iota,\beta}+\frac{1}{n+1}\Psi_{\eta\gamma}^{\iota}g^{\eta\gamma}\tau^{\alpha\beta}\tau^{\lambda\kappa}g_{\lambda\kappa}g_{\alpha\iota,\beta}
$$
$$
+2S_{\eta}^{\kappa\beta}g_{,\beta}^{\eta\lambda}g_{\lambda\kappa}-\Psi_{\alpha\eta}^{\kappa}\tau^{\alpha\beta}g_{\lambda\kappa}g_{,\beta}^{\lambda\eta} \tag{41}
$$

Finally, *in absence of the skew-symmetric part, we get the differential equations for the transport coefficients valid for n > 2*:

$$B_{\mu\nu} = T_{\mu\nu} - \frac{1}{n-2}\, g_{\mu\nu} T = W^{(S)}_{\mu\nu} \tag{42}$$

with $T_{\mu\nu}$ and T given by Eq. (40) and Eq. (41), respectively.

• Property of Tensor $W^{(S)}_{\mu\nu}$

Tensor $B_{\mu\nu}$ satisfies the Bianchi identity for symmetric connection which, written in the linearised form, reads

$$L^{\mu\lambda}\frac{\partial B_{\lambda\nu}}{\partial X^{\mu}} - \frac{1}{2} L^{\mu\lambda}\frac{\partial B_{\mu\lambda}}{\partial X^{\nu}} \equiv 0 \tag{43}$$

The validity of identity (43) may also be checked by direct inspection. Hence, also tensor $W^{(S)}_{\mu\nu}(h)$ satisfies the same identity

$$L^{\mu\lambda}\frac{\partial W^{(S)}_{\lambda\nu}}{\partial X^{\mu}} - \frac{1}{2} L^{\mu\lambda}\frac{\partial W^{(S)}_{\mu\lambda}}{\partial X^{\nu}} \equiv 0 \tag{44}$$

• Observations

By direct inspection, we may check the validity of the following important identities:

$$\Psi^{\lambda}_{\lambda\kappa} = \Psi^{\lambda}_{\kappa\lambda} = 0 \quad ; \quad S^{\lambda\mu}_{\lambda} = -\frac{(n-1)}{2}\Psi^{\mu}_{\lambda\kappa} g^{\lambda\kappa} \tag{45}$$

3.1 *Onsager's Region*

The transport coefficients tend to Onsager's matrix as the thermodynamic system approaches thermodynamic equilibrium. The thermodynamic region where the thermodynamic forces are linearly connected to the conjugate thermodynamic fluxes is referred to as the *linear region of thermodynamics* or *Onsager's region* [14, 15]. Hence, as the thermodynamic forces go to zero, the metric $g_{\mu\nu}$ tends to Onsager matrix $L_{\mu\nu}$ (or, equivalently, the perturbation $h_{\mu\nu}$ of the metric tensor tends to zero):

$$\lim_{X^{\lambda}\to 0} g_{\mu\nu} = L_{\mu\nu} \tag{46}$$

Condition (46) is referred to as *Onsager's condition.*

3.2 *Near the Onsager Region*

Let us compute the first nonlinear contributions of Eq. (42). In this case, since $\Psi^{\lambda}_{\mu\nu}$ is already of the first order in $h_{\mu\nu}$, to obtain equations valid up to the third order, we should develop quantities $g^{\mu\nu}$, g, σ^2, $\tau^{\lambda\kappa}$, etc. up to the second order. Hence, by setting

$$g_{\mu\nu} \simeq L_{\mu\nu} + h_{\mu\nu} \quad ; \quad \sigma \simeq \sigma_{(L)}(1 + h_{\mu\nu}\tau^{\mu\nu}_{(L)}) \tag{47}$$

with $\sigma_{(L)} \equiv L_{\mu\nu}X^{\mu}X^{\nu}$ and $\tau^{\mu\nu}_{(L)} \equiv X^{\mu}X^{\nu}/\sigma_{(L)}$, we get

$$\begin{aligned}
g^{\mu\nu} &\simeq L^{\mu\nu} - h^{\mu\nu} + h^{\mu\lambda}h^{\nu\kappa}L_{\lambda\kappa} + O(h^3)\\
\tau^{\mu\nu} &\simeq \tau^{\mu\nu}_{(L)}\Big(1 - h_{\lambda\kappa}\tau^{\lambda\kappa}_{(L)} + (h_{\lambda\kappa}\tau^{\lambda\kappa}_{(L)})^2\Big) + O(h^3)\\
\sigma^2 &\simeq \sigma_{(L)}\Big(1 + 2h_{\mu\nu}\tau^{\mu\nu}_{(L)} + (h_{\mu\nu}\tau^{\mu\nu}_{(L)})^2\Big) + O(h^3)\\
\frac{1}{\sigma} &\simeq \frac{1}{\sigma_{(L)}}\Big(1 - h_{\lambda\kappa}\tau^{\lambda\kappa}_{(L)} + (h_{\lambda\kappa}\tau^{\lambda\kappa}_{(L)})^2\Big) + O(h^3)\\
\frac{1}{\sigma^2} &\simeq \frac{1}{\sigma^2_{(L)}}\Big(1 - 2h_{\lambda\kappa}\tau^{\lambda\kappa}_{(L)} + 3(h_{\lambda\kappa}\tau^{\lambda\kappa}_{(L)})^2\Big) + O(h^3)\\
g &\simeq L\Big(1 + L^{\lambda\kappa}h_{\lambda\kappa} + \frac{1}{2}\big((L^{\lambda\kappa}h_{\lambda\kappa})^2 - L^{\lambda\alpha}L^{\kappa\beta}h_{\alpha\kappa}h_{\beta\lambda}\big)\Big) + O(h^3)\\
g^{1/2} &\simeq L^{1/2}\Big(1 + \frac{1}{2}L^{\lambda\kappa}h_{\lambda\kappa} + \frac{1}{8}(L^{\lambda\kappa}h_{\lambda\kappa})^2 - \frac{1}{4}L^{\lambda\alpha}L^{\kappa\beta}h_{\alpha\kappa}h_{\beta\lambda}\Big) + O(h^3)\\
g^{-1/2} &\simeq L^{-1/2}\Big(1 - \frac{1}{2}L^{\lambda\kappa}h_{\lambda\kappa} + \frac{1}{8}(L^{\lambda\kappa}h_{\lambda\kappa})^2 + \frac{1}{4}L^{\lambda\alpha}L^{\kappa\beta}h_{\alpha\kappa}h_{\beta\lambda}\Big) + O(h^3)
\end{aligned} \tag{48}$$

with L denoting the determinant of Onsager's matrix.

4 Two-Dimensional Transport Equations

In two dimensions, the curvature tensor $B_{\mu\nu\lambda\kappa}$ has only one component, since all nonzero components may be obtained from B_{0101}. Equivalently, the curvature tensor may be written in terms of the scalar B

$$B_{\lambda\mu\kappa\nu} = \frac{1}{2}B\left(g_{\lambda\kappa}g_{\mu\nu} - g_{\lambda\nu}g_{\mu\kappa}\right) \tag{49}$$

So, B alone completely characterises the local geometry. From Eq. (49), we find the expressions for $B_{\mu\nu} \equiv B_{\lambda\mu\kappa\nu}g^{\lambda\kappa}$ and $B \equiv B_{\lambda\mu\kappa\nu}g^{\lambda\kappa}g^{\mu\nu}$. We get

$$B_{\mu\nu} - \frac{1}{2}Bg_{\mu\nu} \equiv 0 \tag{50}$$

Hence, Eq. (42) is meaningless in two dimensions (see also Refs. [22, 23]). It is easy to convince ourselves that, in analogy with the works for $1+1$ gravity [19, 24], also in our case the only non-trivial version of Eq. (42) for $n=2$ has to read

$$B = -T \tag{51}$$

with

$$\begin{aligned}
T = \quad & 2S_{\eta}^{\kappa\beta} g^{\eta\lambda}_{,\beta} g_{\lambda\kappa} \\
& - \Psi^{\kappa}_{\alpha\eta} \tau^{\alpha\beta} g_{\lambda\kappa} g^{\lambda\eta}_{,\beta} - \Psi^{\eta}_{\alpha\gamma} \tau^{\alpha\beta} \tau^{\lambda\kappa} g^{\iota\gamma} g_{\lambda\kappa} g_{\eta\iota,\beta} \\
& + \frac{1}{3} g_{\eta\alpha} g^{-1/2} (g^{1/2} \Psi^{\alpha}_{\lambda\kappa} g^{\lambda\kappa} \tau^{\eta\beta})_{,\beta} \\
& + \frac{1}{3} \Psi^{\iota}_{\eta\gamma} g^{\eta\gamma} \tau^{\alpha\beta} \tau^{\lambda\kappa} g_{\lambda\kappa} g_{\alpha\iota,\beta} \\
\Gamma^{\mu}_{\alpha\beta} = \quad & \begin{Bmatrix} \mu \\ \alpha\beta \end{Bmatrix} + \frac{1}{2\sigma} X^{\mu} X^{\eta} g_{\alpha\beta,\eta} \\
& - \frac{1}{6\sigma} \left(\delta^{\mu}_{\alpha} X^{\nu} X^{\eta} g_{\beta\nu,\eta} + \delta^{\mu}_{\beta} X^{\nu} X^{\eta} g_{\alpha\nu,\eta} \right)
\end{aligned} \tag{52}$$

It is useful to recall the well-known result from differential geometry; *all two-dimensional manifolds are conformally flat*. Hence, the transport coefficients out of Onsager's region can always be brought into the form

$$g_{\mu\nu} = L_{\mu\nu} \exp \phi(X) \tag{53}$$

with ϕ denoting a scalar field depending on the thermodynamic forces. By plugging Eq. (53) into Eqs. (51) and (52), we get the PDE which has to be solved for the conformal field ϕ. In this case, the Onsager condition requires $\phi(0) = 0$.

Concerning the action, we adopt the expression proposed in literature [20]. This action reads

$$I = \int \mathcal{N}(B+T)\sqrt{g}\, d^2X \tag{54}$$

where $\mathcal{N}$ is an auxiliary scalar field (analogous to the *dilaton field* [21]), which plays the role of a Lagrangian multiplier. Notice that in this formalism, the dynamical fields present in the action (54) are the dilatation field and the transport coefficients. In this case, the affine connection does not play the role of an independent field (it is a dynamical variable only when $n > 2$) and it intervenes in the dynamics through the second expression of Eq. (52). By varying this action with respect to $\mathcal{N}$ we get Eq. (51), while variation with respect to the transport coefficients yields the PDE for $\mathcal{N}$. The PDE for the transport coefficients is decoupled from that for the dilaton field. However, as we will see in the next work, this will not be the case when the skew-symmetric part of the transport coefficients is different from zero.

Let us now determine the nonlinear partial differential equation satisfied by a conformal factor Λ of the metric $g_{\mu\nu}$. A conformal manifold is a manifold equipped with an equivalence class of metric tensors, in which two metrics $g_{\mu\nu}$ and $\tilde{g}_{\mu\nu}$ are equivalent if and only if

$$g_{\mu\nu} = \Lambda(X)\tilde{g}_{\mu\nu} \tag{55}$$

where $\Lambda(x)$ is a real-valued smooth function defined on the manifold referred to as *conformal factor*. An equivalence class of such metrics is known as a conformal metric or conformal class. Thus, a conformal metric may be regarded as a metric that is only defined *up to scale*. A conformal metric is conformally flat if there is a metric representing it that is flat, i.e.,

$$g_{\mu\nu} = \Lambda(X)L_{\mu\nu} \tag{56}$$

Often conformal metrics are treated by selecting a metric in the conformal class, and applying only *conformally invariant* constructions to the chosen metric. From Eq. (56), we get

$$\begin{aligned} &g^{\mu\nu} = \frac{1}{\Lambda}L^{\mu\nu};\ g = \Lambda^n L;\ L_{\mu\nu}X^\nu = \Lambda^{-1}X_\mu;\ L^{\mu\nu}X_\nu = \Lambda x^\mu \\ &\Gamma^\lambda_{\mu\nu} = \frac{1}{2\Lambda}\left(\Lambda_{,\nu}\delta^\lambda_\mu + \Lambda_{,\mu}\delta^\lambda_\nu - \Lambda_{,\kappa}L^{\kappa\lambda}_{\mu\nu}\right) + \frac{X^\lambda}{2\sigma}L_{\mu\nu}X^\kappa\Lambda_{,\kappa} \\ &\quad - \frac{\Lambda_{,\kappa}X^\kappa}{2(n+1)\Lambda\sigma}\left(X_\nu\delta^\lambda_\mu + X_\mu\delta^\lambda_\nu\right) \end{aligned} \tag{57}$$

For the two-dimensional case, we have

$$\begin{aligned} T &= \frac{2}{9\Lambda\sigma}\Lambda_{,\lambda}X^\lambda + \frac{2}{9\Lambda\sigma}\Lambda_{,\lambda,\kappa}X^\lambda X^\kappa + \frac{4}{9\Lambda^2\sigma}\Lambda_{,\lambda}\Lambda_{,\kappa}X^\lambda X^\kappa \\ B &= \frac{\Lambda_{,\lambda,\kappa}L^{\lambda\kappa}}{\Lambda^2} - \frac{\Lambda_{,\lambda}\Lambda_{,\kappa}L^{\lambda\kappa}}{\Lambda^3} - \frac{2\Lambda_{,\lambda}X^\lambda}{3\Lambda\sigma} - \frac{2\Lambda_{,\lambda,\kappa}X^\lambda X^\kappa}{3\Lambda\sigma} \\ &\quad + \frac{5\Lambda_{,\lambda}\Lambda_{,\kappa}X^\lambda X^\kappa}{9\Lambda^2\sigma} \end{aligned} \tag{58}$$

where Eqs. (28), (41), and (57) have been taken into account. From Eq. (51), we get

$$\begin{aligned} B + T = \sigma_{(L)}&\left(\Lambda\Lambda_{,\mu,\nu} - \Lambda_{,\mu}\Lambda_{,\nu}\right)L^{\mu\nu} - \\ &\left(\frac{4}{9}\Lambda\Lambda_{,\mu,\nu} - \Lambda_{,\mu}\Lambda_{,\nu}\right)X^\mu X^\nu - \frac{4}{9}\Lambda\Lambda_{,\mu}X^\mu = 0 \end{aligned} \tag{59}$$

Now, by setting $\Lambda(X) = \exp(\phi(X))$, Eq. (59) reads [36]

$$L^{\mu\nu}\frac{\partial^2\phi}{\partial X^\mu \partial X^\nu} - \frac{4}{9\sigma_{(L)}}X^\mu X^\nu \frac{\partial^2\phi}{\partial X^\mu \partial X^\nu} - \frac{4}{9\sigma_{(L)}}X^\mu \frac{\partial\phi}{\partial X^\mu} + \frac{5}{9\sigma_{(L)}}\left(X^\mu \frac{\partial\phi}{\partial X^\mu}\right)^2 = 0 \tag{60}$$

Equation (60) has to be solved for the conformal factor ϕ. By introducing the differential operators, invariant under TCT [11],

$$\mathcal{O} \equiv X^\mu \frac{\partial}{\partial X^\mu}, \quad \Box^2 \equiv L^{\mu\nu}\frac{\partial^2}{\partial X^\mu \partial X^\nu} \tag{61}$$

Equation (60) can be cast into a manifestly TCT-covariant form

$$\left(9\sigma_{(L)}\Box^2 - 4\mathcal{O}^2\right)\phi + 5\left(\mathcal{O}\phi\right)^2 = 0 \tag{62}$$

Let us now perform the following coordinate transformation[3]:

$$X'^\lambda = A^\lambda_\kappa X^\kappa, \quad \text{with } A^\mu_\nu \text{ such that } A^\alpha_\lambda L^{\lambda\kappa} A^\beta_\kappa = \mathrm{I}^{\alpha\beta} \tag{63}$$

with $\mathrm{I}^{\alpha\beta}$ denoting the Identity matrix. Notice that, since the matrix $L_{\mu\nu}$ is a positive-definite matrix, there exists always a matrix A^μ_ν, which satisfies condition (63). Finally, we get

$$\left(9\sigma'_{(L)}\Box'^2 - 4\mathcal{O}'^2\right)\phi' + 5\left(\mathcal{O}'\phi'\right)^2 = 0 \quad \text{with}$$
$$\sigma'_{(L)} = X'^{1^2} + X'^{2^2}, \quad \Box'^2 \equiv \frac{\partial^2}{\partial X'^{1^2}} + \frac{\partial^2}{\partial X'^{2^2}} \tag{64}$$

Equation (62) (or, equivalently, Eq. (64)) is, in Thermodynamical Field Theory (TFT), analogous to Liouville's equation in Riemannian (or pseudo-Riemannian) geometry [36].

5 Linearised Transport Equations

When the transport coefficients are close to Onsager's matrix, we may set

$$g_{\mu\nu} \simeq L_{\mu\nu} + h_{\mu\nu} \tag{65}$$

with $h_{\mu\nu}$ considered as a small perturbation of the transport matrix coefficients. We also introduce a small parameter ε of the order of σ^{-1} (considered as a small quantity).

[3] Note that linear transformations of coordinates are allowed because this class of transformations belongs to the TCT-group [11].

The linearised Transport Equations are obtained by discarding systematically in the following calculations:
(i) All the terms of order $hh,\ hhh,\ldots$;
(ii) All the terms of order ε^2, of order σ^{-2} or of higher order.

5.1 *Linearised Transport Equations for n > 2*

At the dominant order in $h_{\mu\nu}$, we get for $n > 2$

$$B_{\mu\nu}=T_{\mu\nu}-\frac{1}{2}L_{\mu\nu}\left(\Psi^{\beta}_{\lambda\kappa}L^{\lambda\kappa}+\frac{2}{(n+1)(n-2)}\Psi^{\lambda}_{\mu\nu}L^{\mu\nu}L_{\lambda\kappa}\tau^{\kappa\beta}_{(L)}\right)_{,\beta}=W^{(S)}_{\mu\nu} \tag{66}$$

where

$$B_{\mu\nu}\simeq\frac{\partial\Gamma^{\lambda}_{\lambda\mu}}{\partial X^{\nu}}-\frac{\partial\Gamma^{\lambda}_{\mu\nu}}{\partial X^{\lambda}}\qquad\text{hence}$$

$$\begin{aligned}B_{\mu\nu}\simeq\frac{1}{2}\Big(&L^{\lambda\kappa}\frac{\partial^2 h_{\mu\nu}}{\partial X^{\lambda}X^{\kappa}}+L^{\lambda\kappa}\frac{\partial^2 h_{\lambda\kappa}}{\partial X^{\mu}X^{\nu}}-L^{\lambda\kappa}\frac{\partial^2 h_{\kappa\mu}}{\partial X^{\lambda}X^{\nu}}-L^{\lambda\kappa}\frac{\partial^2 h_{\kappa\nu}}{\partial X^{\lambda}X^{\mu}}\\&-\frac{\partial}{\partial X^{\lambda}}\big(\tau^{\lambda\kappa}\frac{\partial h_{\mu\nu}}{\partial X^{\kappa}}\big)+\frac{1}{n+1}\frac{\partial}{\partial X^{\mu}}\big(\tau^{\lambda\kappa}\frac{\partial h_{\nu\kappa}}{\partial X^{\lambda}}\big)+\frac{1}{n+1}\frac{\partial}{\partial X^{\nu}}\big(\tau^{\lambda\kappa}\frac{\partial h_{\mu\kappa}}{\partial X^{\lambda}}\big)\Big)\end{aligned} \tag{67}$$

or

$$\begin{aligned}B_{\mu\nu}&=B^{(0)}_{\mu\nu}+B^{(1)}_{\mu\nu}\qquad\text{with}\\B^{(0)}_{\mu\nu}&=\frac{1}{2}\Big(L^{\lambda\kappa}\frac{\partial^2 h_{\mu\nu}}{\partial X^{\lambda}X^{\kappa}}+L^{\lambda\kappa}\frac{\partial^2 h_{\lambda\kappa}}{\partial X^{\mu}X^{\nu}}-L^{\lambda\kappa}\frac{\partial^2 h_{\kappa\mu}}{\partial X^{\lambda}X^{\nu}}-L^{\lambda\kappa}\frac{\partial^2 h_{\kappa\nu}}{\partial X^{\lambda}X^{\mu}}\Big)\\B^{(1)}_{\mu\nu}&=\frac{1}{2}\Big(-\frac{\partial}{\partial X^{\lambda}}\big(\tau^{\lambda\kappa}\frac{\partial h_{\mu\nu}}{\partial X^{\kappa}}\big)+\frac{1}{n+1}\frac{\partial}{\partial X^{\mu}}\big(\tau^{\lambda\kappa}\frac{\partial h_{\nu\kappa}}{\partial X^{\lambda}}\big)+\frac{1}{n+1}\frac{\partial}{\partial X^{\nu}}\big(\tau^{\lambda\kappa}\frac{\partial h_{\mu\kappa}}{\partial X^{\lambda}}\big)\Big)\end{aligned} \tag{68}$$

We also have for $n > 2$

$$\begin{aligned}T_{\mu\nu}=&-\Psi^{\beta}_{\mu\nu,\beta}+\frac{1}{2}\Psi^{\beta}_{\lambda\kappa,\beta}L^{\lambda\kappa}L_{\mu\nu}-\frac{1}{2}\Big((\Psi^{\lambda}_{\kappa\nu}L_{\lambda\mu}+\Psi^{\lambda}_{\kappa\mu}L_{\lambda\nu})\tau^{\kappa\beta}_{(L)}\Big)_{,\beta}\\&+\frac{1}{2(n+1)}\Big(\Psi^{\lambda}_{\eta\alpha}\tau^{\kappa\beta}_{(L)}+\Psi^{\kappa}_{\eta\alpha}\tau^{\lambda\beta}_{(L)}\Big)_{,\beta}L^{\eta\alpha}L_{\lambda\mu}L_{\kappa\nu}\\\Psi^{\mu}_{\lambda\kappa}=&\ \frac{1}{2}\tau^{\mu\eta}_{(L)}h_{\lambda\kappa,\eta}-\frac{1}{2(n+1)}\tau^{\eta\alpha}_{(L)}h_{\lambda\eta,\alpha}\delta^{\mu}_{\kappa}-\frac{1}{2(n+1)}\tau^{\eta\alpha}_{(L)}h_{\kappa\eta,\alpha}\delta^{\mu}_{\lambda}\\T=&\ T_{\mu\nu}L^{\mu\nu}=\Big(\frac{n-2}{2}\Psi^{\beta}_{\mu\nu}L^{\mu\nu}+\frac{1}{n+1}\Psi^{\lambda}_{\mu\nu}L^{\mu\nu}L_{\lambda\kappa}\tau^{\kappa\beta}_{(L)}\Big)_{,\beta}\\B=&-\Big(\Psi^{\beta}_{\mu\nu}L^{\mu\nu}+\frac{2}{(n+1)(n-2)}\Psi^{\lambda}_{\mu\nu}L^{\mu\nu}L_{\lambda\kappa}\tau^{\kappa\beta}_{(L)}\Big)_{,\beta}\end{aligned} \tag{69}$$

with

$$\tau_{(L)}^{\mu\nu} = \frac{X^\mu X^\nu}{\sigma_{(L)}} \quad ; \quad \sigma_{(L)} = L_{\mu\nu} X^\mu X^\nu \tag{70}$$

After a little algebra, we find that identity (44) (or Eq. (43)) implies

$$\frac{\partial T^{\mu\nu}}{\partial X^\nu} = 0 \tag{71}$$

This *conservation law* is consistent with the fact that our PDE (and the Lagrangian) are invariant under the TCT. Hence, for the Noether theorem, this invariance is associated to a *conserved current* and, so, a conserved *source tensor* [34].

As mentioned above, one way to get the approximate solution of Eq. (66) is to introduce a parameter ε of the order of σ^{-1}, which we consider to be a small quantity. By setting

$$h_{\mu\nu} \simeq h_{\mu\nu}^{(0)} + \varepsilon h_{\mu\nu}^{(1)} \quad \text{with} \quad \varepsilon \sim O(\sigma^{-1}) \tag{72}$$

the linearised Transport Equations for $n > 2$ read

$$\begin{aligned}
&B_{\mu\nu}^{(0)}(h^{(0)}) = 0 \\
&B_{\mu\nu}^{(0)}(\epsilon h^{(1)}) = T_{\mu\nu}(h^{(0)}) - B_{\mu\nu}^{(1)}(h^{(0)}) - \frac{1}{2} L_{\mu\nu} \left((\Psi_{\lambda\kappa}^{\beta}(h^{(0)}) L^{\lambda\kappa} \right)_{,\beta} \\
&\quad - \frac{1}{(n+1)(n-2)} L_{\mu\nu} \left(\Psi_{\lambda\kappa}^{\alpha}(h^{(0)}) L^{\lambda\kappa} L_{\alpha\eta} \tau_{(L)}^{\eta\beta} \right)_{,\beta} \equiv W_{\mu\nu}^{(S)}(h^{(0)})
\end{aligned} \tag{73}$$

Note that $W_{\mu\nu}^{(S)}(h^{(0)}) \to 0$ as $h^{(0)} \to 0$.

5.2 *Examples of Simplification of the Linearised Transport Equations*

It is worth mentioning that in several cases, the second PDE of system (73) simplifies significantly. Indeed, we have already noticed that tensor $W_{\mu\nu}^{(S)}$ satisfies the identity

$$L^{\mu\lambda} \frac{\partial W_{\lambda\nu}^{(S)}}{\partial X^\mu} - \frac{1}{2} L^{\mu\lambda} \frac{\partial W_{\mu\lambda}^{(S)}}{\partial X^\nu} \equiv 0 \tag{74}$$

Now, let us suppose to have solved the following Poisson PDE:

$$L^{\lambda\kappa} \frac{\partial^2 h_{\mu\nu}^{(1)}}{\partial X^\lambda \partial X^\kappa} = W_{\mu\nu}^{(S)} \qquad \text{with} \quad h_{\mu\nu}^{(1)} \mid_{\partial\Omega} = 0 \tag{75}$$

with $\partial\Omega$ denoting the boundary.[4] From Eqs. (74) and (75), we get

$$L^{\lambda\kappa}\frac{\partial^2\mathcal{R}_\nu(x)}{\partial X^\lambda\partial X^\kappa}=0\quad \text{with}\quad \mathcal{R}_\nu(x)\equiv\frac{1}{2}L^{\mu\lambda}\frac{\partial h_{\mu\lambda}}{\partial X^\nu}-L^{\mu\lambda}\frac{\partial h_{\lambda\nu}}{\partial X^\mu} \tag{76}$$

Hence, if it happens, for example, that $\mathcal{R}_\nu(x)\mid_{\partial\Omega}=0+\mathcal{O}(\varepsilon^2)$ we also have $\mathcal{R}_\nu(x)=0+\mathcal{O}(\varepsilon^2)$ throughout the space. In other words, if it happens, for example, that the derivative of the perturbation $h^{(1)}_{\mu\nu}$ vanishes at the boundary

$$\frac{\partial h^{(1)}_{\mu\nu}}{\partial X^\lambda}\Big|_{\partial\Omega}=0 \tag{77}$$

the second PDE of system (73) reduces to a Poisson PDE and the Transport Equations to be solved reduce to

$$\begin{aligned}&B^{(0)}_{\mu\nu}(h^{(0)})=0\\&L^{\lambda\kappa}\frac{\partial^2\varepsilon h^{(1)}_{\mu\nu}(X)}{\partial X^\lambda\partial X^\kappa}=W^{(S)}_{\mu\nu}(X)\end{aligned} \tag{78}$$

There is another important case where the second PDE of system (73) reduces to a Poisson PDE. This happens when the perturbation takes the form

$$h_{\mu\nu}(X)=L_{\mu\nu}h(X) \tag{79}$$

with $h(X)$ indicating a scalar field. As we shall see in Sect. 7, this is exactly what happens for the two-dimensional case (see Eq. (53)). PDEs (78) should be solved with the boundary conditions specified in the Annex.

5.3 Linearised Transport Equation for n = 2

As seen in Sect. 4, for $n=2$ the PDE to be solved is

$$B=-T \tag{80}$$

Hence, the linearised Transport Equation reads

$$\begin{aligned}&B=0\qquad \text{for the homogeneous case}\\&B=-T_{(L)}\quad \text{for the inhomogeneous case}\end{aligned} \tag{81}$$

[4] Note that the boundary conditions have already been satisfied at zero order, and for this reason, $\bar{h}^{(1)}_{\mu\nu}$ should vanish at the boundary.

with

$$T_{(L)} = \frac{1}{3} L_{\eta\alpha} L^{\lambda\kappa} \left(\Psi^{\alpha}_{\lambda\kappa} \tau^{\eta\beta}_{(L)} \right)_{,\beta} \tag{82}$$

We have already mentioned that for $n = 2$, the solution of Eq. (81) can always be brought into the form (see Eq. (53) in Sect. 4)

$$g_{\mu\nu} = L_{\mu\nu} f(x) \tag{83}$$

By setting

$$h_{\mu\nu} \simeq h^{(0)}_{\mu\nu} + \varepsilon h^{(1)}_{\mu\nu} \quad \text{with} \quad \varepsilon \sim O(\sigma^{-1}) \tag{84}$$

we get the linearised Transport Equations for $n = 2$ as

$$\begin{aligned} &B^{(0)}(h^{(0)}) = B^{(0)}_{\mu\nu} L^{\mu\nu}(h^{(0)}) = 0 \\ &B^{(0)}(\epsilon h^{(1)}) = B^{(0)}_{\mu\nu} L^{\mu\nu}(\epsilon h^{(1)}) = -B^{(1)}_{\mu\nu} L^{\mu\nu}(h^{(0)}) - T_{(L)}(h^{(0)}) = W^{(S)}(x) \end{aligned} \tag{85}$$

The analytic solution of system (85) (or system (81)) can be found in Sect. 7.

6 TFT Gauge Invariance

In field theories, different configurations of the unobservable fields can result in identical observable quantities. A transformation from one such field configuration to another is called a *gauge transformation*; the lack of change in the measurable quantities, despite the field being transformed, is a property called *gauge invariance*. In this section, we shall clarify the physical meaning of the gauge invariance in the Thermodynamical Field Theory. To carry out this task, we need first to recall some fundamental theorems concerning the solution of the differential equations (73). After this, in Sect. 6.2 we provide the physical interpretation of the gauge invariance in the TFT.

6.1 *Basic Theorems for the PDEs* $B^{(0)}_{\mu\nu}(h) = W^{(S)}_{\mu\nu}$

Let us consider the PDE $B^{(0)}_{\mu\nu}(h) = W^{(S)}_{\mu\nu}$ where the source $W^{(S)}_{\mu\nu}$ may be either different from zero or absent. By direct inspection, we find that if $h_{\mu\nu}(X)$ is a solution of $B^{(0)}_{\mu\nu}(h) = W^{(S)}_{\mu\nu}$, then so will be

$$\widehat{h}_{\mu\nu}(X) = h_{\mu\nu}(X) + \frac{\partial u_{\nu}(X)}{\partial X^{\mu}} + \frac{\partial u_{\mu}(X)}{\partial X^{\nu}} \tag{86}$$

where $u_\mu(X)$ are n small but otherwise arbitrary functions of X^μ. Hence, tensor $B^{(0)}_{\mu\nu}(h)$ is unaffected by gauge transformations (86). Thanks to this gauge invariance, we have the following theorem [42].

Theorem 1 *If one knows a specific solution $\bar{h}_{\mu\nu}$ to the linearised equations (73) for a given $T_{\mu\nu}$, one can obtain another solution that describes precisely the same physical situation by the change of gauge (86), in which u_μ are arbitrary but small functions.*

So, if we are able to find a particular solution of Eq. (73), say $\bar{h}_{\mu\nu}(X)$, all the other solutions $\widehat{h}_{\mu\nu}(X)$ can be found by adding to the particular solution $\bar{h}_{\mu\nu}(X)$ the tensor $\frac{\partial u_\nu(X)}{\partial X^\mu} + \frac{\partial u_\mu(X)}{\partial X^\nu}$. In addition, $\widehat{h}_{\mu\nu}(X)$ and $h_{\mu\nu}(X)$ possess the *same physical meaning*.

We also have the following.

Theorem 2 *If one knows a specific solution $\bar{h}_{\mu\nu}$ to the second equation of system (73) for a given $W_{\mu\nu}$, it is always possible to choose u_μ such that the new solution $\widehat{h}_{\mu\nu}(X)$ satisfies the gauge*

$$\frac{1}{2}L^{\lambda\kappa}\frac{\partial\widehat{h}_{\lambda\kappa}}{\partial X^\nu} = L^{\lambda\kappa}\frac{\partial\widehat{h}_{\lambda\nu}}{\partial X^\kappa} \quad \text{with}$$
$$\widehat{h}_{\mu\nu}(X) = \bar{h}_{\mu\nu}(X) + \frac{\partial u_\nu(X)}{\partial X^\mu} + \frac{\partial u_\mu(X)}{\partial X^\nu} \tag{87}$$

Indeed, $\widehat{h}_{\mu\nu}(X)$ manifestly satisfies the second equation of system (73), and the gauge condition [i.e., the first equation of Eq. (87)] is satisfied by choosing u_ν such that

$$L^{\lambda\kappa}\frac{\partial^2 u_\nu(X)}{\partial X^\lambda \partial X^\kappa} = \frac{1}{2}L^{\lambda\kappa}\frac{\bar{h}_{\lambda\kappa}(X)}{\partial X^\nu} - L^{\lambda\kappa}\frac{\bar{h}_{\kappa\nu}(X)}{\partial X^\lambda} \tag{88}$$

Note that, thanks to Eq. (88), $\widehat{h}_{\mu\nu}(X)$ satisfies the second PDE of system (73) because it satisfies *simultaneously* the gauge condition (87) and the following Poisson's PDE:

$$L^{\lambda\kappa}\frac{\partial^2\widehat{h}_{\mu\nu}(X)}{\partial X^\lambda \partial X^\kappa} = W^{(S)}_{\mu\nu}(X) \tag{89}$$

In conclusion, if we know a specific solution $\bar{h}_{\mu\nu}(X)$, thanks to Eqs. (87)–(88), we shall also able to get the expression for $\widehat{h}_{\mu\nu}(X)$ satisfying simultaneously the gauge condition and Poisson's PDE (89).

Incidentally, we also have the following theorem [42].

Theorem 3 *By performing the following change of variables $X^\mu \to X'^\mu$*

$$X'^\mu = X^\mu + L^{\mu\nu}u_\nu(X) \tag{90}$$

where u_ν is the solution of Eq. (88), *the transformed tensor of the unknown $h_{\mu\nu}(X)$ is a solution of Poisson's PDE* (89).

Indeed, by direct inspection, we can check that we find exactly the same PDE for the transformed tensor $h'_{\mu\nu}(X')$, obtained by $h_{\mu\nu}$ after the coordinate transformation $X^{\mu} \rightarrow X'^{\mu}$. The only precaution to be taken is to remain within the limits of validity of the *weak-field approximation* and, therefore, nonlinear terms of the type hh, hu, uu, and of higher order must be neglected.

6.2 A Note on the Physical Meaning of the Gauge Invariance

Let us consider a group of transformations of the field variables that leaves unchanged the basic physical observable. This group of transformations is called *gauge transformations*, and a theory where *all* the basic observables are unchanged under a gauge transformation of the field variables is referred to as a *gauge-invariant theory*. In Electrodynamics, for example, the Lorentz transformation and the Coulomb transformation are both gauge transformations since they do not affect the values of the electrodynamic observables, i.e., the values of the electric and the magnetic fields. In other words, the experimentalist is not able to detect the gauge-transformation choice, with any kind of system, and he is unable to notice any difference between two different gauge choices.

Another example is the theory of the General Relativity (GRT) since transformations (86) leave unchanged the physical observables, i.e., the Ricci tensor, the Einstein tensor, and the Riemannian curvature tensor. Hence, in GRT, transformations (86) may be regarded as the gauge transformations.

Thus, *the concept of gauge invariance is intimately related to one of unchanged physical observables*. This means that, before starting calculations, we should firstly identify all the physical observables linked to these transformations and, successively, check whether the values of these observables may be affected by a field-variables transformation.

The following example will make clear the concept. Let us suppose (absurdly, of course) that, in classical Electrodynamics, the physical observables are not only the electric field **E** and the magnetic field **B**, but also the scalar potential ϕ and the vector potential **A**. Thus, we suppose that an experimentalist is able to measure, with his instruments, also the numerical values of these two variables (in addition to the electromagnetic fields **E** and **B**). The electromagnetic fields **E** and **B** will still remain unaffected under Lorentz's or Coulomb's transformations. However, the question is: *May we still consider this New Electrodynamics as a gauge-invariant theory?* The answer is No. The only thing that has been changed is the fact that, in this *new Electrodynamics*, the experimentalist is now able to measure the scalar potential ϕ and the potential vector **A** (in addition to the electric and the magnetic fields). What happens then? If we perform calculations by using the Lorentz transformation, as to the electric field and the magnetic field, the experimentalist will confirm the good agreement between the theoretical predictions and the experimental data. However, in general, he will find a discrepancy between the experimental data and the values of the scalar potential ϕ_L and of the potential vector $\mathbf{A}_L$ established by the Lorentz transformation. For the experimentalist, the only way to overcome this impasse is to know the mathematical expression linking the scalar potential ϕ_L and the potential vector $\mathbf{A}_L$ with the scalar potential $\phi_{Exp.}$ and the potential vector $\mathbf{A}_{Exp.}$ measured in a laboratory. In this case, the *new Electrodynamics* loses its status of *gauge-invariant theory*.

Let us now consider another comparison. Let us compare the Thermodynamical Field Theory (TFT) with the General Relativity Theory (GRT). Here, there is a clear divergence between the TFT and the GRT. Indeed, as mentioned above, the physical observables in the GRT are the Ricci tensor, the Einstein tensor, and the Riemannian curvature tensor. Transformations (86) leave unchanged these physical observables. Hence in GRT, *without loss of generality*, we may suppose that there exists a specific solution of the PDE $B^{(0)}_{\mu\nu}(h^{(0)}) = 0$. Note that it does matter if, in reality, we do not know the mathematical expression of this solution; the proof of its existence is just sufficient. For **Theorem 2**, we may imagine to perform transformation (86), with u_ν satisfying Eq. (88), such that the new unknown reduces to the PDE (89) (and it satisfies, at the same time, the gauge condition). All of this is consistent with the General Covariance Principle (GCP), which allows choosing the coordinate system as we like such as the coordinate transformation (90) with u_ν satisfying the PDE (88), where the second PDE of system (73) reduces to Eq. (89) (ref. to **Theorem 3** and [42, 43]). In conclusion, in GRT we do not need to know the mathematical expression of a specific solution of the second equation of system (73), and we may start calculations by solving directly Eq. (89).

The case of TFT is utterly different. Firstly, we cannot evoke the validity of the General Covariance Principle and, even more importantly, the physical observables are the unknown $h_{\mu\nu}$ (indeed, the $h_{\mu\nu}$ are the transport coefficients). Hence, in analogy with what we said concerning the case of the *new Electrodynamics*, if we want to communicate with experimentalists we are compelled to find, firstly, a specific solution of Eq. (73) (either analytically or numerically) and, successively, obtain the general solution by applying Eq. (86). In conclusion, the TFT does not possess the status of *gauge-invariant theory* even though **Theorem 1** and **Theorem 2** remain valid and are very useful for performing calculations.

7 Solution of the Linearised Equations

As seen in Sect. 6.1, to get a concrete expression of a solution of Eq. (73), firstly we have to be able to find a specific solution for this PDE. Successively, according to **Theorem 1**, all the other solutions can be obtained by Eq. (86). **Theorem 2** allows getting the solution satisfying Eq. (89) by solving Eq. (88).

- **Solution of the Transport Equations for the Two-Dimensional Case**

For the two-dimensional case, the PDE to be solved reads

$$\begin{aligned} &B^{(0)}(h^{(0)}) = 0 \\ &B^{(0)}(\epsilon h^{(1)}) = -B^{(1)}(h^{(0)}) - T_{(L)}(h^{(0)}) = W^{(S)}(h^{(0)}) \end{aligned} \tag{91}$$

We start by solving the homogeneous differential equation $B^{(0)}(h^{(0)}) = 0$. Since our task is to find a specific solution, we look for a solution of the form (83) [4].

$$\bar{h}^{(0)}_{\mu\nu}(x) = L_{\mu\nu}\bar{h}^{0)}(x) \tag{92}$$

We have

$$\begin{aligned} B^{(0)}(\bar{h}^{(0)}) = \frac{1}{2}L^{\mu\nu}\Big(& L_{\mu\nu}L^{\lambda\kappa}\frac{\partial^2\bar{h}^{(0)}}{\partial X^\lambda X^\kappa} \\ & + L^{\lambda\kappa}L_{\lambda\kappa}\frac{\partial^2\bar{h}^{(0)}}{\partial X^\mu X^\nu} - L^{\lambda\kappa}L_{\kappa\mu}\frac{\partial^2\bar{h}^{0)}}{\partial X^\lambda X^\nu} - L^{\lambda\kappa}L_{\kappa\nu}\frac{\partial^2\bar{h}^{(0)}}{\partial X^\lambda X^\mu}\Big) = 0 \end{aligned} \tag{93}$$

By noticing that the sum of the last three contributions on the r.h.s. of Eq. (93) vanishes identically, we finally get

$$B^{(0)}(\bar{h}^{(0)}) = L^{\lambda\kappa}\frac{\partial^2\bar{h}^{(0)}}{\partial X^\lambda X^\kappa} = 0 \tag{94}$$

The second PDE of system (91) reduces to a Poisson PDE. Indeed, also in this case, we look for a special solution of the form

$$\bar{h}^{(1)}_{\mu\nu}(X) = L_{\mu\nu}\bar{h}^{(1)}(X) \tag{95}$$

By inserting Eq. (95) to the second PDE of system (91), we get

$$L^{\lambda\kappa}\frac{\partial^2\epsilon\bar{h}^{(1)}}{\partial x^\lambda\partial x^\kappa} = \left(\frac{2}{3}\tau^{\alpha\beta}_{(L)}h^{(0)}_{,\alpha} - \frac{1}{6}L_{\eta\alpha}L^{\iota\varrho}\Psi^\alpha_{\iota\varrho}\tau^{\eta\beta}_{(L)}\right)_{,\beta} = W^{(S)}(h^{(0)}) \tag{96}$$

since the contribution

$$L^{\lambda\kappa}L_{\lambda\kappa}\frac{\partial^2\epsilon\bar{h}^{(1)}}{\partial X^\mu X^\nu} - L^{\lambda\kappa}L_{\kappa\mu}\frac{\partial^2\epsilon\bar{h}^{1)}}{\partial X^\lambda X^\nu} - L^{\lambda\kappa}L_{\kappa\nu}\frac{\partial^2\epsilon\bar{h}^{(1)}}{\partial X^\lambda X^\mu} \equiv 0 \tag{97}$$

vanishes identically. In conclusion, for $n = 2$ the PDEs to be solved are

$$\begin{aligned} &L^{\lambda\kappa}\frac{\partial^2\bar{h}^{(0)}(X)}{\partial X^\lambda\partial X^\kappa} = 0 \\ &L^{\lambda\kappa}\frac{\partial^2\epsilon\bar{h}^{(1)}(X)}{\partial X^\lambda\partial X^\kappa} = \frac{1}{6}\left(4\tau^{\alpha\beta}_{(L)}\bar{h}^{(0)}_{,\alpha} - L_{\eta\alpha}L^{\iota\varrho}\Psi^\alpha_{\iota\varrho}\tau^{\eta\beta}_{(L)}\right)_{,\beta} = W^{(S)}(X) \end{aligned} \tag{98}$$

By performing the following orthogonal coordinate transformation

$$X'^{\lambda} = A^{\lambda}_{\kappa} X^{\kappa} \tag{99}$$

Equation (98) reads

$$\begin{aligned} &A^{\alpha}_{\lambda} L^{\lambda\kappa} A^{\beta}_{\kappa} \frac{\partial^2 \bar{h}'^{(0)}(X')}{\partial X'^{\alpha} \partial X'^{\beta}} = 0 \\ &A^{\alpha}_{\lambda} L^{\lambda\kappa} A^{\beta}_{\kappa} \frac{\partial^2 \epsilon \bar{h}'^{(1)}(X')}{\partial X'^{\alpha} \partial X'^{\beta}} = \left(\frac{2}{3}\tau'^{\alpha\beta}_{(L)} \bar{h}'^{(0)}_{,\alpha} - \frac{1}{6} L_{\eta\alpha} L^{\iota\varrho} \Psi'^{\alpha}_{\iota\varrho} \tau'^{\eta\beta}_{(L)}\right)_{,\beta} \end{aligned} \tag{100}$$

Note that linear transformations of coordinates are allowed because this class of transformations belongs to the group TCT [11]. Since the tensor $L^{\mu\nu}$ is a symmetric positive-definite matrix, it is always possible to determine A^{λ}_{κ} such that

$$A^{\alpha}_{\lambda} L^{\lambda\kappa} A^{\beta}_{\kappa} = \mathrm{I}^{\alpha\beta} \tag{101}$$

with $\mathrm{I}^{\alpha\beta}$ denoting the components of the identity matrix. As a consequence, Eq. (100) reads

$$\begin{aligned} &\mathrm{I}^{\alpha\beta} \frac{\partial^2 \bar{h}'^{(0)}(X')}{\partial X'^{\alpha} \partial X'^{\beta}} = 0 \\ &\mathrm{I}^{\alpha\beta} \frac{\partial^2 \epsilon \bar{h}'^{(1)}(X')}{\partial x'^{\alpha} \partial X'^{\beta}} = \left(\frac{2}{3}\tau'^{\alpha\beta}_{(L)} \bar{h}'^{(0)}_{,\alpha} - \frac{1}{6} L_{\eta\alpha} L^{\iota\varrho} \Psi'^{\alpha}_{\iota\varrho} \tau'^{\eta\beta}_{(L)}\right)_{,\beta} = W'^{(S)}(X') \end{aligned} \tag{102}$$

Tensors $\bar{h}^{(i)}_{\mu\nu}(x)$ (with $i = 0, 1$) are obtained by tensors $\bar{h}'^{(i)}_{\mu\nu}(X')$ by means of the coordinate transformation (99), with A^{μ}_{ν} determined by Eq. (101). We conclude this part of our analysis by noticing that the conformal field ϕ, defined in Eq. (53), is determined by $\bar{h}^{(i)}$ (with $i = 0, 1$) through the expression

$$\phi = \ln\bigl(1 + \bar{h}^{0)} + (\epsilon \bar{h}^{(1)}) \cdots\bigr) \simeq \bar{h}^{0)} + (\epsilon \bar{h}^{(1)}) \tag{103}$$

since $\bar{h}^{(i)}$ are non-negative scalar fields. Note that, by setting

$$\phi \simeq \phi_0 + \varepsilon \phi_1, \quad \text{with} \quad \varepsilon \sim O(\sigma^{-1}_{(L)}) \tag{104}$$

we get

$$\Box^2 \phi_0 = 0 \tag{105}$$

$$\Box^2 (\varepsilon \phi_1) = \frac{4}{9\sigma_{(L)}} \mathcal{O}^2 \phi_0 \tag{106}$$

which can be derived also by Eq. (102). Equation (102), subject to the appropriate boundary conditions, admits solutions that can be obtained analytically. Let us consider the first equation of system (102). This is a Laplacian PDE. According to the arguments illustrated in Appendix, we have to solve the Laplacian equations in the first quadrant by imposing that the solutions vanish on the axes, and they are constant on the arc of the circle of radius R_0. Successively, the entire solution, valid for all quadrants, is obtained by applying, the Schwartz principle [44]. In Appendix, we report the solution of the Laplacian equation subjected to appropriate boundary conditions, for the case of two independent thermodynamic forces. We have [4]

$$\bar{h}'^{(0)}_{\mu\nu}(X'^1, X'^2) = \frac{2}{\pi} L_{\mu\nu} \arctan\left[\frac{4R_0^2 \mid X'^1 X'^2 \mid}{R_0^4 - (X'^{12} + X'^{22})^2}\right] \tag{107}$$

where the new variables $\{X'^\mu\}$ (with $\mu = 1, 2$) are linked to the old ones $\{X^\mu\}$ by the (constant) 2×2 matrix A^λ_η, which satisfies the relation

$$X'^\mu = A^\mu_\nu X^\nu \qquad \text{with} \qquad A^\alpha_\lambda L^{\lambda\kappa} A^\beta_\kappa = \mathrm{I}^{\alpha\beta} \tag{108}$$

The value of constant χ and the expression of the radius R_0 are determined by the thermodynamic system under consideration and by the specific problem to be solved. An example of calculation can be found in [36]. It is worth mentioning that the method illustrated in Appendix applies also for obtaining the solution for n independent thermodynamic forces.

Let us now find the solution of the inhomogeneous problem. In this case, the second PDE of system (73) reduces to a Poisson PDE. Appendix reports the analytic solution of Poisson's PDE for a two-dimensional thermodynamic space. In polar coordinates ρ, θ, we have

$$\begin{aligned} \epsilon \bar{h}'^{(1)}_{\mu\nu}(\rho, \theta) = &- L_{\mu\nu} R_0^2 \sum_{n=1}^{\infty} \Big[\frac{\sin 2(n-1)\theta}{4(n-1)} \Big(\rho^{2(n-1)} \Big(\int_0^\rho t^{-2n+3} \widehat{W}'^{(S)}_n(t) dt - \widehat{\alpha}_n\Big) \\ &- \rho^{-2(n-1)} \int_0^\rho t^{2n-1} \widehat{W}'^{(S)}_n(t) dt\Big)\Big] \end{aligned} \tag{109}$$

with

$$\widehat{\alpha}_n = \int_0^1 \left(t^{-2n+3} - t^{2n-1}\right) \widehat{W}'^{(S)}_n(t) dt \qquad \text{and}$$

$$\widehat{W}'^{(S)}_n(\rho) \equiv \frac{1}{\pi} \int_{-\pi}^{\pi} W'^{(S)}(\rho, \theta) \sin(2(n-1)\theta) \, d\theta$$

where ρ and θ are linked to the new variables $\{X'^\mu\}$ by the usual relations $\rho = (X'^{12} + X'^{22})^{1/2}$ and $\theta = \arctan(X'^2/X'^1)$. Figure 4 illustrates solution $\bar{h}'^{(1)}_{11}(\rho, \theta)$ in polar coordinates ρ and θ in case of $W'(S)(X') \sim \varepsilon \bar{h}'^{(0)}(x')$ and $\chi = 1$.

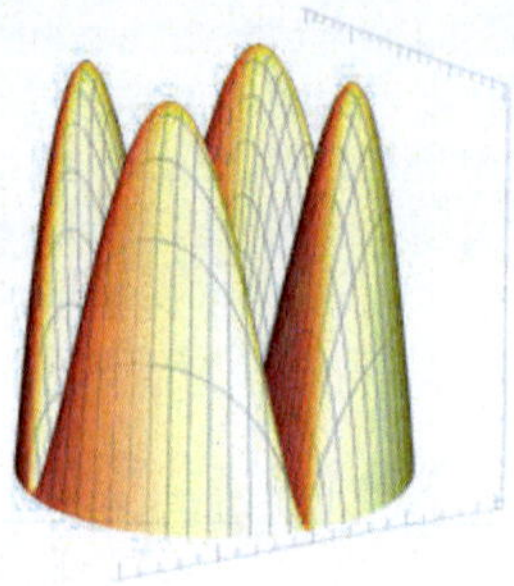

Fig. 4 $\bar{h}'^{(1)}_{11}(\rho, \theta)$ in polar coordinates ρ and θ in case of $W'^{(S)}(X') \sim \varepsilon \bar{h}'^{(0)}(X')$ and $\chi = 1$

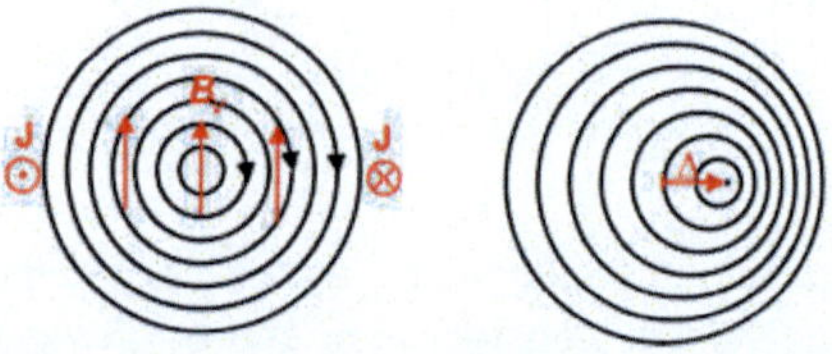

Fig. 5 **The Shafranov shift**. In tokamak-plasmas, the plasma pressure leads to an outward shift Δ of the centre of the magnetic flux surfaces. **J** indicates the direction of the electric current that flows inside the plasma. Note that the poloidal magnetic field increases and the magnetic pressure can, then, balance the outward force [34]

8 Testing the Validity of the PDE (64)—Computation of Heat Loss in L-mode, Collisional FTU-Plasma

The aim of this section is to test the validity of the PDE (64). For this purpose, we compare the theoretical predictions with the experimental data provided by the EUROfusion Consortium in Frascati (Rome-Italy) for FTU-plasmas.[5] We started by comparing the theoretical predictions of Eq. (64) subjected to the correct boundary conditions, with the experimental data for FTU-plasmas, in a fully collisional regime. So, in the first phase, experiments have been performed in a zone of the Tokamak where the turbulent effects are almost *frozen*. In our calculations, we have also taken into account the Shafranov-shift (which is not negligible in FTU-plasmas). The physical explanation of the Shafranov-shift is briefly sketched in Fig. 5.

As to the boundary conditions, these have been obtained in the following way: **(a)** First of all, we have to satisfy the Onsager condition. Hence, the solution should vanish at the origin of the axes, i.e., $\phi(\mathbf{0}) = 0$;

[5] Here, we shall not enter in describing technical details [45], since all this is out of scope of the present work. Our aim is to show the good agreement between the proposed approach and experiments. The interested reader can find a detailed description of the comparison between theory and the experimental data in FTU-plasmas in our article submitted for publication in a review specialised in the field of thermonuclear fusion.

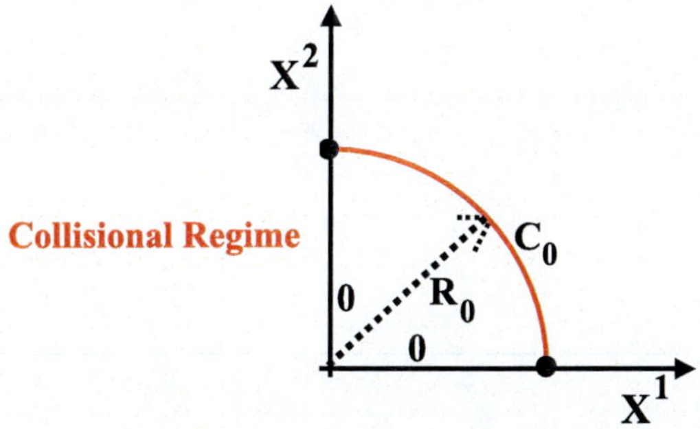

Fig. 6 **Boundary conditions for ϕ for collisional FTU-plasmas.** The validity of Onsager's relations is ensured by imposing that ϕ vanishes along the axes. In addition, we have to impose that there are no privileged directions for very large values of the thermodynamic forces. This condition is satisfied by imposing, in the first quadrant, $\phi(r = R_0, \theta) = c_0 \neq 0$

(b) Experimental evidence shows that, in the pure collisional regime, the pure effects (such as Fourier's law and Fick's law) are very robust laws. So, we have to impose $\phi(X^1 = 0, X^2) = \phi(X^1, X^2 = 0) = 0$;
(c) There are no privileged directions when the thermodynamic forces tend to infinity (or for very large values of the thermodynamic forces). In other words, $\phi(r = R_0, \theta) = const. \equiv c_0 \neq 0$ on the arc of a circle of radius R_0 (with R_0 very large). Here, (r, θ) denotes the polar coordinates: $r = (X^{1^2} + X^{2^2})^{1/2}$ and $\theta = \arctan(X^2/X^1)$.

The boundary conditions, in the case of FTU-plasmas in a fully collisional regime, are depicted in Fig. 6.

Now, we are able to solve the PDE (64) in the first quadrant. After having obtained the solution in the first quadrant, successively we shall be able to reconstruct the entire solution which is valid for the whole circle by using the Schwartz principle [44]. Parameters R_0 and c_0 have been determined as follows.

The scaling parameter R_0 is determined such that
(i) a solution of the TFT equation exists everywhere in the physical system, hence it cannot be too small;
(ii) the solution area is maximised in the space of the thermodynamic forces, i.e., R_0 defines the minimal circle enclosing the solution area—see Fig. 7.

The Dirichlet boundary condition c_0 is determined such that the thermodynamic forces X^1 and X^2, solutions of the system, maximise the electron heat loss. It is numerically found to be approximately equal to $c_0 \simeq -4.5$.

Figure 8 shows a comparison between experimental data for fully collisional FTU-plasmas and the theoretical predictions of Eq. (64), subjected to the boundary conditions illustrated in Fig. 6. On the vertical axis, we have the (surface magnetic-averaged) radial electron heat flux, and on the horizontal axis the minor radius of the tokamak. The lowest dashed profile corresponds to the Onsager theory (i.e., the neoclassical theory) and the bold line to the *Thermodynamical Field Theory* (TFT)

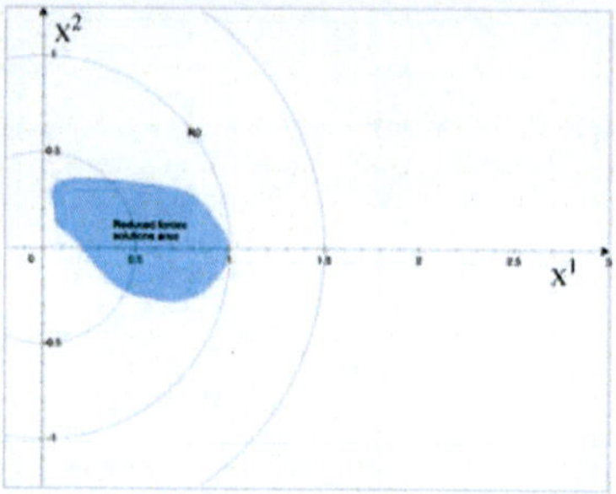

Fig. 7 **Determination of the value** R_0. R_0 defines the minimal circle enclosing the solution area. Here, the solution area is maximised in the thermodynamic forces space, spanned by X^1 and X^2

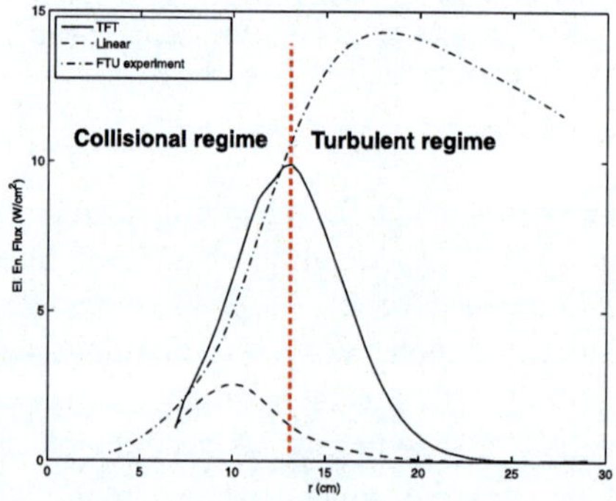

Fig. 8 **Electron heat loss in FTU-plasmas versus the minor radius of the Tokamak**. The highest dashed line is the experimental profile. These data have been provided by Marinucci from the ENEA C.R.-EUROfusion in Frascati [46]. The bold line is the theoretical profile obtained by the nonlinear theory satisfying the TCP (TFT) and the lowest dashed profile corresponds to the theoretical prediction obtained by Onsager's theory (i.e., by the neoclassical theory)

satisfying the TCP, respectively. The highest profile is the experimental data provided by the ENEA C.R.-EUROfusion. As we can see, the TCP principle is well satisfied in the core of the plasma where plasma is in the fully collisional transport regime. Towards the edge of the tokamak, transport is dominated by turbulence.

9 Conclusions

A non-Riemannian geometry has been constructed out of the components of the affine connection which has been determined by imposing the validity of the *General Evolution Criterion* for non-equilibrium systems relaxing towards a steady state. Relaxation expresses an intrinsic physical property of a thermodynamic system. The affine connection, on the other hand, is an intrinsic property of geometry allowing to perform derivatives and determine the equation for the shortest path. It is spontaneous to argue that a correct thermodynamical-geometrical theory should correlate these two properties: relaxation of a system with the affine connection. It is important to recall that the General Evolution Criterion is valid for systems, even far, from equilibrium and even in turbulent regimes. More specifically, this theorem has been derived only from the balance equations for mass, energy, and momentum *without assuming the validity of the Onsager reciprocity relations and without neglecting any terms, including the terms leading to turbulence in hydrodynamic systems*. Successively, Glansdorff and Prigogine showed the validity of this theorem also for plasmas governed by the plasma-dynamic PDEs. Incidentally, if we assume that the transport coefficients are a small perturbation of the Onsager matrix and in the limit $\sigma >> 1$, all terms leading to turbulence disappear and the General Evolution Criterion is trivially satisfied. In this limit case, we obtain the closure relations reported in ref. [11].

Action (31) (*or action* (27)), *takes into account all the terms of the balance equations, including those leading to turbulence, and its validity range coincides with that of the General Evolution Criterion*. The action principle leads, for $n > 2$, to the PDEs (42) and, for $n = 2$, to Eq. (64), respectively. To get these equations, we did not neglect any terms and we did not require that the transport coefficients are close to the Onsager matrix. Moreover, we have noted that for FTU-plasmas, turbulence appears in a region of the Tokamak where the values of the dimensionless entropy production are of order $\mathcal{O}(1)$. Hence, in this work, we did not assume that $\sigma >> 1$. Successively, we have applied Eq. (64) to FTU-plasmas in a collisional regime. This regime requires that the *pure Onsager laws* (i.e., the Fourier law and the Fick law) are *very robust laws* and for this, we are bound to impose as boundary conditions that along the thermodynamic axes the transport coefficients must coincide with the Onsager ones. It is worth noticing that variable ϕ is not a perturbation. It is our opinion that it is a great success that in the collision regime the theoretical predictions, resulting from a PDEs so different from the standard equations that we are used to seeing in the literature, are in very good agreement with the experimental data. Since the PDE (64) has been derived without neglecting any term present in the dynamic equations (i.e., the energy, mass, and momentum balance equations), it is quite natural to propose Eq. (64) as a good candidate also for describing transport in two-dimensional turbulent systems. So in our opinion, we may analyse the electron heat loss for FTU-plasmas even in the turbulent zone if we specify the appropriate boundary conditions. In the turbulent zone, the system is (very) far from thermodynamic equilibrium. Thus, in this zone we have to release the very strict condition that along the (thermodynamic) axes the solution must coincide with the Onsager relations. Indeed, in a turbulent regime the Onsager regression hypothesis for microscopic fluctuations of small non-equilibrium disturbances is violated [12]. However, we have to maintain the hypothesis that for large values of the thermodynamic forces, the space tends to flatten towards a constant metric (that no longer coincides with Onsager's metric). The boundary conditions, in the case of Tokamak-plasmas in the collisional regime (first circle) with the ones in the turbulent regime (i.e., in the annulus) are depicted in Fig. 9. We are currently analysing the stability of the constant solution of Eq. (64) in the annulus region. Concretely, we have to determine the con-

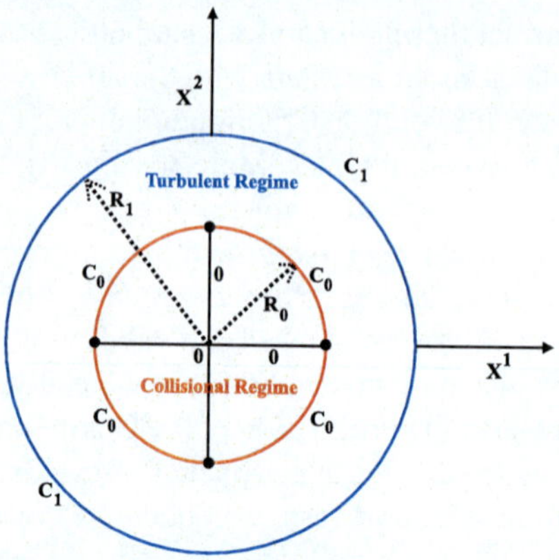

Fig. 9 Boundary conditions for ϕ for FTU-plasmas in collisional and turbulent regimes. In the turbulent zone, plasmas are far from equilibrium. So, in the annulus the Onsager relations are no longer valid. Of course, we have always to impose that there are no privileged directions for very large values (say for $r = R_1$) of the thermodynamic forces. This condition is satisfied by imposing, in the first quadrant, $\phi(r = R_1, \theta) = c_1 \neq 0$. By combining these boundary conditions with those specified for FTU-plasmas in the collision regime, we are finally able to find the solution of Eq. (64), valid in the collision zone as well as in the turbulent one

ditions where the constant solution of Eq. (64) loses its stability towards a chaotic one. This task may be accomplished by applying, for example, the mathematical methods reported in Ref. [47]. The ultimate aim of our work is to apply our approach to the *Divertor Tokamak Test facility* (DTT) to be built in Italy and to ITER.

We conclude with some comments about the validity of Eq. (1). It is known that the most general flux-force transport relations take the form

$$J_\mu(\mathbf{r}, t) = \int_\Omega d\mathbf{r}' \int_0^t dt' \mathcal{G}_{\mu\nu}[X(\mathbf{r}', t')] X'^\nu(\mathbf{r} - \mathbf{r}', t - t') \tag{110}$$

with Ω denoting the volume occupied by the system. The space-time-dependent coefficients $\mathcal{G}_{\mu\nu}$ are called *nonlocal transport coefficients*: they should not be confused with coefficients $\varpi_{\mu\nu}$ (they do not have the same dimension). The nonlocal and non-Markovian Eq. (110) expresses the fact that the flux at a given point ($\mathbf{r}$, t) could be influenced by the values of the forces in its spatial environment and by its history. Whenever the spatial and temporal ranges of influence are sufficiently small, the delocalisation and the retardation of the forces can be neglected under the integral

$$\begin{aligned} &\mathcal{G}_{\mu\nu}[X(\mathbf{r}', t')] X'^\nu(\mathbf{r} - \mathbf{r}', t - t') \\ &= 2\varpi_{\mu\nu}[X(\mathbf{r}, t)] X^\nu(\mathbf{r}, t) \delta(\mathbf{r} - \mathbf{r}') \delta(t - t') \end{aligned} \tag{111}$$

with δ denoting Dirac's delta function. In this case, the transport equations reduce to

$$J_\mu(\mathbf{r}, t) = \varpi_{\mu\nu}[X(\mathbf{r}, t)]X^\nu(\mathbf{r}, t) \tag{112}$$

In the vast majority of cases studied at present in transport theory, it is assumed that the transport equations are of the form of Eq. (112). However, equations of the form (110) may be met when we deal with anomalous transport processes such as transport in turbulent tokamak plasmas—see, for example, Ref. [40]. Hence, Eq. (111) establishes, in some sort, the limit of validity of Eq. (1) and, in this case, the fluxes should be evaluated by using Eq. (110). Nonetheless, we would like to stress the following. Hydrodynamic turbulence is normally studied through the Navier-Stokes equations, supported by the conservation equations for the mass and energy (the so-called *mass-energy balance equations*). The set of hydrodynamic equations are closed through relations of the form (112) where, for Newtonian fluids, $\tau_{\mu\nu}$ depends only on the thermodynamical quantities, and not on their gradients. The experimental data are in excellent agreement with the numerical simulations—see, for example, Ref. [48]. For non-Newtonian fluids, turbulence is still analysed by closing the balance equations with equations of the form (112) where the viscosity coefficients depend not only on the thermodynamic quantities but also on their gradients—see, for example, Ref. [49]. Also in this case, the experimental data are in excellent agreement with the numerical simulations. Even transport phenomena in Tokamak-plasmas in the weak-collisional regime are analysed by closing the balance equations with equations of the type (112)—see, for example, Ref. [1]. This is for saying that Eq. (112) are *very robust* equations and their validity goes well beyond the collisional, or the weak-collisional, regime. This case is very similar to what happens for the Onsager reciprocity relations: even if, according to the non-equilibrium statistical physics and the kinetic theory, these relations should have been valid only in the vicinity of the thermodynamic equilibrium in reality their validity goes well beyond the thermodynamic equilibrium, up to be valid even in turbulent hydrodynamic regimes.

In conclusion, before further complicating the mathematical formalism, it is the author's opinion that it is still worth analysing the turbulence in Tokamak-plasmas by closing the balance equations with local equations of the type (112) and comparing a posteriori the theoretical predictions with the experimental data.

By passing, there is another important point which is worthwhile mentioning. In this manuscript, the thermodynamic quantities (number density, temperature, pressure, etc.) are evaluated at the local equilibrium state. This is not inconsistent with the fact that the arbitrary state of a thermodynamic system is close to, but not in a state of local equilibrium. Indeed, as known, it is always possible to construct a representation in such a way that the thermodynamic quantities evaluated with a distribution function close to a Maxwellian do coincide exactly with those evaluated at the local equilibrium state—see, for example, the textbook [50].

Data Availability Statement

The data that support the findings of this study are available from the corresponding author upon reasonable request.

Acknowledgements I pay tribute to my friend, Prof. Slava Belyi. I remember fondly, and emotionally, the fruitful discussions, at times animated, on issues concerning Non-equilibrium Thermodynamics at the home of Prof. Ilya Prigogine. Professor I. Prigogine often invited us to dinner at his home, and I remember that the discussions soon ended in converging on topics concerning the dynamics of systems out of equilibrium, the role of fluctuations, and the formulation of the sub-dynamics. I would like also to pay tribute to my colleague and friend Prof. Enrique Tirapegui, co-author of several manuscripts of this series of works. I am indebted to Dr. M. Marinucci from the ENEA—Frascati (Rome-Italy) for having provided the experimental data for the FTU-plasmas.

Appendix

• Specification of the Boundary Conditions

The purpose of this section is to specify the boundary conditions for the first equation of system (102), i.e., for the Laplace PDE

$$\mathrm{I}^{\lambda\kappa}\frac{\partial^2 f(x)}{\partial X^\lambda \partial X^\kappa} = 0 \tag{113}$$

with $\mathrm{I}^{\lambda\kappa}$ denoting the components of the Identity matrix. This task will be accomplished by taking into account the Onsager theory and experimental evidence. More specifically, we should require the validity of the following conditions:

(a) The solution must coincide with the Onsager matrix when the system approaches equilibrium. We refer to this condition as *Onsager's condition*;
(b) Experimental evidence shows that the pure effects (such as Fourier's law and Fick's law) are very robust laws. Hence, for the unidimensional case (i.e., for $n = 1$), we impose $g_{11} = L_{11}$ (or $h_{11} = 0$);
(c) There are no privileged directions when the thermodynamic forces tend to infinity (or for very large values of the thermodynamic forces);
(d) For isotropic substances, the solution must be invariant under permutation among the (dimensionless) thermodynamic forces. Hence, for isotropic materials, the solution should be invariant with respect to the permutation of the axes X^i;
(e) The solution holding throughout the space may be obtained by using the Schwartz principle [44];
(f) The boundary conditions for the n-dimensional case may be derived from the knowledge of the solution of Eq. (113) for the $(n-1)$-dimensional case.

The analysis of the two-dimensional case will make clear this approach. Once this case is solved, we shall be able to specify the boundary conditions for a three-dimensional thermodynamic space and so on.

(i) First of all we have to satisfy the Onsager condition. Hence, the solution should vanish at the origin of the axes:

$$f(\mathbf{0}) = 0 \tag{114}$$

Fig. 10 The two-dimensional boundary conditions

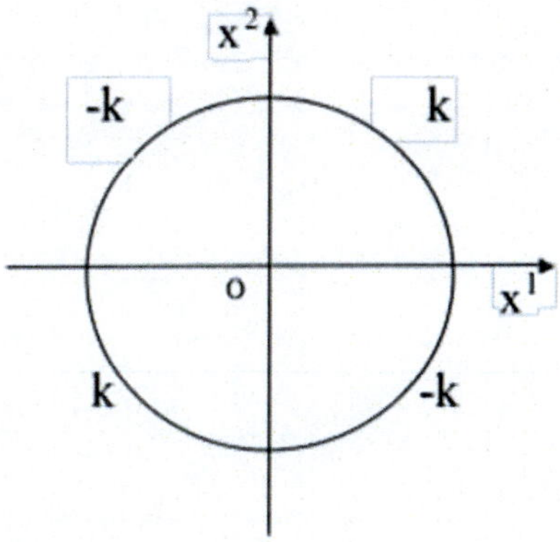

As a consequence, in two dimensions, we should have $f(r=0,\theta)=0$, with $r=({X^1}^2+{X^2}^2)^{1/2}$ and $\theta=\arctan(X^2/X^1)$, respectively.

(ii) From condition **(b)**, we have that the perturbation of the transport coefficients is zero on the axes $X^1=0$ and $X^2=0$, and condition **(c)** requires that the solution should be a constant different from zero, say with value $k\neq 0$, on the arc of a circle of radius R_0 (with R_0 very large). Hence, we should have

$$f(\mathbf{X})|_{(\theta=0,r)} = f(\mathbf{X})|_{(\theta=\pi/2,r)} = 0 \text{ and } f(\mathbf{X})|_{(0<\theta<\pi/2,r=R_0)} = k \neq 0 \quad (115)$$

(iii) The solution should be invariant with respect to the permutation of the axes X^1 and X^2;

(iv) After having obtained the solution in the first quadrant, successively we shall be able to reconstruct the entire solution which is valid for the whole circle by using the Schwartz principle [44].

Note that, according to the previous boundary condition **(b)**, the first derivatives of the solution have discontinuity points. However, the C^2 smoothness inside the domain is automatically assured by Weyl's lemma [51]. Hence, the solutions are of class C^2 inside the circle, except at the boundary where they are at least of class C^0.

By taking into account conditions **(i)–(iv)**, it is easy to convince ourselves that, for $n=2$, the correct boundary condition reads [29, 32]

$$f(R_0,\theta)=\begin{cases} k & \text{if } 0<\theta<\frac{\pi}{2} \\ -k & \text{if } \frac{\pi}{2}<\theta<\pi \\ k & \text{if } -\pi<\theta<-\frac{\pi}{2} \\ -k & \text{if } -\frac{\pi}{2}<\theta<0 \end{cases} \quad (116)$$

• **Solution of Eq. (113) for the Two-Dimensional Case**

The solution of this problem can be found in [29, 32]. Here, we shall solve Eq. (113) subject to the boundary conditions depicted in Fig. 10.

As known, the solution of Eq. (113) with boundary conditions (116) can be written in general as (see, for example [52] or [53]):

$$f(r,\theta) = \frac{a_0}{2} + \sum_{n=1}^{\infty} (a_n \cos n\theta + b_n \sin n\theta)\left(\frac{r}{R_0}\right)^n \tag{117}$$

with

$$\begin{aligned}
a_0 &= 2\int_{-\pi}^{\pi} f(R_0,\theta)\, d\theta \\
a_n &= \frac{1}{\pi}\int_{-\pi}^{\pi} f(R_0,\theta)\cos(n\theta)\, d\theta \qquad \text{for} \quad \mathrm{n} \geq 1 \\
b_n &= \frac{1}{\pi}\int_{-\pi}^{\pi} f(R_0,\theta)\sin(n\theta)\, d\theta \qquad \text{for} \quad \mathrm{n} \geq 1
\end{aligned} \tag{118}$$

Integrals (118) can be computed. We get

$$\begin{aligned}
a_n &= 0 & (\mathrm{n} = 0, 1, 2\cdots) \\
b_n &= -k\frac{8\cos(\frac{n\pi}{2})\sin(\frac{n\pi}{4})^2}{n\pi} & (\mathrm{n} = 1, 2\cdots)
\end{aligned} \tag{119}$$

Therefore, solution $f(r,\theta)$ can be written as

$$f(r,\theta) = -\frac{8k}{\pi}\sum_{n=1}^{\infty}\frac{\cos(\frac{n\pi}{2})\sin(\frac{n\pi}{4})^2}{n}\sin(n\theta)\left(\frac{r}{R_0}\right)^n \tag{120}$$

Solution (120) can be brought into the form

$$f(r,\theta) = \frac{4k}{\pi}\sum_{n=1}^{\infty}\frac{\sin(2(n-1)\theta)}{2n-1}\left(\frac{r}{R_0}\right)^n \tag{121}$$

The sum in Eq. (121) can be evaluated [54], and we find the compact expression

$$f(r,\theta) = \frac{2k}{\pi}\arctan\left[\frac{2\rho^2\sin(2\theta)}{1-\rho^4}\right] \qquad \text{where} \quad \rho \equiv \frac{r}{R_0} \leq 1 \tag{122}$$

or, in coordinates x^1 and x^2:

$$f(X^1,X^2) = \frac{2k}{\pi}\arctan\left[\frac{4R_0^2X^1X^2}{R_0^4 - ({X^1}^2 + {X^2}^2)^2}\right] \tag{123}$$

Solution (123) is valid only in the quadrants $X^1X^2 > 0$. The general solution, valid in all quadrants, is obtained by using Schwartz's principle:

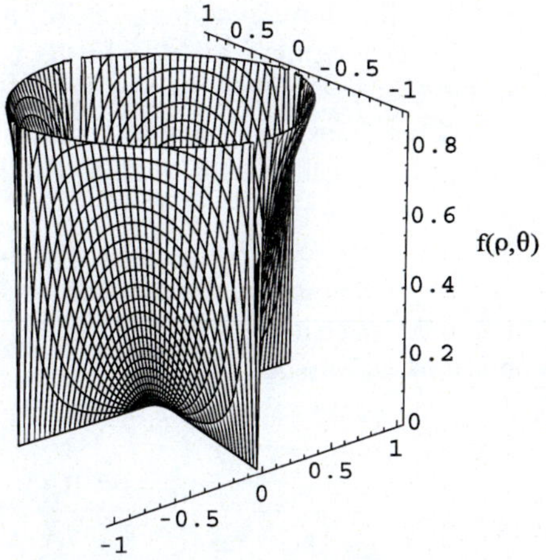

Fig. 11 Solution (124) (with $k = 1$) in polar coordinates

$$f(X^1, X^2) = \frac{2k}{\pi} \arctan\left[\frac{4R_0^2 \mid X^1 X^2 \mid}{R_0^4 - ({X^1}^2 + {X^2}^2)^2}\right] \tag{124}$$

In our original problem, we found that constant $k = 1$ [45]. Solution (124) is illustrated in Fig. 11.

• Boundary Conditions for the Three-Dimensional Space

The boundary conditions for a three-dimensional space are derived directly from the solution of Eq. (113) in two dimensions. Indeed, let us consider the first octant of the space. In the planes $X^3 = 0$, $X^2 = 0$, and $X^1 = 0$, we should re-obtain the expressions for the transport coefficients which we have been found by solving the two-dimensional case. So, on the planes $X^1 = 0$, $X^2 = 0$, and $X^3 = 0$, solution $f(X^1, X^2, X^3)$ should satisfy the boundary conditions

$$f(x^1 = 0, X^2, X^3) = f_1(X^2, X^3) = \frac{2k}{\pi} \arctan\left[\frac{4R_0^2 X^2 X^3}{R_0^4 - ({X^2}^2 + {X^3}^2)^2}\right]$$
$$f(X^1, X^2 = 0, X^3) = f_2(X^1, X^3) = \frac{2k}{\pi} \arctan\left[\frac{4R_0^2 X^1 X^3}{R_0^4 - ({X^1}^2 + {X^3}^2)^2}\right]$$
$$f(X^1, X^2, X^3 = 0) = f_3(X^1, X^2) = \frac{2k}{\pi} \arctan\left[\frac{4R_0^2 X^1 X^2}{R_0^4 - ({X^1}^2 + {X^2}^2)^2}\right], \tag{125}$$

respectively. In addition, the above condition **(d)** is satisfied by imposing that the solution $f(r, \theta, \phi)$ (with r, θ, and ϕ denoting the radial coordinate, the azimuth, and the zenith angle, respectively) is *constant* on the spherical cap of radius R_0, centred at the origin of the axes and located in the first octant:

$$f(r = R_0, \theta, \phi) = k \neq 0 \tag{126}$$

In this way, the above conditions **(a)**–**(d)** have been satisfied and we have obtained a well-posed Dirichlet's problem in the first octant. Now, we are in a position to solve the Laplace PDE in the first octant, subject to the above-derived Dirichlet's boundary conditions, by using standard methods of mathematical physics.[6] As for the two-dimensional case, the solution holding throughout the space may be obtained by using the Schwartz principle. Of course, the above method is not limited to the three-dimensional case, and it can naturally be extended for getting the boundary conditions for an n-dimensional space, once the solution of Eq. (113) for a $(n-1)$-dimensional space is obtained.

• Solutions of Poisson's PDE

In this section, we solve the Poisson PDE subjected to the appropriate boundary conditions,

$$\mathrm{I}^{\lambda\kappa}\frac{\partial h^{(1)}(X)}{\partial X^{\lambda}\partial X^{\kappa}} = W^{(S)}(X) \tag{127}$$

with source $W^{(S)}(X)$ given by the right-hand side of the second equation in system (102). According to the general procedure, we have, firstly, to find a particular solution of the Poisson PDE which should be solved with all homogeneous boundary conditions. The individual conditions must retain their type (Dirichlet, Neumann, or Robin type) in this sub-problem. Successively, we have to add any solution of the homogeneous Laplace equation with the non-homogeneous boundary conditions. Also in this case, the individual conditions must retain their type (Dirichlet, Neumann, or Robin type) in the sub-problem. The complete solution of the Poisson equation is the sum of the solution of the two sub-problems: the solution of the *Poisson sub-problem* plus the solution of the *Laplace sub-problem*. Since the boundary conditions have already been satisfied when we solved the PDE at the dominant order (i.e., the equations for $h^{(0)}_{\mu\nu}$), the only task that we have to accomplish is to find the particular solution of the inhomogeneous Poisson PDE subject to homogeneous boundary conditions. The solution to the homogeneous equation allows us to obtain a system of basis functions that satisfy the given boundary conditions.

For a two-dimensional space (i.e., in case of two independent thermodynamic forces), Eq. (121) suggests the following expression for the solution of the Poisson equation

$$h^{(1)}(r,\theta) = \sum_{n=1}^{\infty} a_n(r)\sin\left(2(n-1)\theta\right) \tag{128}$$

Equation (128) satisfies the homogeneous boundary conditions at $\theta = 0$ and $\theta = \pi/2$. Now, we substitute Eq. (128) into the Poisson equation (127), written in polar coordinates, i.e.:

$$\frac{1}{r}\frac{\partial}{\partial r}\left(r\frac{\partial h^{(1)}}{\partial r}\right) + \frac{1}{r^2}\frac{\partial^2 h^{(1)}}{\partial \theta^2} = W^{(S)}(r,\theta) \tag{129}$$

[6] See, for example, the reference books [53, 55, 56].

and we solve the equation by using the orthogonality relations for the sine functions. We have also to take into account that, at equilibrium (i.e., at $\rho = 0$), we have to re-obtain the Onsager matrix and we have also to satisfy the homogeneous condition at $\rho = 1$. Finally, after simple calculations, we get the following ordinary differential equation for $a_n(r)$, subject to the following conditions:

$$\begin{aligned}
&\rho^2 a_n''(\rho) + \rho a_n'(\rho) - 4(n-1)^2 a_n(\rho) = R_0^2 \rho^2 \widehat{W}_n^{(S)}(\rho) \\
&a_n(0) = 0 \ ; \ a_n(1) = 0 \qquad \text{where} \\
&\widehat{W}_n^{(S)}(\rho) \equiv \frac{1}{\pi} \int_{-\pi}^{\pi} W^{(S)}(\rho, \theta) \sin(2(n-1)\theta)\, d\theta
\end{aligned} \tag{130}$$

with $\rho \equiv r/R_0 \leq 1$ and the suffix "prime" denoting the derivative with respect to ρ, respectively. Equation (130) corresponds to a *non-homogeneous Euler equation of the 2ndorder.* By using standard methods of integration (see, for example, [57]), by imposing the Onsager and the homogeneous conditions, after simple calculations we finally get

$$\begin{aligned}
a_n(\rho) &= -\frac{R_0^2}{4(n-1)} \Big(\rho^{-2(n-1)} \int_0^{\rho} t^{2n-1} \widehat{W}_n^{(S)}(t) dt \\
&\quad - \rho^{2(n-1)} \int_0^{\rho} t^{-2n+3} \widehat{W}_n^{(S)}(t) dt \Big) + \alpha_n \rho^{2(n-1)} \quad \text{with} \\
\alpha_n &= -\frac{R_0^2}{4(n-1)} \int_0^1 \left(t^{-2n+3} - t^{2n-1} \right) \widehat{W}_n^{(S)}(t) dt
\end{aligned} \tag{131}$$

Solution (131) satisfies the Onsager condition since[7]

$$\lim_{\rho \to 0} \rho^{-2(n-1)} \int_0^{\rho} t^{2n-1} \widehat{W}_n^{(S)}(t) dt = 0 \tag{132}$$

Hence,

$$\begin{aligned}
h^{(1)}(\rho, \theta) &= R_0^2 \sum_{n=1}^{\infty} \Big[\frac{\sin 2(n-1)\theta}{4(n-1)} \Big(\rho^{2(n-1)} \Big(\int_0^{\rho} t^{-2n+3} \widehat{W}_n^{(S)}(t) dt - \widehat{\alpha}_n \Big) \\
&\quad - \rho^{-2(n-1)} \int_0^{\rho} t^{2n-1} \widehat{W}_n^{(S)}(t) dt \Big) \Big] \quad \text{with} \\
\widehat{\alpha}_n &= \int_0^1 \left(t^{-2n+3} - t^{2n-1} \right) \widehat{W}_n^{(S)}(t) dt \quad \text{and} \\
\widehat{W}_n^{(S)}(\rho) &\equiv \frac{1}{\pi} \int_{-\pi}^{\pi} W^{(S)}(\rho, \theta) \sin(2(n-1)\theta)\, d\theta
\end{aligned} \tag{133}$$

---

[7] The indeterminate form can be solved by using Hospital's rule and $\lim_{\rho \to 0} \widehat{W}_n^{(S)}(\rho) = 0$.

It is also easy to convince ourselves that such a procedure also applies for solving the Poisson PDE in case of n independent thermodynamic forces.

References

1. R. Balescu, *Transport Processes in Plasmas: Neoclassical Transport*, vol. 2 (Elsevier, Amsterdam, 1988)
2. J.-A. Carrillo, T. Goudon, P. Lafitte, F. Vecil, Numerical schemes of diffusion asymptotics and moment closures for kinetic equations. J. Sci. Comput. **36**, 113–149 (2008). https://hal.archives-ouvertes.fr/hal-00768401. ISSN: 0885-7474, EISSN: 1573-7691
3. T. Morita, K. Kazuo Hiroike, A new approach to the theory of classical fluids. I. Prog. Theor. Phys. **23**, 1003–1027 (1960)
4. G. Sonnino, *A Field Theory Approach to Thermodynamics of Irreversible Processes*. Thèse d'Habilitation à Diriger des Recherches - H.D.R. (Institut Nonlinéaire de Nice (INLN), 2002)
5. G. Sonnino, Nuovo Cimento Soc. Ital. Fis. B **115**, 1057 (2000)
6. G. Sonnino, Nuovo Cimento B **118**(10), 1175 (2003)
7. G. Sonnino, Thermodynamic field theory: an approach to thermodynamics of irreversible processes, in *Proceedings of the 9th International Workshop on Instabilities and Nonequilibrium Structures*, ed. by Viña del Mar (Kluwer, Dordrecht, 2001), p. 291
8. G. Sonnino, Int. J. Quantum Chem. **98**, 191 (2004)
9. G. Sonnino, J. Evslin, Int. J. Quantum Chem. **107**, 968 (2007)
10. G. Sonnino, J. Evslin, Phys. Lett. A **365**, 364 (2007)
11. G. Sonnino, Phys. Rev. E **79**, 051126 (2009)
12. L. Onsager, Phys. Rev. **37**, 405 (1931)
13. L. Onsager, Phys. Rev. **38**, 2265 (1931)
14. I. Prigogine, *Etude Thermodynamique des Phénomènes Irréversibles*, Thèse d'Aggrégation de l'Enseignement Supérieur de l'Université Libre de Bruxelles (U.L.B.) (1947)
15. I. Prigogine, *Thermodynamics of Irreversible Processes* (Wiley, 1954), p. 42
16. S.R. De Groot, P. Mazur, *Non-equilibrium Thermodynamics* (Dover Publications, Inc., New York, 1984)
17. J.E. Verschaffelt, The thermomechanical effect. Bull. Classe Sci. Acad. Roy. Belg. **37**, 853 (1951)
18. R.O. Davies, Physica **18**, 182 (1952)
19. R. Jackiw, *Liouville field theory: a two-dimensional model for gravity ?*, in *Quantum Theory of Gravity*, ed. by S.M. Christensen (Adam Higler, Bristol, 1984)
20. R. Jackiw, Nucl. Phys. B **252**, 343 (1985)
21. M. Cavaglià, Phys. Rev. D **59**, 084011 (1999)
22. J.D. Brown, *Lower Dimensional Gravity* (World Scientific, Singapore, 1988)
23. P. Collas, D. Klein, *The Dirac Equation in Curved Spacetime—A Gide for Calculations*. Springer Briefs in Physics (Netherlands, 2019). ISSN 2191-5423
24. C. Teitelboim, The Hamiltonian structure of two-dimensional space-time and its relation with the conformal anomaly, in *Quantum Theory of Gravity*, ed. by S.M. Christensen (Adam Higler, Bristol, 1984)
25. P. Peeters, G. Sonnino, Il Nuovo Cimento B **115**, 1083 (2000)
26. G. Sonnino, P. Peeters, Chaos Interdiscip. J. Nonlinear Sci. **14**(3), 910 (2004). https://doi.org/10.1063/1.1774411
27. G. Sonnino, Nuovo Cimento B **118**(10), 1155 (2003)
28. G. Sonnino, J. Math. Phys. (JMP) **51**, 122703 (2010)
29. G. Sonnino, J. Evslin, A. Sonnino, Entropy **17**(11), 7567 (2015). https://doi.org/10.3390/e17117567

30. P. Glansdorff, I. Prigogine, *Thermodynamic Theory of Structure, Stability and Fluctuations* (Wiley, London, New York, 1971)
31. P. Glansdorff, I. Prigogine, Sur Les Propriétés Différentielles de La Production d'Entropie. Physica **20**, 773 (1954)
32. G. Sonnino, A. Sonnino, J. Therm. Catal. **5**, 129 (2014)
33. G. Sonnino, A. Sonnino, The thermodynamic coordinate transformations and the thermodynamic covariance principle, in *Current Perspectives on Chemical Sciences*, vol. 2 (Book Publisher International, India, United Kingdom, 2020). https://doi.org/10.9734/bpi/cpcs/v2
34. G. Sonnino, J. Evslin, A. Sonnino, G. Steinbrecher, E. Tirapegui, Phys. Rev. E **94**, 042103 (2016)
35. G. Sonnino, Geometry and symmetry in non-equilibrium thermodynamic systems. Am. Inst. Phys. (AIP) Conf. Proc. **1853**, 030002 (2017). https://doi.org/10.1063/1.498535. Workshop: *MaxEnt 2016—36 International Workshop on Bayesian Inference and Maximum Entropy Methods in Science and Engineering*, Ghent (Belgium), 10–15 July 2016
36. G. Sonnino, P. Nardone, P. Peeters, E. Tirapegui, Chaos **30**(6), 063110 (2020). https://doi.org/10.1063/5.0006213
37. L.P. Eisenhart, *Non-Riemannian Geometry*, vol. VIII (American Mathematical Society, Colloqium Publications, New York, 1927)
38. D. Fitts, *NonEquilibrium Thermodynamics. A Phenomenological Theory of Irreversible Processes in Fluid Systems* (McGraw-Hill Book Company, Inc., New York, 1962)
39. C. Vidal, G. Dewel, P. Borckmans, *Au-delà de l'équilibre* (Hermann Éditeurs Des Sciences et des Arts, Paris, 1994)
40. R. Balescu, *Aspects in Anomalous Transport in Plasmas*. Series in Plasma Physics (Institute of Physics, Bristol, 2005), p. 417
41. F.L. Hinton, R.D. Hazeltine, Rev. Mod. Phys. **48**, 239 (1976)
42. C.W. Misner, K.S. Thorne, J.H. Wheeler, *Gravitation* (W. H. Freeman, San Francisco, 1973). 978-0-7167-0344-0
43. S. Weinberg, *Gravitation and Cosmology. Principle and Applications of the General Theory of Relativity* (Wiley, New York, Chichester, Brisbane, Toronto, Singapore, 1972)
44. R. Courant, D. Hilbert, *Method of Mathematical Physics*, vol. 1 (Wiley Classics Editions Published in 1989) (Wiley, 1937)
45. G. Sonnino, P. Peeters, Phys. Plasmas **15**, 062309/1–062309/23 (2008)
46. M. Marinucci (private communication, 2015)
47. J. Awrejcewicz, V.A. Krysko, I.V. Papkova, A.V. Krysko, *Deterministic Chaos in One-Dimension Continuous Systems*. Series on Nonlinear Science (World Scientific, Singapore, 2016)
48. W. Kollmann, *Navier-Stokes Turbulence* (Springer Nature Switzerland, 2019)
49. J.L. Lumley, Turbulence in Non-Newtonian Fluids. Phys. Fluids **7**, 335 (1964)
50. R. Balescu, *Transport Processes in Plasmas: Classical Transport*, vol. 1 (Elsevier, Amsterdam, 1988)
51. H. Weyl, The method of orthogonal projections in potential theory. Duke Math. J. **7**, 411–444 (1940). See Lemma 2, p. 415
52. E.C. Zachmanoglou, Dale W. Thoe, *Introduction to Partial Differential Equations with Applications* (Dover Inc., New York, 1976)
53. P.M. Morse, H. Feshbach, *Methods of Theoretical Physics* (Mc Graw-Hill Book, Inc., 1953)
54. I.S. Gradshteyn, I.M. Ryzhik, *Table of Integrals, Series, and Products*, 7th edn. (Academic Press, Amsterdam, Boston, Heidelberg, London, New York, Oxford, Paris, San Diego, San Francisco, Singapore, Sydney, Tokyo, 2007). ISBN-13: 978-0-12-373637-6. ISBN-10: 0-12-373637-4
55. E.T. Wittaker, G.N. Watson, *A Course of Modern Analysis* (Cambridge University Press, 1927)
56. R. Courant, D. Hilbert, *Method of Mathematical Physics*, vol. 2 (Wiley Classics Editions Published in 1989) (Wiley, 1937)
57. W. Walter, *Ordinary Differential Equations* (Springer, New York Inc., 1998). ISBN 978-1-4612-5

Simulating Deterministic Dynamics by Drawing Coloured Balls at Random in Urns

Léon Brenig

Abstract In this work, we illustrate by detailed examples an equivalence relation established in a previous publication between deterministic nonlinear differential dynamical systems and random Markovian processes.

This equivalence is shown here to differ from the equivalence relation between Langevin noisy systems or, more generally, stochastic differential equations and their associated deterministic Fokker-Planck partial derivative equations (PDEs). Here, in contrast, we are concerned with a correspondence between random Polya urn processes and deterministic systems governed by nonlinear ordinary differential equations (ODEs). More precisely, this relation is an isomorphism between each random urn process and a corresponding equivalence class of deterministic non-linear differential systems that transform one into the other under quasi-monomial transformations.

The object of the article is to illustrate this relation with several examples: 1. The population dynamics described by Lotka-Volterra equations. 2. The Lorenz system. 3. The asymmetric top with dissipation. 4. The cosmological equations describing the dynamics of Friedmann-Lemaître universes in the presence of N competitive and interacting barotropic fluids.

In all these four examples, we find the conditions under which the dynamical evolution is equivalent to balanced urn processes.

The bridge between the two domains of deterministic dynamical systems and of stochastic processes allows for transferring results from one field to the other. It also provides a new tool for simulating deterministic systems using random urn processes.

To my dear friend Slava who, certainly, would have liked the idea of simulating the Universe evolution by drawing coloured balls at random from a black box.

L. Brenig (✉)
Service de Physique des Systèmes Dynamiques, Faculté des Sciences, Université Libre de Bruxelles, 1050 Brussels, Belgium
e-mail: leon.brenig@ulb.be

L. Brenig et al. (eds.), *Nonequilibrium Thermodynamics and Fluctuation Kinetics*, Fundamental Theories of Physics 208,
https://doi.org/10.1007/978-3-031-04458-8_6

1 Introduction

Since the discovery of the Brownian motion in 1827 by R. Brown and its theoretical explanation by Einstein [1] and Smoluchowski [2], the random processes and, more generally, the mathematical probability theory have made their entry in the description of the physical world. They provide an efficient idealization of the effect of the enormous number of molecules colliding against a small but macroscopic object immersed in a fluid at thermodynamic equilibrium.

Shortly after the articles of Einstein and Smoluchowski, the idea of describing by random variables the short time-scale action of many microscopic degrees of freedom on the large time-scale evolution of macroscopic bodies has been generalized in a systematic way to other systems. The first to do so is Langevin [3] who introduced random variables at the level of the mechanical description of macroscopic systems. He described the action of a multitude of short time-scale collisions of molecules on a macroscopic body as a random time-dependent external force $\eta(t)$. Thus, for a macroscopic particle of mass m moving in a fluid with dissipative friction coefficient μ, P. Langevin modified the Newton equation into

$$m\frac{d\mathbf{v}}{dt} = -\mu\mathbf{v} + \eta(t) \tag{1}$$

where the random force $\eta(t)$ is a Gaussian white noise vector, that is, its average is zero, it is uncorrelated to the random velocity $\mathbf{v}$ and its auto-correlations are delta correlated:

$$< \eta_i(t)\eta_j(t') >= 2\mu k_B T \delta_{ij}\delta(t-t') \; ; \; i, j = 1, 2, 3 \tag{2}$$

in which k_B is the Boltzmann constant and T is the absolute temperature. Remark that Eq. (2) represents the simplest form of the so-called fluctuation-dissipation relations that have been and still are at the focus of intense research in Statistical Physics, and particularly in the work of V.V. (Slava) Belyi [4].

Very soon, the introduction of random forces made its way into classical mechanics in order to describe other mechanical systems immersed in fluids or in other environments made of many microscopic degrees of freedom. When these degrees of freedom interact with the immersed macroscopic system on time scales that are much smaller than the characteristic time scale of the purely mechanical system, these interactions can be approximated by Gaussian white noises with similar statistical properties as given in Eq. (2).

However, noise does not affect only mechanical systems. It also exists in electric circuits where the unavoidable thermal noise is generated by the random motion of the electrons in the conductors. This noise affects the potential and appears as a Gaussian white noise in the differential equations of the circuits. As an example, the RLC circuit obeys the following differential equation:

$$L\frac{d^2q(t)}{dt^2} + R\frac{dq(t)}{dt} + \frac{1}{C}q(t) = V + \xi(t) \tag{3}$$

where $q(t)$ is the charge at a given point of the circuit, L is the inductance, R is the resistance and C is the capacitance. V is the potential source and $\xi(t)$ is a Gaussian white noise.

More generally, noise is introduced in many dynamical systems governed by systems of ordinary differential equations in other domains of science like atmospheric dynamics, chemical reactions, population dynamics or economy. Noise of various origins may affect source terms or coefficients of these ODEs. In all these cases, the random variable called noise represents an idealization of the effect of a multitude of deterministic variables interacting with the described system on very short time scales. Mathematically, the introduction of noise in deterministic differential systems transforms them into so-called stochastic differential equations [5]. The functions that are solutions of these equations are themselves random processes. The extension of a deterministic system of ODEs to a system of stochastic differential equations (SDEs) is not unique. This is related to the non-differentiability of the Gaussian white noise process. Among the infinity of such extensions, two of them are the most used, the Itô and the Stratonovich forms, and obey their own generalized calculus rules [5].

The main concern for us at this point is the following fundamental property. To each system of stochastic differential equations is associated a deterministic partial differential equation, the Fokker-Planck equation. This equation governs the probability distribution of the random variables that are solutions of the stochastic differential equations.

More precisely, any N-dimensional deterministic system of ODEs:

$$\frac{dx_i(t)}{dt} = F_i(\mathbf{x}(\mathrm{t})) \; ; i = 1, \ldots, N \tag{4}$$

can be randomized into a system of SDEs:

$$dx_i(t) = F_i(\mathbf{x}(\mathrm{t}))dt + \sum_{j=1}^{M} G_{ij}(\mathbf{x}(\mathrm{t}))dW_{j;t} \tag{5}$$

in which M may differ from N, and where the F_i and the G_{ij} are functions which, respectively, represent drift and diffusion contributions to the dynamics. The symbols $dW_{j;t}$ denote the Itô or Stratonovich version of the differential associated to the j-component of the N-dimensional white noise vector. The dependent variables $x_i(t)$ are random variables and their probability distribution, $p(x_1, \ldots, x_N; t)$, obeys a deterministic partial derivative equation, the Fokker-Planck equation (FPE). This equation takes different form according to the choice of the Itô or the Stratonovich differential calculus. This choice depends on the physical nature of the modeled system [5].

In the Itô framework, the Fokker-Planck equation associated to Eq. (5) reads

$$\frac{\partial}{\partial t}p(\mathbf{x};t) = -\sum_{i=1}^{N}\frac{\partial}{\partial x_i}[F_i(\mathbf{x})p(\mathbf{x};t)] + \sum_{i=1}^{N}\sum_{j=1}^{N}\frac{\partial^2}{\partial x_i \partial x_j}[D_{ij}(\mathbf{x})p(\mathbf{x};t)] \quad (6)$$

with

$$D_{ij}(\mathbf{x}) = \frac{1}{2}\sum_{k=1}^{M} G_{ik}(\mathbf{x})G_{kj}^{T}(\mathbf{x})$$

while in the Stratonovich sense, the Fokker-Planck equation associated to (5) corresponds to

$$\frac{\partial}{\partial t}p(\mathbf{x};t) = -\sum_{i=1}^{N}\frac{\partial}{\partial x_i}[F_i(\mathbf{x})p(\mathbf{x};t)] + \frac{1}{2}\sum_{k=1}^{M}\sum_{i=1}^{N}\frac{\partial}{\partial x_i}\{G_{ik}(\mathbf{x})\sum_{j=1}^{N}\frac{\partial}{\partial x_j}[G_{kj}^{T}(\mathbf{x})p(\mathbf{x};t)]\} \quad (7)$$

Let us insist on the deterministic character of the two above partial derivative equations. Once an initial probability distribution is given, these equations provide their solution at later times with certainty. The Fokker-Planck equation represents a deterministic process of drift with diffusion. As a conclusion, the above considerations show the existence of an equivalence relation between the set of Stochastic Differential Equations (SDE) and the set of Fokker-Planck Equations (FPE). We can summarize this by the expression SDE ≅ FPE.

The object of this article is to show and illustrate the existence of another kind of equivalence relation between a random process and a deterministic evolution than the above-described one. More precisely, as we show in the sequel, there is an equivalence relation between the so-called urn processes and a set of nonlinear dynamical systems governed by ODEs, the quasi-polynomial dynamical systems. Denoting the set of urn processes by Urn Process and the set of quasi-polynomial systems by QP systems, we can summarize this new equivalence relation by Urn Process ≅ QP Systems.

In Sect. 2, we introduce the deterministic quasi-polynomial systems and their canonical forms. Section 3 introduces the urn processes. In Sect. 4, the equivalence Urn Process ≅ QP Systems is established. In Sect. 5, four examples of such equivalence are presented. Conclusions and perspectives are drawn in Sect. 6.

2 The Quasi-polynomial Dynamical Systems and Their Canonical Forms

The class of quasi-polynomial systems [6] corresponds to all the deterministic ODEs systems that can be cast in the form

$$\frac{dx_i}{dt} = x_i \sum_{j=1}^{N} A_{ij} \prod_{k=1}^{n} x_k^{B_{jk}} \qquad \text{for } i = 1, \ldots, n, \tag{8}$$

where A is any constant $n \times N$ matrix, B is any constant $N \times n$ matrix, n and N are any integers and do not need to be equal.

This class includes all the systems with right-hand sides involving linear and polynomial functions of the dependent variables $x_1(t), \ldots, x_n(t)$. However, it extends further to systems with non-polynomial nonlinearities like sums of monomials with non-integer exponents—this is the origin of the name "quasi-polynomial". Nonlinear systems with functions in their right-hand side that are themselves solutions of other polynomial differential systems belong also to the QP systems class. Practically, this class contains many systems relevant to the different domains of Physics and, more generally, to the scientific fields that are using mathematical modelling. Paradigmatic systems like the Lorenz or the Rössler models belong to that class.

QP systems of equations are covariant under the group of quasi-monomial transformations:

$$x_i = \prod_{k=1}^{n} \tilde{x}_k^{C_{ik}} \qquad \text{for } i = 1, \ldots, n \tag{9}$$

for any invertible matrix C belonging to $GL(n, R)$. These nonlinear transformations form a group isomorphic to $GL(n, R)$. They amount to linear transformations between the logarithms of the variables x_i and $\tilde{x}_i$. Being covariant means that Eq. (8) keeps the same QP form when rewritten in terms of the new variables $\tilde{x}_i$:

$$\dot{\tilde{x}}_i = \tilde{x}_i \sum_{j=1}^{N} \tilde{A}_{ij} \prod_{k=1}^{n} \tilde{x}_k^{\tilde{B}_{jk}} \qquad \text{for } i = 1, \ldots, n \tag{10}$$

with

$$\tilde{A} = C^{-1} A \tag{11}$$

and

$$\tilde{B} = BC. \tag{12}$$

In the two previous equations, the product is the matrix product. From these equations, it is clear that

$$\tilde{B}\tilde{A} = BA. \tag{13}$$

In other words, the $N \times N$ matrix $M = BA$ is an invariant of transformations (9). This means that all the QP systems with matrices A and B such that their products BA are all equal to a given $N \times N$ matrix M belong to the same equivalence class. Moreover, since the transformations (9) are diffeomorphisms all the systems belonging to that equivalence class are dynamically equivalent.

Now, each equivalence class of QP systems, labelled by a given matrix M, contains two remarkable systems. We called them the canonical forms of the QP systems belonging to that equivalence class. One of them is the Lotka-Volterra canonical form. It is obtained by considering all the quasi-monomials $\prod_{k=1}^{n} x_k^{B_{jk}}$ appearing in the QP system (8) as new dynamical variables:

$$u_j = \prod_{k=1}^{n} x_k^{B_{jk}}; \ j = 1, \ldots, N. \tag{14}$$

Then, it is easily checked that the ODEs governing these new variables are

$$\frac{du_j}{dt} = u_j \sum_{k=1}^{N} M_{jk} u_k; \ j = 1, \ldots, N \tag{15}$$

with $M = BA$ which, precisely, is the invariant matrix labelling the equivalence class. System (15) can also be seen as a N-dimensional QP system with $A = M$ and $B = I$ where I is the $N \times N$ identity matrix. In fact, this type of systems is well known under the name of Lotka-Volterra equations. The nonlinearity of these equations is the simplest that can exist: quadratic nonlinearity. The Lotka-Volterra systems are the subject of an extensive literature and all the results reported in it can apply to all the QP systems belonging to the equivalence class labelled by the matrix M.

The second canonical form related to the same equivalence class is obtained from the Lotka-Volterra system (15) by just making a transformation of the form (9) with $C = M$:

$$u_k = \prod_{j=1}^{N} \tilde{u}_j^{M_{kj}}; \ k = 1, \ldots, N. \tag{16}$$

This transformation implies that matrix M is invertible. If this is the case, following Eqs. (11) and (12), the QP system in the new variables $\tilde{u}_j$ has the new matrices $\tilde{A} = C^{-1}A = M^{-1}M = I$ and $\tilde{B} = BC = IM = M$ and reads

$$\frac{d\tilde{u_j}}{dt} = \tilde{u_j} \prod_{k=1}^{N} \tilde{u}_k^{M_{jk}}; \ j = 1, \ldots, N. \tag{17}$$

This is the second canonical form. Though, it is equivalent to the Lotka-Volterra system by the transformation (16), its fame is not comparable with that of the Lotka-Volterra equations. To my knowledge, it only appears in the context of the QP systems theory that is reported here, and in the theory of the urn processes as we shall see below. In absence of well-established name, we called this canonical form, the Monomial system. This name reflects its peculiarity of having right-hand sides that are monomials or, more generally, quasi-monomials of the dependent variables.

Before ending this section, let us remark that the transformation of the general QP system form (8) into the Lotka-Volterra and the monomial forms is always possible even if the matrices B and M are not invertible. This is shown in reference [6] and references therein. Essentially, this comes from the fact that when these matrices are of non-maximal rank r, the QP system or its Lotka-Volterra canonical form reduces to a $r-$dimensional QP system or to a r-dimensional Lotka-Volterra system. The smaller matrices B_{red} and M_{red} of these reduced systems are of maximal rank r and can be inverted.

3 Urn Processes

A urn process [7] requires four items: a box (the so-called urn), objects (e.g. balls) differing by only one characteristics (e.g. colour), an infinite reservoir of such objects and a prescribed set of replacement rules of the objects in the urn. Here, we shall consider balls with N possible colours. The replacement rules make the composition of the urn evolve at each discrete step. Each such step consists in

1. Picking a ball at random in the urn, with equal chance for all balls present in the urn.
2. The colour of the picked ball is noted and the ball is reintroduced in the urn.
3. Depending on the colour of the drawn ball, prescribed numbers of balls of each colour are taken from the reservoir and transferred to the urn: If the ball drawn from the urn is of colour i then one has to transfer M_{ij} balls of colour j from the reservoir into the urn, with $i, j \in \{1, \dots, N\}$. These natural numbers M_{ij} form a $N \times N$ matrix, the replacement matrix M.

Some entries of M can be negative, in which case the balls of the corresponding colours are transferred from the urn to the reservoir at each step in which these entries are required. If some of the diagonal entries are negative, conditions must be imposed in order to avoid blocking the process. The distribution of balls of each colour in the urn at a given step n forms its composition vector at that step, U_n and the sequence $(U_n;\ n \geq 0)$ represents the evolution of the urn process, where the initial composition U_0 of the urn is given.

We shall consider urn processes for which the total number of balls transferred at each step is constant. This implies that

$$\sum_{j=1}^{N} M_{ij} = \sigma. \tag{18}$$

The parameter σ is called the balance, and urn processes fulfilling the above condition are called balanced urn processes. In the sequel, we shall only consider balanced urns.

4 Equivalence Between Balanced Urn Processes and the QP Differential systems

As mentioned in the Introduction, there is a link between the Monomial system that plays the rôle of a canonical form in the QP systems theory on the one hand, and the balanced urn processes, on the other hand. This result obtained by Ph. Flajolet, Ph. Dumas and V. Puyhaubert in their seminal paper [8] constitutes the first step in establishing an equivalence between the two domains of random urn processes and deterministic dynamical systems.

Let us briefly summarize their finding.

A balanced urn with balls of N colours is considered with a given replacement matrix M. The initial composition vector of the urn $U_0 = (u_{10}, \ldots, u_{N0})$ is given.

Let us define the history of length n of the urn process as a succession of n replacement steps of the urn's content starting from the initial composition vector U_0. In other words, this is a trajectory of the urn's content in the N-dimensional space of composition vectors. Since the urn process is balanced, the probability of a history of length n to be realized is uniform. As a result, it is easy to determine the probability of finding in the urn a composition vector $U = (u_1, \ldots, u_N)$ at step n after starting from the initial composition vector U_0 . It is, clearly, given by the ratio of the number of histories of length n starting at U_0 and ending at U over the total number of possible histories of length n starting at U_0. This can be stated as follows:

$$P(U_n = U \mid U_0) = \frac{\left[x_1^{u_1} \ldots x_N^{u_N} z^n\right] H(x_1, \ldots, x_N, u_{10}, \ldots, u_{N0}, z)}{[z^n]\, H(1, \ldots, 1, u_{10}, \ldots, u_{N0}, z)}, \tag{19}$$

where the notation $[x^m]\, S(x)$ for a power series $S(x)$ corresponds to the coefficient of x^m in that series and where the counting generating function is defined as

$$H(x_1, \ldots, x_N, u_{10}, \ldots, u_{N0}, z) \equiv \sum_{n:0}^{\infty} \sum_{u_1:0}^{\infty} \cdots \sum_{u_N=0}^{\infty} H_n(u_{10}, \ldots, u_{N0}, u_1, \ldots, u_N) x_1^{u_1} \ldots x_N^{u_N} \frac{z^n}{n!}, \tag{20}$$

where $H_n(u_{10}, \ldots, u_{N0}, u_1, \ldots, u_N)$ counts the number of histories connecting in exactly n steps the initial composition vector $U_0 = (u_{10}, \ldots, u_{N0})$ to the given composition vector $U = (u_1, \ldots, u_N)$. One can see that the probability given by (19) is known once the counting generating function $H(x_1, \ldots, x_N, u_{10}, \ldots, u_{N0}, z)$ is known. Hence, the central goal to be reached is to determine that function. This is what Flajolet, Dumas and Puyhaubert succeeded to do. They showed that

$$H(x_1, \ldots, x_N, u_{10}, \ldots, u_{N0}, z) = [X_1(z)]^{u_{10}} \ldots [X_N(z)]^{u_{N0}}, \tag{21}$$

where the functions $X_1(z), \ldots, X_N(z)$ are the components of the solution vector at time $t = z$ of the system of ODEs:

$$\frac{dX_i}{dt} = X_i \prod_{j=1}^{N} X_j^{M_{ij}} \tag{22}$$

with initial conditions $X_1(0) = x_1, \ldots, X_N(0) = x_N$.

Amazingly, this result immediately shows a link with the theory of QP systems exposed in the previous section. It is quite obvious that the above system of ODEs is exactly the monomial system that appeared in Sect. 2. There it was shown to be one of the two canonical systems characterizing an equivalence class of QP systems with matrices A and B such that $BA = M$.

Now, considering the properties derived in Sect. 2, if one identifies this matrix M with the replacement matrix of a balanced urn process, one gets the following result:

A balanced urn process with replacement matrix M is not only equivalent to a monomial differential system like (22) *but, also, to all the QP systems that are equivalent to it under the quasi-monomial transformations* (9).

As announced in the Introduction, this theorem establishes an equivalence relation between a class of random processes, the balanced urn processes and a class of deterministic differential systems, the QP systems. More precisely, this equivalence is an isomorphism between the set of balanced urn processes and the *quotient* of the set of QP systems with respect to the group QM of quasi-monomial transformations (9). This can be expressed shortly as Urn Processes≅ QP Systems/QM.

This theorem has already been reported in a previous article [9]. In the sequel of the present article, we illustrate this result with four detailed examples.

5 Examples

5.1 Lotka-Volterra Systems

Lotka-Volterra systems have been first introduced in order to model the evolution of populations of animal and plant species in interaction in a given territory. They are of the general form:

$$\frac{dx_i}{dt} = \lambda_i x_i + x_i \sum_{j=1}^{n} N_{ij} x_j \qquad \text{for } i = 1, \ldots, n \tag{23}$$

where matrix N is the interaction matrix between species. It is also called the community matrix.

In order to make the connection with the QP systems as given by Eq. 8, one has to increase the dimension of the above system from n to $n + 1$. Since it is a sum of linear and quadratic terms, we can write it in the QP format:

$$\frac{dx_i}{dt} = x_i \sum_{j=1}^{n+1} A_{ij} \prod_{k=1}^{n} x_k^{B_{jk}} \qquad \text{for } i = 1, \ldots, n \tag{24}$$

with

$$A_{ij} = N_{ij} \ ; \ i, j = 1, \ldots, n \text{ and } A_{i\,n+1} = \lambda_i \tag{25}$$

$$B_{jk} = \delta_{jk}; \ j, k = 1, \ldots, n \text{ and } B_{n+1\,k} = 0. \tag{26}$$

Introducing the new variables:

$$u_j = \prod_{k=1}^{n} x_k^{B_{jk}}; \ j = 1, \ldots, n+1 \tag{27}$$

one gets the $(n+1)$-dimensional form of (23):

$$\frac{du_j}{dt} = u_j \sum_{k=1}^{N} M_{jk} u_k; \ j = 1, \ldots, n+1 \tag{28}$$

with matrix M given by

$$M_{jk} = N_{jk} \ ; \ j, k = 1, \ldots, n \tag{29}$$

$$M_{j\,n+1} = \lambda_j \ ; \ j = 1, \ldots, n$$

$$M_{n+1\,k} = 0 \ ; \ k = 1, \ldots, n+1$$

Using the equivalence theorem of Sect. 4, one, thus, can say that the population dynamical equations of form (23) are equivalent with balanced urn processes. However, the balance constraint imposes conditions on the interaction matrix N:

$$\sum_{l=1}^{n+1} M_{jl} = \sum_{l=1}^{n} N_{jl} + \lambda_j = \sigma \ ; \ j = 1, \ldots, n+1. \tag{30}$$

For $j = n + 1$, Eq. (29) impose

$$\sum_{l=1}^{n+1} M_{n+1\,l} = 0. \tag{31}$$

Consequently, one must have

$$\sigma = 0 \tag{32}$$

and Eq. (30) becomes

$$\sum_{l=1}^{n} N_{jl} = -\lambda_j \ ; \ j = 1, \ldots, n. \tag{33}$$

Hence, due to the balance condition, the (n+1)-colour urn process with the above-described matrix M as replacement matrix is equivalent to the following population dynamical Lotka-Volterra system:

$$\frac{dx_i}{dt} = x_i \sum_{j=1}^{n} N_{ij}(x_j - 1) \qquad \text{for } i = 1, \ldots, n. \tag{34}$$

Remarkably, for $n = 3$, this system has been shown by A. Arneodo, P. Coullet and C. Tresser to have a chaotic attractor in a domain of the parameter space [10]. What are the effects of such a chaotic regime on the urn process? This question remains open.

Also, another result is that all the QP systems that belong to the equivalence class characterized by the matrix M given by (29) have a chaotic attractor for parameters values derived from the parameter values for which Eq. (34) has a strange attractor (for $n = 3$).

5.2 *The Lorenz System*

The Lorenz system:

$$\frac{dx_1}{dt} = \sigma(x_2 - x_1) \tag{35}$$

$$\frac{dx_2}{dt} = \rho x_1 - x_2 - x_1 x_3$$

$$\frac{dx_3}{dt} = -\beta x_3 + x_1 x_2$$

is a simplification of the hydrodynamical equations for an incompressible thermal convection flow in the presence of gravity.

As is well known, it possesses chaotic solutions in a certain domain of the parameter space. In other words, for these values of the parameters, the trajectories are asymptotically attracted towards a strange attractor.

This system can be cast into the QP format (8) for $n = 3$ and $N = 5$, with the following matrices A and B:

$$A = \begin{bmatrix} \sigma & 0 & 0 & 0 & -\sigma \\ 0 & \rho & -1 & 0 & -1 \\ 0 & 0 & 0 & 1 & -\beta \end{bmatrix} \tag{36}$$

$$B = \begin{bmatrix} -1 & 1 & 0 \\ 1 & -1 & 0 \\ 1 & -1 & 1 \\ 1 & 1 & -1 \\ 0 & 0 & 0 \end{bmatrix}. \tag{37}$$

The matrix M of the associated Lotka-Volterra and monomial canonical systems is

$$M = BA = \begin{pmatrix} -\sigma & \rho & -1 & 0 & \sigma - 1 \\ \sigma & -\rho & 1 & 0 & 1 - \sigma \\ \sigma & -\rho & 1 & 1 & 1 - \sigma - \beta \\ \sigma & \rho & -1 & -1 & -1 - \sigma + \beta \\ 0 & 0 & 0 & 0 & 0 \end{pmatrix}. \tag{38}$$

At this point, one should mention the work of Kozlov and Vakulenko [11]. In their article, they find a 10-dimensional Lotka-Volterra system with linear term of form (23) that would be dynamically equivalent to the Lorenz system. This means that the corresponding Lotka-Volterra without linear term is 11 dimensional. In contrast, we find here a five-dimensional Lotka-Volterra without linear term that is equivalent to the Lorenz system.

Now, let us come to the equivalence of the Lorenz system with a random urn process. This equivalence is realized if the matrix M fulfills the following balance condition:

$$\sum_{j=1}^{5} M_{ij} = \sigma\, ;\; i = 1, \ldots, 5. \tag{39}$$

However, since the last row of that matrix is zero, $\sigma = 0$, which leads to

$$\sum_{j=1}^{5} M_{ij} = 0\, ;\; i = 1, \ldots, 5. \tag{40}$$

These conditions are satisfied by the Lorenz system's parameters for arbitrary σ, $\rho = 2$ and $\beta = 1$. Notice that unfortunately the same symbol sigma is used for the balance parameter and the first parameter of the Lorentz system. They are of course independent.

5.3 Asymmetric Top with Dissipation

The dynamics of an asymmetric top with dissipation satisfies the following system of ODEs:

$$\frac{dx_1}{dt} = \lambda_1 x_1 + \alpha x_2 x_3 \tag{41}$$

$$\frac{dx_2}{dt} = \lambda_2 x_2 + \beta x_1 x_3$$

$$\frac{dx_3}{dt} = \lambda_3 x_3 + \gamma x_1 x_2$$

.

The coefficients α, β, γ are the inertia momenta while $\lambda_1, \lambda_2, \lambda_3$ are the dissipation coefficients.

Several operations must be performed in order to put this system in the QP form (8). First, the dependent variables are transformed by $u_i = e^{-\lambda_i t} x_i$ for $i = 1, 2, 3$. Next, a supplementary dependent variable $u_4 = e^{-t}$ is added, and its ODE is added to system (41). This transforms system (41) into

$$\frac{du_1}{dt} = \alpha u_2 u_3 u_4^{\mu_1} \tag{42}$$

$$\frac{du_2}{dt} = \beta u_1 u_3 u_4^{\mu_2}$$

$$\frac{du_3}{dt} = \gamma u_2 u_1 u_4^{\mu_3}$$

$$\frac{du_4}{dt} = -u_4$$

with the definitions $\mu_1 = \lambda_1 - \lambda_2 - \lambda_3$, $\mu_2 = \lambda_2 - \lambda_1 - \lambda_3$, $\mu_3 = \lambda_3 - \lambda_1 - \lambda_2$.

Now, system (42) can be written in the QP form (8) with $n = N = 4$ and matrices:

$$A = \begin{pmatrix} \alpha & 0 & 0 & 0 \\ 0 & \beta & 0 & 0 \\ 0 & 0 & \gamma & 0 \\ 0 & 0 & 0 & -1 \end{pmatrix} \tag{43}$$

$$B = \begin{pmatrix} -1 & 1 & 1 & \mu_1 \\ 1 & -1 & 1 & \mu_2 \\ 1 & 1 & -1 & \mu_3 \\ 0 & 0 & 0 & 0 \end{pmatrix}. \tag{44}$$

The matrix M of the two canonical forms is obtained as follows:

$$M = BA = \begin{pmatrix} -\alpha & \beta & \gamma & -\mu_1 \\ \alpha & -\beta & \gamma & -\mu_2 \\ \alpha & \beta & -\gamma & -\mu_3 \\ 0 & 0 & 0 & 0 \end{pmatrix}. \tag{45}$$

Finally, in order to be equivalent to a balanced urn process, the parameters must fulfill the balance condition for the matrix M:

$$\sum_{j=1}^{4} M_{ij} = \sigma \; ; \; i = 1, \ldots, 4. \tag{46}$$

Again, here, due to the last row containing only zeroes, one has $\sigma = 0$. Hence, the balance condition is now

$$\sum_{j=1}^{5} M_{ij} = 0 \; ; \; i = 1, \ldots, 5. \tag{47}$$

Using these conditions along with the definitions of μ_1, μ_2, μ_3, one gets

$$\alpha = -\lambda_1, \; \beta = -\lambda_2, \; \gamma = -\lambda_3. \tag{48}$$

These conditions are compatible with the physical meaning of inertia momenta and dissipation coefficients of the parameters. Consequently, with these relations, the deterministic asymmetric top with dissipation is equivalent to a balanced urn random process.

5.4 Cosmological Dynamics

In a remarkable article, J. Perez and co-authors [12] showed that a good model for cosmological expansion is given by a Lotka-Volterra system. More precisely, they demonstrated that the dynamics of a homogeneous and isotropic Friedmann-Lemaître (FL) expanding universe containing an arbitrary number of interacting cosmological fluids is well described by a Lotka-Volterra system. Among these fluids filling the universe, there are the baryonic matter, radiation, dark matter and dark energy. In this picture, the fluids are viewed as species that compete for feeding upon their unique energy resource, curvature, while interacting one with the others.

Let us briefly sketch the general framework. The universe in which these fluids evolve obeys the Einstein equations with a non-vanishing cosmological constant Λ. Among the solutions of the Einstein equation, the FL metric is chosen. It describes an expanding homogeneous, isotropic universe. This metric involves the scale factor $a(t)$ where t is the synchronous time. The fluids filling this universe are assumed to be barotropic. That is, each fluid of index i obeys an equation of state of the kind $\rho_i = \omega_i p_i$ where ρ_i is the energy density of the fluid, ω_i is its barotropic index with $-1 \leq \omega_i \leq 1$, and p_i is the pressure. Instead of using the synchronous time t, a better adapted variable is considered, $\lambda = \ln a(t)$. The two variables t and λ are related by $d\lambda = H\, dt$ where H is the Hubble constant, $H = \frac{1}{a}\frac{da}{dt}$.

In terms of these quantities, the system of ODEs describing the evolution in λ of the energy densities associated to the different fields, and also to the dark energy Λ and to the curvature, are given by

$$\frac{dx_i}{d\lambda} = r_i x_i + x_i \sum_{j=1}^{N} A_{ij} x_j \qquad \text{for } i = 1, \ldots, N, \tag{49}$$

where

$$x_i = \frac{8\pi G \rho_i}{3H^2} \tag{50}$$

$$r_i = -1 - 3\omega_i$$

$$A_{ij} = 1 + 3\omega_j + \varepsilon_{ij}$$

with $\varepsilon_{ji} = -\varepsilon_{ij}$ represents the coupling constant between the two fluids i and j, while G is the universal constant of gravity. The antisymmetry of the tensor ε_{ij} ensures the condition of energy balance $\sum_{i=1}^{N} \sum_{j=1}^{N} \varepsilon_{ij} x_i x_j = 0$.

Now, let us see whether this deterministic dynamical system is equivalent to a possible random urn process. The N-dimensional system of ODEs (49), clearly, is of the Lotka-Volterra type with linear term and, thus, is similar to Eq. (23). Following the same reasoning leading from Eqs. (23) to (28), system (49) is reshaped in a $(N + 1)$-dimensional Lotka-Volterra system without linear term as in Eq. (28).

$$\frac{du_j}{dt} = u_j \sum_{k=1}^{N+1} M_{jk} u_k; \;\; j = 1, \ldots, N + 1. \tag{51}$$

The matrix M appearing in this system is given by

$$M_{jk} = A_{jk}\,; \;\; j, k = 1, \ldots, N \tag{52}$$

$$M_{j\,N+1} = r_j\ ;\ j = 1, \ldots, N$$

$$M_{N+1\,k} = 0\ ;\ k = 1, \ldots, N+1$$

.

As explained in Sect. 2, system (51) is a canonical form for an equivalence class of QP systems. Its matrix M is also the matrix of the second canonical form, the Monomial system. According to the equivalence theorem between balanced urn processes and deterministic QP systems stated at the end of Sect. 4, this matrix is also the replacement matrix of a balanced urn. However, this is true only if the balance condition of the urn process is fulfilled. This condition imposes constraints on the matrix M, as we see now.

Indeed, the balance of the urn imposes

$$\sum_{l=1}^{N+1} M_{jl} = \sum_{l=1}^{N} A_{jl} + r_j = \sigma\ ;\ j = 1, \ldots, N+1. \tag{53}$$

However, for $j = N+1$, Eq. (53) along with Eq. (52) gives

$$\sigma = 0. \tag{54}$$

Inserting the two last equations of (50) in (53) along with $\sigma = 0$ leads to

$$\sum_{j=1}^{N} \varepsilon_{ij} = 1 - N - 3 \sum_{j=1; j \neq i}^{N} \omega_j\ ;\ i = 1, \ldots, N \tag{55}$$

So, with the above conditions, one can say that there exists a balanced urn process, a random process, that is equivalent to the cosmological evolution described by the deterministic system (49).

6 Conclusive Remarks and Perspectives

In the Introduction, we have compared two types of equivalence between random processes and deterministic equations, the equivalence SDE $\cong$ FPE and the equivalence Urn Processes $\cong$ QP Systems/QM. One should stress the difference between the two relations. In the first equivalence, the physical system appears in the random side of the equivalence, SDE. While, in the second equivalence, the physics appears in the deterministic side of the equivalence, QP Systems/QM. Furthermore, in the first equivalence, the physical model is modified by the introduction of random external forces, while in the second equivalence the physical system is unchanged.

Let us also remark an important point that did not seem to bother us when we treated the examples. The entries of the replacement matrix of an urn process must be integers. This is, of course, due to the discrete nature of balls. However, in the examples we discussed in Sect. 5, there was no physical reason for having only integer values of the parameters. In all the considered cases, the population dynamics, the Lorenz system, the asymmetric top or the cosmological evolution, Physics does not limit the parameters to integer values. Integer parameters would be non-generic in these contexts. So, the question is, is it possible to extend the concept of balanced urn process to real entries in the replacement matrix? The answer is positive and is developed in many articles among which the following [13–15]. The essential point in what concerns us is that the extension to real values of the entries of the replacement matrix does not affect the main result: The isomorphism between the class of balanced urn processes and the quotient of the class of QP differential systems by the group of quasi-monomial transformations is maintained. This will be the object of a next publication.

Another extension of the balanced urn process is possible. The entries of the replacement matrix could be themselves random variables with given probabilities. Hence, the randomness of the urn process is, in some way, increased. Instead of limiting the randomicity to the act of drawing a ball of a given colour from the urn, in this generalization the replacement of balls required by the colour of the picked ball also becomes random. Such an extension has been considered in several works. In one of them, the Flajolet-Dubois-Puyhaubert theorem is extended [16]. This result proves that the equivalence between urn processes and QP differential systems extends to urns with random replacement matrices. Work is underway on this subject.

As a practical perspective, the property of equivalence discussed in this article could be exploited in order to develop a new approach to simulate numerically deterministic dynamical systems. It could, indeed, prove to be faster and easier to draw virtual coloured balls on computer than run Runge-Kutta or other integration schemes to solve the differential equations of the deterministic description.

A last reflection should be made about the equivalence between random processes such as balanced urn processes, and deterministic differential systems such as the QP systems. This equivalence is not without philosophical consequences. It is particularly striking in the example we have considered about the equivalence of the deterministic cosmological evolution equations with a large urn process. Indeed, this seems to mean that the evolution of the universe is at the same time deterministic and random. *Se non è vero, è bene trovato*!

References

1. A. Einstein, Zur Theorie der Brownschen Bewegung. Annalen der Physik **324**, 371–381 (1906)
2. M. Smoluchowski, Zur kinetischen Theorie der Brownschen Molekularbewegung und der Suspensionen. Annalen der Physik **326**, 756–780 (1906)
3. P. Langevin, Sur la théorie du mouvement brownien. C. R. Acad. Sci. Paris. **146**, 530–533 (1908)
4. V.V. Belyi, Fluctuation-dissipation relation and quality factor for slow processes. Phys. Rev. E **69**, 017104 (2004)
5. B. Øksendal, *Stochastic Differential Equation: An introduction with Applications.* (Springer, 2013)
6. L. Brenig, Reducing nonlinear dynamical systems to canonical forms. Phil. Trans. R. Soc. A **376**, 20170384 (2018)
7. H.M. Mahmoud, Pólya Urn Models. (Chapman & Hall/CRC, 2008)
8. Ph. Flajolet, Ph. Dumas, V. Puyhaubert, Some exactly solvable models of urn process theory, Fourth Colloquium on Mathematics and Computer Science Algorithms, Trees, Combinatorics and Probabilities, 2006, Nancy, France. Discrete Mathematics and Theoretical Computer Science, AG, pp. 59–118 (DMTCS Proceedings, 2006). HAL Open Science: hal-01184710
9. L. Brenig, I. Gléria, T.M. Rocha Filho, A. Figueiredo, B. Hernàndez-Bermejo, Equivalence between nonlinear dynamical systems and urn processes. J. Phys. A: Math. Theor. **51**, 485101 (2018)
10. A. Arneodo, P. Coullet, C. Tresser, Strange attractors in Volterra equations for species in competitions. Phys. Lett. **79A**, 259–263 (1980)
11. V. Kozlov, S. Vakulenko, On chaos in Lotka-Volterra systems: an analytical approach. Nonlinearity **26**, 2299–2314 (2013)
12. J. Perez, A. Füzfa, T. Carletti, L. Mélot, L. Guedezounme, The Jungle Universe: coupled cosmological models in a Lotka–Volterra framework. Gen. Relativ. Gravit. **46**, 1753 (2014). arXiv: 1306.1037v2
13. R. Pemantle, A survey of random processes with reinforcement. Probab. Surv. **4**, 1–79 (2007)
14. A. Dasgupta, K. Maulik, Strong laws for urn models with balanced replacement matrices. Electron. J. Probab. **16**, 1723–1749 (2011)
15. C. Mailler, J.-F. Marckert, Measure-valued Polya urn processes. Electron. J. Probab. **22**, 1–33 (2017)
16. B. Morcrette, H.M. Mahmoud, Exactly solvable balanced tenable urns with random entries via the analytic methodology. Discrete Mathe. Theor. Comput. Sci. (DMTCS) Proceed. AQ, 219–232 (2012)

Langton's Ant as an Elementary Turing Machine

Jean Pierre Boon

Abstract The automaton known as 'Langton's ant' describes the step by step motion of an elementary object (the ant) on a square lattice. Yet despite the simplicity of its rule, the automaton exhibits amazing dynamical behaviour leading to a propagation phase where the particle (the ant) dynamics produce a regular periodic pattern (called 'highway'). The dynamical pattern so created by the ant on the highway produces a recurring mechanism for an elementary Turing machine (Boon in J Stat Phys 102:355, 2001 [1]).

1 Introduction

The automaton known as 'Langton's ant' was devised by Langton [2] and its dynamics are illustrated in Fig. 1; it has been a recurring theme in the mathematical and physical literature [3]. There are two reasons. The first is of physical relevance: the automaton known as Langton's ant (which I describe below) offers a paradigm of complexity out of simplicity. The second reason is mathematical: despite the simplicity of the basic algorithm, the analytical description of the spatio-temporal dynamics generated by the automaton exhibits propagation dynamics which obeys a general difference equation and the speed of the ant in the highway ($c = \sqrt{2}/52$) follows exactly from the equation whose continuous limit gives a propagation-dispersion equation [4].

This paper is dedicated to the memory of my friend Slava Belyi.

J. P. Boon (✉)
Physics Department CP 231, Universite Libre de Bruxelles, Brussels, Belgium
e-mail: Jean-Pierre.Boon@ulb.be

L. Brenig et al. (eds.), *Nonequilibrium Thermodynamics and Fluctuation Kinetics*, Fundamental Theories of Physics 208,
https://doi.org/10.1007/978-3-031-04458-8_7

2 The Automaton Process

The basic process governing the automaton dynamics follows a simple rule. The automaton universe is the square lattice with checker board parity, therefore, defining H sites and V sites. A particle moves from site to site (by one lattice unit length) in the direction given by an indicator. One may think of the indicator as a 'spin' (up or down) defining the state of the site. When the particle arrives at a site with spin up (down), it is scattered to the right (left) making an angle of $+\pi/2$ $(-\pi/2)$ with respect to its incoming velocity vector. But the particle modifies the state of the visited site (up $\Longleftrightarrow$ down) so that on its next visit, the particle is deflected in the direction opposite to the scattering direction of its former visit. Thus, the particle entering from below an H site with spin-up is scattered East, and on its next visit to that same site (now with spin down), if it arrives from above, it will be scattered East again, while if it arrives from below, it will be scattered West. Similar reasoning shows how the particle is scattered North or South on V sites.

At the initial time, all sites are in the same state (all spins up or all spins down), and the position and velocity direction of the particle are fixed, but arbitrary. So if we paint the sites black or white according to their spin state, we start initially with say an all white universe. Then as the particle moves, the visited sites turn alternately black and white depending on whether they are visited an odd or even number of times. This colour coding offers a way to observe the evolution of the automaton universe. The particle starts exploring the universe by first creating centrally symmetric transient patterns (see figures in Ref. [3]), then after about 10 000 time steps (9977 to be precise), it leaves a seemingly 'random territory' to enter a 'highway' (see Fig. 1). Note that this 'disordered phase' is not what a random walk would produce: the automaton is deterministic and its rules create correlations between successive states of the substrate, so also between successive positions of the particle. The power spectrum computed from the particle position time correlation function measured over the first 9977 time steps goes like $\sim\nu^{-\zeta}$ with $\zeta \simeq 4/3$. In the ordered phase ('highway'), the power spectrum shows a peak at $\nu = 1/104$ with harmonics. In the highway, the trajectory shows a periodic pattern where the particle travels with a constant propagation speed.[1] Here, I show analytically that the propagation speed is $c = \sqrt{2}/52$ (in lattice units) as measured in automaton simulations [3].

Because of the complexity of the dynamics on the square lattice, Grosfils et al. [5] developed a one-dimensional version of the automaton for which they provided a complete mathematical analysis which is also applicable to the two-dimensional triangular lattice. One of their main results is the mean-field equation describing the microscopic dynamics of the particle subject to the more general condition that the spins at the initial time are randomly distributed on the lattice. The equation reads, for the one-dimensional lattice

$$f(r+1,t) = q\, f(r,t-1) + (1-q)\, f(r,t-3)\,, \tag{1}$$

[1] A theorem by Bunimovich and Troubetzkoy [6] demonstrates that the automaton fulfils the conditions for unboundedness of the trajectory of the particle.

and, for the two-dimensional triangular lattice

$$f(r+1,t) = q(1-q)\, f(r,t-2) + [q^2 + (1-q)^2]\, f(r,t-8) \\ + (1-q)q\, f(r,t-14)$$

where $f(r,t)$ is the single particle distribution function, i.e. the probability that the particle visits site r *for the first time* at time t, and q is the probability that the immediately previously visited site along the propagation strip (the highway) has initially spin up, i.e. the probability that the particle be scattered, in the one-dimensional case, along the direction of its velocity vector when arriving at the scattering site at $r-1$, and in the two-dimensional triangular case, along the direction forming clockwise an angle of $+2\pi/3$ with respect to the incoming velocity vector of the particle. Equations (1) and (2) express the probability of the first visit to a site along the propagation strip in terms of the probability of an earlier visit to the previous site along the strip.[2] The equations were shown to yield exact solutions for propagative behaviour in the two classes of models considered by Grosfils et al. [5].

Equations (1) and (2) are particular cases of the following general equation

$$\sum_{j=0}^{n} f(r+\rho,t) = \sum_{j=0}^{n} p_j(q)\, f(r,t-\tau_j) \;\; ; \;\; \tau_j = (1+\alpha j)\, m\, \tau \;\; , \qquad (2)$$

with $\sum_{j=0}^{n} p_j = 1$ ($n \le r/\rho$) and where $f(r,t)$ is the *first visit* distribution function. Here ρ denotes the elementary space increment of the dynamics along the propagation strip; p_j is the probability that the particle propagates from r to $r+\rho$ in τ_j time steps, i.e. τ_j is the time delay between two successive first visits on the strip (more precisely on the one-dimensional edge of the strip) for the path with probability p_j, and m is the corresponding minimum number of automaton time steps ($\tau_0 = m\tau$, where τ is the automaton time step; $\tau = 1$). The sum is over all possible time delays, weighted by the probability p_j (a polynomial function of q). α denotes the number of lattice unit lengths in an 'elementary loop', i.e. the minimum number of lattice unit lengths necessary to return to a site.[3] Equation (2) implies the assumption that first visits occur after a finite number of recurrences (n finite in Eq. (2)), i.e. a finite number of possible paths (not identical loops) between two successive first visits; this defines a general class of automata which includes the 1-D, 2-D triangular and square lattice models.

Now from the expectation value of the time delay, computed with (2)

$$E[\tau(q)] = \sum_{j=0}^{n} \tau_j\, p_j(q) = (1+\alpha\,\langle j\rangle)\, m\, \tau \;\; ; \;\; \langle j\rangle = \sum_{j=0}^{n} j\, p_j(q)\, , \qquad (3)$$

---

[2] In the two-dimensional case, the equation describes the one-dimensional propagation motion along the edge of the strip.

[3] An interesting equation follows from the continuous limit of (2) as discussed in [4].

one obtains immediately the average propagation speed of the particle: $c(q) = \rho/E[\tau(q)]$.

It is straightforward to verify that Eqs. (1) and (2) are particular cases of the general equation (2): for 1-D: $\alpha = 2,\ m = 1,\ \rho = 1,\ n = 1$, with $p_0 = q,\ p_1 = 1 - q$; for 2-D (triangular lattice): $\alpha = 3,\ m = 2,\ \rho = 1,\ n = 2$, with $p_0 = q\,(1-q),\ p_1 = q^2 + (1-q)^2,\ p_2 = (1-q)\,q$. The corresponding propagation speeds are then readily obtained from (3); for the one-dimensional case one finds $c(q) = 1/\langle\tau(q)\rangle = [1 + 2(1-q)]^{-1} = 1/(3-2q)$, and for the triangular lattice: $\langle\tau(q)\rangle = [1 + 3(q^2 + (1-q)^2 + 2q(1-q))] \times 2$, so that $c = 1/8$. These results are in exact agreement with those obtained in [5].

For the 2-D square lattice: $\alpha = 4,\ m = 2 \times 4,\ \rho = 2\sqrt{2}$. The value of ρ is easily checked by inspection of the highway path shown in the upper box of Fig. 1: it is the length of the elementary increment along the edge of the propagation strip. Correspondingly, m is 2×4 (the minimum number of time steps necessary to move one elementary space increment must be counted on each edge of the strip). For the square lattice, one does not know the value of n, but from the structure of the p_j's for the 1-D and 2-D triangular lattices given above, one can infer that $n = 6$, with $p_0 = p_6 = q^2\,(1-q)^2,\ p_1 = p_5 = q\,(1-q)\,[q^2 + (1-q)^2],\ p_2 = p_4 = p_0 + p_1,\ p_3 = [q^2 + (1-q)^2]^2$. However, the precise expressions are unimportant for the automaton describing Langton's ant, because all sites are initially in the same spin state; so $q = 1$, and only one p_j is non-zero: $p_3 = 1$. Equation (2) then reads

$$f(r + 2\sqrt{2}, t) = f(r, t - \tau_3) \ ; \ \tau_3 = (1 + 4 \times 3)\,2 \times 4 = 104\,, \qquad (4)$$

which describes the dynamics of the particle in the highway. This result shows that a displacement of length $2\sqrt{2}$ along the edge of the strip is performed in 104 automaton time steps. Consequently the propagation speed of Langton's ant in the highway is $c = \rho/\tau_3 = 2\sqrt{2}/104 = \sqrt{2}/52$.

Although the initial condition with all spins in the same state may appear as a particular configuration, it should not be considered as a non-typical one, in the sense that it produces propagation. In the 1-D and 2-D triangular lattices, propagation always occurs regardless of the initial spin configuration [5]. In the square lattice, propagation only occurs with all spins initially up or down (or periodically distributed[4]) is related to the fact that the scattering angle here is $\pm\pi/2$, which can be conjectured as an indication of criticality (at angles smaller than $\pi/2$, propagation is never observed).

[4] Propagating patterns with different modes of propagation depending on the periodicity of the spin distribution are discussed in [7, 8].

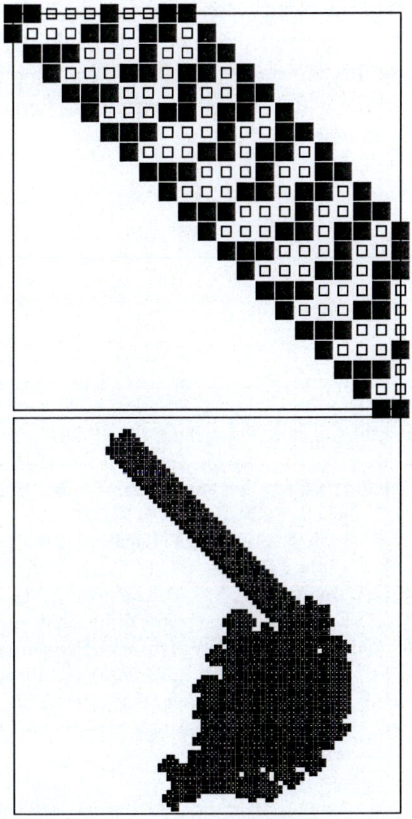

Fig. 1 Langton's ant propagation highway: the lower panel shows the area of the square lattice covered by the ant during its first phase looking apparently as a randomly covered area where from it emerges after 9977 steps into a 'highway' shown as a blowup in the upper figure and where it propagates describing a regular periodic pattern. The colour code of the black and white squares indicates whether the 'spin' is left up or down (left or right) after the last passage of the ant

3 Concluding Comments

The origin of particle propagation in 1-D and 2-D triangular lattices was shown to be a 'blocking mechanism' [5], and the question was raised as to whether such a mechanism also exits in the square lattice. Although the precise blocking mechanism has yet to be identified, that the same general equation, Eq. (2), describes propagation in 1-D, 2-D triangular and square lattices suggests that a similar blocking mechanism is responsible for the construction of Langton's highway.

Grosfils et al. [5] developed extensively this type of automaton in the case of the triangular lattice and in their conclusions they discuss an interesting 'reorganisation corollary' stating that *All sites located on one edge of the propagation strip are in the initial state of the sites on the other edge, shifted upstream by one lattice unit length*, a feature clearly illustrated in Fig. 1. The same corollary applies trivially to the spin states (up and down spins interchanged as 0's and 1's) on the edges of the highway of Langton's ant. From this observation, it follows that the particle dynamics functions as a control operator which transcribes and shifts the string of characters (0's and

1's) of the input tape (on one edge) to the output tape (on the other edge), i.e. (i) the control operator plays the role of the EXCHANGE gate in Feynman's model of a quantum computer [9], and (ii) can then be interpreted as the controller of a Turing machine which shifts the string of characters of the input tape (on one edge) to the output tape (on the other edge).

Acknowledgements I acknowledge fruitful discussions with Patrick Grosfils and James F. Lutsko.

References

1. The main body of the present paper appeared originally in: J.P. Boon, J. Stat. Phys. **102**, 355 (2001)
2. C.G. Langton, Phys. D **22**, 120 (1986)
3. D. Gale, *Tracking the Automatic Ant* (Springer, New York, 1998); S.E. Troubetskoy, Lewis-Parker lecture: the ant. Alabama J. Math. **21**, 3 (1997); A. Gajardo, A. Moreira, E. Goles, Discrete Appl. Math. **117**, 41 (2003)
4. J.P. Boon, P. Grosfils, J.F. Lutsko, J.Stat. Phys. **113**, 527 (2003)
5. P. Grosfils, J.P. Boon, E.G.D. Cohen, L.A. Bunimovich, J. Stat. Phys. **97**, 575 (1999)
6. L.A. Bunimovich, S.E. Troubetzkoy, J. Stat. Phys. **67**, 289 (1992); ibid. **74**, 1 (1994)
7. E.G.D. Cohen, New types of diffusion in lattice gas cellular automata, in *Microscopic Simulations of Complex Hydrodynamic Phenomena*, ed. by M. Mareschal, B. L. Holian. NATO ASI Series B: Physics, vol. 292 (1992), p. 145; and references in this publication
8. H.F. Meng, E.G.D. Cohen, Phys. Rev. E **50**, 2482 (1994)
9. A.J.G. Hey, R.W. Allen (eds.), *Feynman Lectures on Computation* (Addison-Wesley Publ., New York, 1996)

Kinetic Theory of Far-from-Equilibrium Processes

Phase Transitions in Active Matter Systems

Subir K. Das

Abstract This article reviews various aspects of phase transitions in active matter systems. Scaling phenomena in steady states as well as far-from-steady-states have been considered. The focus has been on systems where particles align their velocities along their neighbors. Such dynamic interactions are known to facilitate clustering. Wherever necessary, results and discussions are provided from relevant passive matter systems. Comparison between the two should help understand the influence of activity at a quantitative level.

1 Introduction

Active Matter systems [1–3] consist of self-propelling particles. These constituents perform motion by continuously drawing energy from the environment. Such systems are inherently away from equilibrium and pose challenging questions concerning nonequilibrium statistical physics [1–26].

Fascinating structures and dynamics are commonly observed in active matter systems. Some pictures in this connection are shown in Fig. 1. These are related to (Fig. 1a) a flock of birds [27], (Fig. 1b) a school of fish [28], (Fig. 1c) a herd of sheeps [29], and (Fig. 1d) a colony of bacteria [30]. In the first three systems, interesting structure in the density field is clearly visible. These snapshots also provide an indication about structure formation in directionalities of particle motion. A more complex and interesting pattern in the velocity field can be observed in Fig. 1d. This is related to the movement of bacteria in a colony.

In many active matter systems, interesting patterns form due to the alignment of velocities of the constituents. In a way, this is similar to pattern formation in systems of inelastically colliding granular particles [31–34]. In the latter system, after each collision, the particles move more parallel to each other. Such directional

S. K. Das (✉)
Theoretical Science Unit and School of Advanced Material, Jawaharlal Nehru Centre for Advanced Scientific Research, Jakkur P.O., Bangalore 560064, India
e-mail: das@jncasr.ac.in

L. Brenig et al. (eds.), *Nonequilibrium Thermodynamics and Fluctuation Kinetics*, Fundamental Theories of Physics 208,
https://doi.org/10.1007/978-3-031-04458-8_8

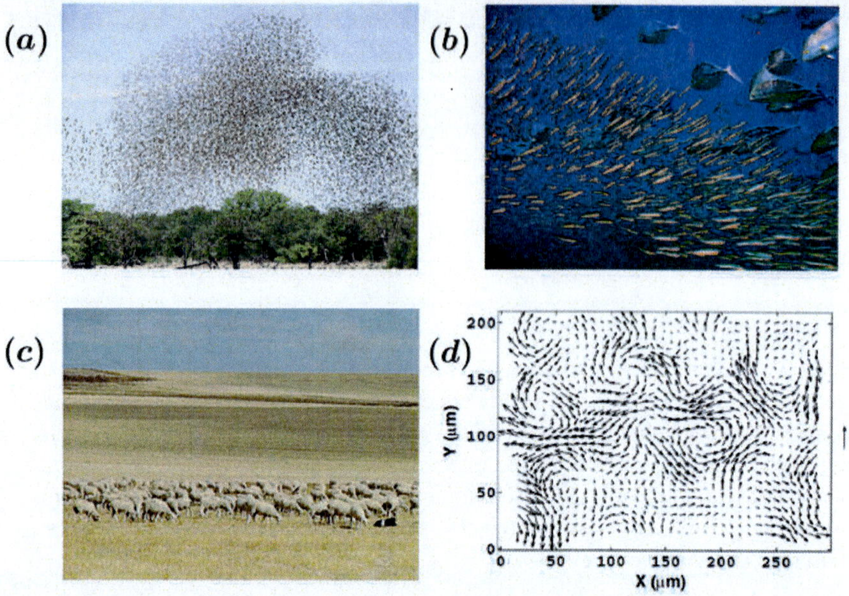

Fig. 1 Depiction of pattern formation in various active matter systems: **a** a flock of birds forming interesting dynamic pattern (*Source* Wikipedia); **b** a school of fish showing clustering and velocity alignment (*Source* wikipedia); **c** a herd of sheeps exhibiting clustering and directionality (*Source* Wikipedia); **d** velocity field in a colony of bacteria forming vortex-like topological defects. Reproduced with permission from C. Dombrowski et al., Phys. Rev. Lett. **93**, 098103 (2004). https://doi.org/10.1103/PhysRevLett.93.098103. For details on the sources and permissions for **a–c**, see text and cited references there

parallelization gives rise to instabilities in density and velocity fields that lead to structure and dynamics similar to phase transitions.

Phase transitions in active matters have received much attention recently [5–7, 9–17, 19, 21–24]. In addition to the formation and evolution of the above mentioned structures [7, 12, 15, 16, 24], belonging to the domain of coarsening phenomena, there exists serious interest in critical phenomena [5, 6, 9–11, 17] as well. Note that steady state in the active case is the counterpart of equilibrium in the passive situation. Critical phenomena in active matter systems are, thus, associated with nonequilibrium transitions in the steady state context. An important objective in this domain, like in the passive case [35–41], has been the identification and understanding of universality. A brief discussion concerning this is provided in the next section, primarily in the passive context.

It is important to note that phase transitions in active matter systems are not only related to the above mentioned alignment interactions. Fascinating structure and dynamics are observed even when constituents perform movements of different kinds, e.g., exhibit Brownian motion [23]. However, in this article, our focus will be on the former. Note, however, that typical active matter systems do not contain thermodynamically large number of particles. This way, in reality, it is expected that there will be rounding off of various anomalies. Nevertheless, it is an important theoretical challenge to figure out, say, via the application of scaling principles, the character of the transitions in the infinite size limit of the system.

2 Phase Transition: An Elementary Discussion

In Fig. 2a, we show a schematic phase diagram of a normal chemical system [39]. Along various lines, two different phases coexist with each other, e.g., solid–liquid coexistence occurs along the broken curve. At the triple point, all the three phases coexist with each other. The vapor–liquid coexistence curve terminates at a critical point. Across the coexistence curves, the density changes discontinuously, e.g., when a system moves from the liquid phase to the vapor phase. This jump tends to vanish as the state point gets closer to criticality. At and beyond the critical point, the system can be brought from one phase to the other without encountering a discontinuity in density.

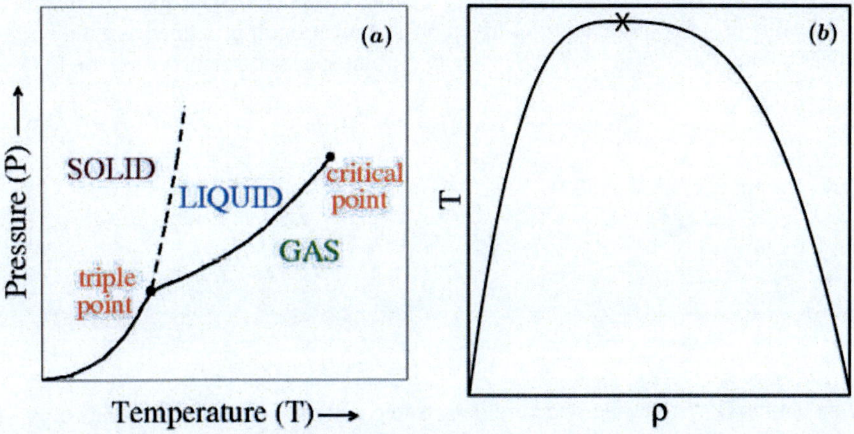

Fig. 2 **a** Schematic phase diagram of a normal chemical system in pressure (P) versus temperature (T) plane. Various phases, viz., solid, liquid, and vapor, are marked. The coexistence curves, along which two different phases coexist, being in equilibrium, are shown by lines. The triple point marks the location of the coexistence of all three phases. The vapor–liquid coexistence curve terminates at the critical point. **b** Coexistence curve for a vapor–liquid transition is schematically shown in temperature versus density plane. The critical point is marked by a cross (×). The left branch of the curve depicts the variation of density in the vapor phase and the right branch shows the same in the liquid phase, with the change of temperature

2.1 Critical Phenomena

The vapor–liquid coexistence curve is shown schematically in Fig. 2b in the temperature versus density plane [39]. The vapor and liquid branches meet each other at the critical point. Various thermodynamic and transport properties exhibit anomalous behavior in the critical vicinity. Below we note these for some of the thermodynamic quantities. The order parameter ψ ($= \rho_l - \rho_v$, ρ_l and ρ_v being the densities along the liquid and vapor branches, respectively), isothermal compressibility (κ_T), isochoric specific heat (C_v), and correlation length (ξ) exhibit the behavior [37, 39, 40]

$$\psi \sim \epsilon^{\beta}, \tag{1}$$

$$\kappa_T \sim \epsilon^{-\gamma}, \tag{2}$$

$$C_v \sim \epsilon^{-\alpha}, \tag{3}$$

and

$$\xi \sim \epsilon^{-\nu}. \tag{4}$$

Here, ϵ is the deviation of temperature (T) at the considered state point from the critical value (T_c) for a temperature-driven phase transition. It is defined as a dimensionless quantity as $\epsilon = |T - T_c|/T_c$. The two-point space correlation function [37]

$$C(r) = \langle \psi(\vec{r})\psi(\vec{0})\rangle - \langle \psi(\vec{r})\rangle\langle \psi(\vec{0})\rangle, \tag{5}$$

follows the functional form [37]

$$C(r) \sim r^{-p} e^{-r/\xi} \tag{6}$$

with

$$p = d - 2 + \bar{\eta}. \tag{7}$$

For an unbiased system, the structure is isotropic and the correlation depends only on the scalar separation (r) between two points, not on the vector distance ($\vec{r}$). In addition, at the critical point, there may exist anomaly in the order parameter with the variation of the relevant external field. For athermal systems, role analogous to temperature can be played by density [37, 39, 40].

In the above equations, β, γ, α, ν, and $\bar{\eta}$ are critical exponents. Other than that for the order parameter, the forms remain valid for an approach to T_c from either side. Typically, the values of the exponents are universal, i.e., they do not depend upon the type of materials. In fact, the universality is so robust that the values of the exponents are independent of the type of transition, e.g., for a para- to ferromagnetic transition across the Curie point or a liquid–liquid transition having a consolute point one observes the same value of an exponent as in a vapor–liquid transition.

Such universality applies to certain amplitude ratios as well. However, depending upon the type of transition, certain definitions may change, e.g., for a magnetic transition, magnetic susceptibility and for a liquid–liquid transition, concentration susceptibility, take the role of compressibility in a vapor–liquid transition [40]. Of course, based on the type of order parameter, e.g., scalar or vector, space dimension, and range of interaction, character of transition and universality may change. For scalar order parameter, if the range of interaction is sufficiently short, typical vapor–liquid transitions, para-to-ferromagnetic transitions, etc., are accurately described by the nearest-neighbor Ising model in terms of the values of the exponents as well as certain amplitude ratios [37, 40]. For this model, the above listed exponents in $d = 3$ are [37, 40]: $\beta \simeq 0.325$, $\gamma \simeq 1.239$, $\alpha \simeq 0.11$, $\nu \simeq 0.629$ and $\bar{\eta} \simeq 0.036$.

Coexistence curves similar to that in Fig. 2 have been obtained for active matter systems as well [3, 5, 6, 9, 19]. In the theoretical and computational literature, standard techniques are being used to estimate the critical point and exponents of relevant singularities for such nonequilibrium transitions. With respect to the existence and characterization of universalities in the active matter domain, however, there exists doubt and lack of consensus [10, 17]. Here the phase transitions are driven by particle motion. In the literature, on passive matter, it is known that differences in transport can alter the universality in dynamics (see below for discussion). In the case of active matter systems, it may be expected that the phase behavior will be different for different self-propulsion rules. It may not be surprising even if the values of the critical exponents for quantities analogous to the ones described above get altered, due to the changes in the self-propulsion rule.

In the equilibrium critical phenomena in passive systems, not only there exists anomaly in thermodynamic quantities, but the dynamics also become atypical. As the critical point is approached, i.e., ξ diverges, the slowest relaxation time, τ, of a system blows up as [38, 40–42]

$$\tau \sim \xi^z. \tag{8}$$

Here, z is a dynamic critical exponent. This is a signature of slow dynamics, a phenomenon often referred to as the critical slowing down. In the dynamical context, other interesting quantities are various transport properties of collective type, e.g., for a vapor–liquid transition, the thermal diffusivity (D_T), shear viscosity (η_v) and bulk viscosity (ζ_v) exhibit the behavior [38, 41, 42]

$$D_T \sim \xi^{-x_D} \tag{9}$$

$$\eta_v \sim \xi^{x_\eta} \tag{10}$$

$$\zeta_v \sim \xi^{x_\zeta}. \tag{11}$$

The exponents x_D, x_η, and x_ζ are positive quantities. Universality in this sub-domain is weaker, e.g., while a magnetic, a solid binary mixture, and a fluid system will give rise to the same static universality class, each of these will belong to different dynamic classes. Despite critical phenomena finding significant importance in the active matter context, this transport-related aspect, as described above, is starting to receive attention only recently [43]. This lack of understanding is interesting, particularly because, as stated above, the transitions in these systems are transport driven.

In addition to the critical phenomena, there exists interest in learning whether scaling picture, analogous to the passive systems [44–56], holds in approach to the steady state in the case of active matter system, following quench of a homogeneous system to a state point inside the miscibility gap. Once again, we provide discussion on this aspect from the perspective of passive matter.

2.2 *Coarsening Phenomena*

Evolution dynamics in systems undergoing phase transitions is characterized by pattern formation in the presence of interesting topological defects [44]. In three-dimensional geometry, for a scalar order parameter, the defects are two-dimensional surfaces or domain walls. For order parameters of more complex types, even more interesting structures arise, e.g., for vector order parameters in $d = 2$ or 3, interesting vortex-like topological defects may appear, with the dimension of the object decided by the competition between space dimension and the number of components in the order parameter. The defect dimension for a n-component vector field, n being unity for the scalar case, in d space dimensions, is given by [44] $d - n$.

In Fig. 3, we have schematically depicted two defect structures, say, related to magnetic transitions. In Fig. 3a, we show a wall that separates a domain of up spins from that of down spins. In the context of phase separation in a binary $(A + B)$ mixture, say, for a liquid–liquid transition [41], the up and down spins may represent A and B particles. In $d = 2$, given that the order parameter here is a scalar quantity, the wall is a line, i.e., one-dimensional object. In Fig. 3b, we have a two-component vector order parameter in $d = 2$. This gives rise to a vortex defect [44] originating from a zero-dimensional object, i.e., a point. Both types of defects can be visualized in the pictures shown in Fig. 1.

In fluid phase separation, one may expect that defects of multiple types in different fields can be simultaneously observed—say, one in the density or concentration field and the other in the flow, i.e., in the velocity field. This picture may be common in passive as well as in active fluids. In the latter case, this is possible even if the underlying basic phase is solid.

Structural information is typically obtained via the calculation of two-point equal time order parameter correlation function [44]:

$$C(\vec{r}, t) = \langle \psi(\vec{r}, t)\psi(\vec{0}, t)\rangle - \langle \psi(\vec{r}, t)\rangle\langle \psi(\vec{0}, t)\rangle. \tag{12}$$

This is similar to the correlation function defined in the critical phenomena context. However, here the order parameter and, thus, the correlation function is a time-dependent quantity, implying the far-from-equilibrium or far-from-steady-state

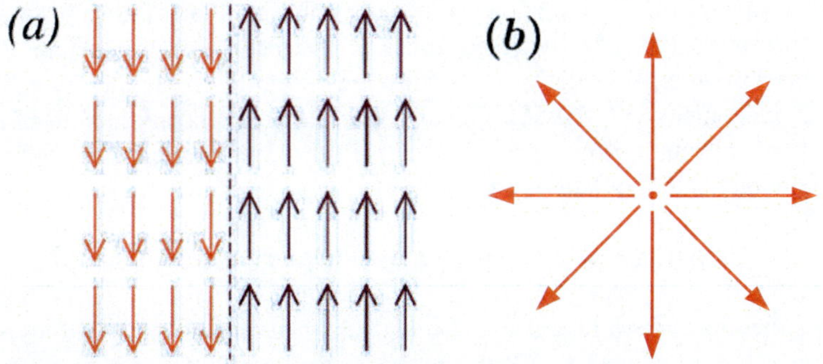

Fig. 3 Schematic diagrams of a few defects in coarsening systems: **a** a line boundary (see the dashed line) between domains for scalar order parameter in space dimension $d = 2$; **b** a vortex formed during the ordering of a two-component vector order parameter in $d = 2$

nature. Furthermore, the discussion on transport phenomena, in the previous subsection, in the context of dynamic critical phenomena, also has strong relevance in the far-from-equilibrium picture that is being described here. Experimentally more relevant quantity is the structure factor, $S(k, t)$, k being the wave number, which is the Fourier transform of $C(r, t)$. The space dimension and order-parameter symmetry is reflected in the behavior of these quantities, e.g., the long wave number tail of $S(k, t)$ is given by [44]

$$S(k, t) \sim k^{-(d+n)}, \tag{13}$$

which is referred to as the Porod tail [44]. This arises from the short distance singularity in $C(r)$, an artifact of scattering from the defects.

These structural quantities exhibit certain scaling properties that arise from self-similar nature of evolution. For non-fractal structures, it is typically observed that [44]

$$C(r, t) \equiv \tilde{C}(r/\ell(t)), \tag{14}$$

and

$$S(k, t) \equiv \ell^d \tilde{S}(k\ell(t)), \tag{15}$$

where $\tilde{C}$ and $\tilde{S}$ are time-independent master functions. In Eqs. (14) and (15), ℓ is the characteristic length scale of the system, i.e., average size of domains under a defect. This quantity grows with time as [44]

$$\ell \sim t^{\delta}. \tag{16}$$

The value of the growth exponent δ depends upon space dimension, mechanism of growth, etc. In simple passive situations, the values of the exponent as well as the full functional form of $C(r,t)$ are known with reasonable accuracy [44, 57]. In active matter systems, the type of activity can certainly influence the overall picture [7, 58].

Another important but relatively less traveled aspect of coarsening dynamics is the aging phenomena [47, 55, 56]. This is concerning the relaxation of an evolving system starting from various ages. Typically, in this sub-domain, one studies the order-parameter autocorrelation function [47]

$$C_{\text{ag}}(t,t_w) = \langle \psi(\vec{r},t)\psi(\vec{r},t_w)\rangle - \langle \psi(\vec{r},t)\rangle\langle \psi(\vec{r},t_w)\rangle. \tag{17}$$

Here, t is the observation time and t_w ($< t$) is the waiting time or age of a system, since the instant of a quench. When plotted versus $t - t_w$, $C_{\text{ag}}(t,t_w)$, in equilibrium situation, for different t_w, overlap with each other. This signifies time translation invariance (TTI) that allows one to obtain better statistics by exploiting time averaging. In the far-from-equilibrium situations, TTI is violated. In the latter case, nevertheless, typically one observes the scaling behavior [47]

$$C_{\text{ag}}(t,t_w) \sim \left(\frac{\ell}{\ell_w}\right)^{-\lambda}, \tag{18}$$

λ being referred to as the aging exponent, and ℓ_w is the value of ℓ for $t = t_w$. In the case of active matter, of course, steady state replaces equilibrium. It is interesting to ask if similar scaling form is obeyed in this case also, for active matter systems.

3 Flocking Transition in Active Matter Systems: A Basic Model

In real life, collections of birds or sheeps or fishes are often seen where the motions of individuals within clusters are along quite the same direction. A minimal model to describe such phenomenon was proposed by Vicsek et al., to be referred to as the Vicsek model (VM) [9] in the following. Within this model, the direction of motion of a constituent particle or object or individual is influenced by the motion of its neighbors. If θ_i represents the direction of motion of the ith particle, at an instant of time t, at the next instant $t + \Delta t$ its direction is updated as

$$\theta_i(t+\Delta t) = \langle \theta_i(t)\rangle_{r_n} + \Delta\theta_i. \tag{19}$$

In Eq. (19), $\langle \theta_i(t)\rangle_{r_n}$ is the average direction of velocity of the neighboring particles lying within a distance r_n from the ith particle. Furthermore, $\Delta\theta_i$ is a noise term value of which lies within $[-\frac{\eta}{2}, \frac{\eta}{2}]$. The particles within the VM move with constant magnitude of velocity, say v_0. The position of a particle, thus, is updated according to

$$\vec{r}_i(t+\Delta t) = \vec{r}_i(t) + \vec{v}_i(t)\Delta t, \tag{20}$$

$\vec{v}_i(t)$ being changed by the modified θ_i values, every next instant of time, keeping the magnitude of the velocity same.

This simple model, often referred to as the Ising model of active matter, provides an interesting phase transition via the competition between density (ρ) of particles and η. The analogy between the two models is not only because of the underlying simplicity, but there also exists similarity in transitions. Note that the Hamiltonian for the Ising model is given by [37, 40]

$$H = -J \sum_{\langle ij \rangle} S_i S_j. \tag{21}$$

For studies related to ordering in ferromagnets, J is positive and S_i or S_j can take values $+1$ and -1. The sign of the majority of the surrounding spins influences the orientation of a spin, if the thermal fluctuation is not very strong. This is similar to the alignment of velocity in the VM. In this sense, the latter, however, is more similar to the XY model or Heisenberg model [44, 59]. In the case of Ising model, the spins or atomic magnetic moments are scalar quantities, as opposed to the continuous cases of XY and Heisenberg models [40, 44]. Like the sum of spins in the magnetic case, the sum of velocities or the normalized average velocity [9], i.e.

$$v_a = \frac{1}{N v_0} \left| \sum_{i=1}^{N} \vec{v}_i \right|, \tag{22}$$

N being the total number of particles in the system, is the order parameter for the transition in the VM. Note that various variants of the VM produce interesting topological defects in the velocity field, like in the XY and Heisenberg magnets, in addition to the vapor–liquid phase separation in the density field. The defect in Fig. 3b has connection with VM or XY model in $d = 2$.

In Fig. 4 we show representative pictures from computer simulations of the VM [9]. In Fig. 4a we show a typical initial configuration, random in positions and velocities. In Fig. 4b we see a late time snapshot that evolved from the initial configuration in Fig. 4a. The picture certainly gives a sense of velocity ordering as well as vapor–liquid-like phase separation in the density field. Late time snapshots from two other sets of system parameters are shown in Fig. 4c, d. Here L is the linear dimension of a system. For a fixed number of particles, L decides the value of ρ. Figure 4 provides an indication of how phase transition can be encountered with the variation in ρ and η.

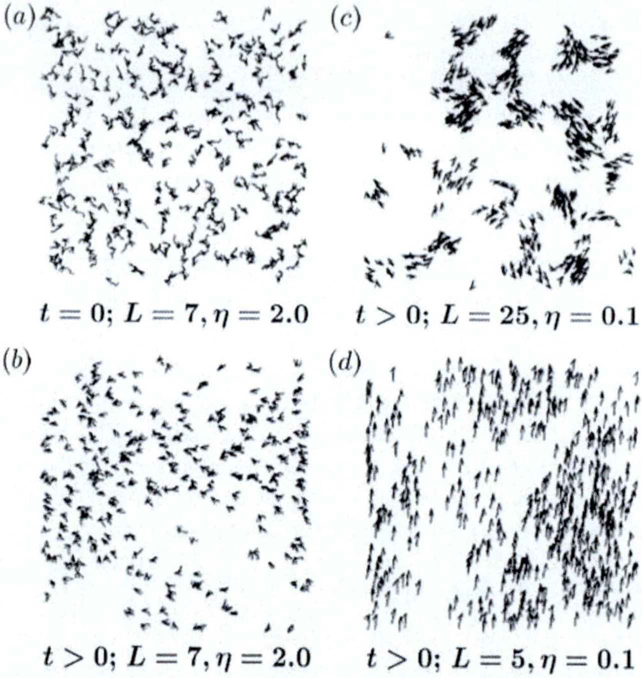

Fig. 4 Snapshots showing approximate positions, directions of motion (see the arrows), and trajectories over the last 20 steps (see continuous curves attached to the arrows) of 300 particles, obtained via computer simulations of the Vicsek model starting from the random initial configuration is shown in part **a**. Late time pictures for a few combinations of density and noise strength are presented in parts **b**–**d**. Reproduced with permission from T. Vicsek et al., Phys. Rev. Lett. **75**, 1226 (1995). https://doi.org/10.1103/PhysRevLett.75.1226

4 Phase Behavior and Critical Aspects in the Vicsek Model

In Fig. 5a we show plots of order parameter as a function of η, that were obtained from the studies of VM [9]. By keeping the value of ρ fixed, results from different system sizes are presented. The general observation is that, with the increase of η, v_a decreases. For larger system sizes, approach of v_a, at large enough η, to zero is clearer. This overall picture is analogous to the critical point behavior in temperature-driven phase transition [9, 37], η playing the role of T, i.e.

$$v_a \sim (\eta_c - \eta)^{\beta}, \tag{23}$$

where η_c is the critical noise strength.

The differences in data sets from systems with different numbers of particles are related to finite-size effects. Such effects are well studied for phase transitions in passive matters [40]. For the latter case, analogous coexistence curves from Monte

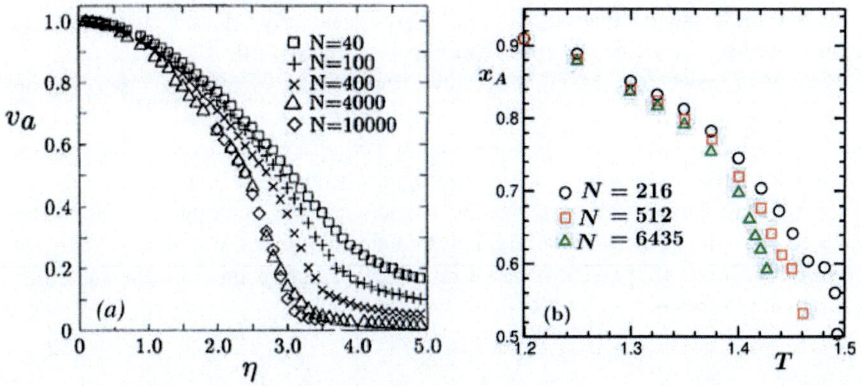

Fig. 5 **a** Variation of order parameter, for the Vicsek model, defined in the text, are shown as a function of noise strength η. The density of particles is kept fixed. Different plots represent results from different system sizes. These data are from simulations in $d = 2$. Reproduced with permission from T. Vicsek et al., Phys. Rev. Lett. **75**, 1226 (1995). https://doi.org/10.1103/PhysRevLett.75.1226. **b** Analogous phase diagrams for a passive binary $(A + B)$ mixture model in $d = 3$. These were obtained from Monte Carlo simulations in a semi-grand-canonical ensemble. Here, x_A represents the concentration of A particles ($x_A = N_A/N$, N_A and N being, respectively, the number of A and total number of particles in the system). These data are borrowed from S. Roy and S.K. Das, J. Chem. Phys. **139**, 064505 (2013)

Carlo simulations of a temperature-driven phase separation in a symmetric binary $(A + B)$ fluid [41], placed in boxes of different sizes, are shown in Fig. 5b. These simulations were performed in a semi-grand-canonical ensemble [41] that allows changes of identities of particles ($A \to B \to A$), thereby inducing fluctuations in the concentration of, say, A species (x_A) in the mixture. The distributions of x_A, obtained by exploiting such fluctuations, at various temperatures, provide crucial information on phase diagram and other critical aspects. The distributions are double-peaked in the coexistence region and single-peaked above the critical point. Below criticality, the locations of the peaks provide points on a coexistence curve. The distributions also allow for the calculations of other quantities, e.g., the susceptibility can be calculated as [40, 41]

$$k_B T \chi = N(\langle x_A^2 \rangle - \langle x_A \rangle^2), \tag{24}$$

where k_B is the Boltzmann constant. For the estimation of the critical point, it is a standard practice to calculate the well-known Binder parameter, i.e., the fourth moment ratio [40],

$$U_L(T) = \frac{(x_A - \langle x_A \rangle)^4}{[(x_A - \langle x_A \rangle)^2]^2}. \tag{25}$$

This dimensionless quantity from different system sizes crosses each other at the critical point. In the case of a vapor–liquid transition, such fluctuations in the density field, leading to the calculations of various experimentally relevant quantities, can be obtained via simulations in grand-canonical ensemble [40].

Choices of smart ensembles in the passive case become possible because of strong universality—very different dynamics lead to the same static class of critical phenomena. In active matter systems, for which dynamics dictate the criticality, it is not possible to choose ensembles at will. Nevertheless, in this case, certain methods can be useful. For systems with fixed size and particle number, such fluctuations can be captured by dividing the simulation box into many sub-boxes. Calculations of v_a in different sub-boxes [17] can provide the desired distribution, and thus the quantities of physical relevance.

5 Transitions in Variants of the Vicsek Model

The original Vicsek model has been combined with passive models to suit studies of various physical pictures. Given that in many biological systems bacteria, typically of the size of colloids, exist in the matrix of bio-polymers, a generic passive model for mixtures of colloids and polymers, referred to as the Asakura–Oosawa (AO) model [60, 61], has been used as the backbone for the studies of effects of Vicsek-like self-propulsion in biologically motivated systems [5, 6].

In the AO model, the colloids and polymers are treated as spheres with radii R_c and R_p, respectively. In the original passive version of the model, colloid–colloid (CC) and colloid–polymer (CP) pair interactions are of hard sphere type. There exists, however, no interaction for the polymer–polymer (PP) pairs. Such a PP interaction has the following justification. Flexible polymers in good solvent conditions take random-walk-like configurations. Upon coarse graining, these can be treated as blobs that can overlap with each other at the free energy cost of $k_B T$.

The above passive model gives rise to phase separation into colloid-rich and polymer-rich phases. Corresponding coexistence curve has been obtained accurately via grand canonical Monte Carlo simulations. For molecular dynamics (MD) [62] studies of this model, however, different interactions [63] were adopted by some authors. When the particles are located r distance apart, the $\alpha\beta$ (= CC or CP) pairs interact via

$$u_{\alpha\beta} = 4\varepsilon_{\alpha\beta}\left[\left(\frac{\sigma_{\alpha\beta}}{r}\right)^{12} - \left(\frac{\sigma_{\alpha\beta}}{r}\right)^{6} + \frac{1}{4}\right]. \tag{26}$$

For PP interactions

$$u_{PP} = 8\varepsilon_{PP}\left[1 - 10\left(\frac{r}{r_{c,PP}}\right)^3 + 15\left(\frac{r}{r_{c,PP}}\right)^4 - 6\left(\frac{r}{r_{c,PP}}\right)\right] \tag{27}$$

was considered. The choices of the system parameters $\varepsilon_{CC} = \varepsilon_{CP} = 1$, $\varepsilon_{PP} = 0.0625$, $\sigma_{CC} = 1$, $\sigma_{CP} = 0.9$, $\sigma_{PP} = 0.8$ and $r_{c,\alpha\beta} = 2^{1/6}\sigma_{\alpha\beta}$, the latter representing cut-off distances in terms of the inter-particle diameters $\sigma_{\alpha\beta}$, produced a phase behavior close to the original version.

Like in the case of the original AO model, the potentials in Eqs. (26) and (27) make the model reasonably insensitive to temperature. Density plays the role of temperature here. For this modified version of the model, the coexistence curve is presented in Fig. 6 on η_C versus η_P plane. Here, η_C and η_P are the packing fractions of colloids and polymers [5, 6, 63]

$$\eta_C = 0.5484\rho_C, \tag{28}$$

$$\eta_P = 0.2808\rho_P, \tag{29}$$

with

$$\rho_\alpha = \frac{N_\alpha}{V}, \tag{30}$$

N_α being the number of particles for species α and V is the system volume.

In the VM, there exists no passive inter-particle interaction. The passive limit of the model, thus, does not exhibit a phase transition. In recent studies [5, 6], colloids in the above mentioned mixtures have been made active via the introduction of the Vicsek activity. Given that the passive counterpart exhibits phase separation, imposition of the self-propulsion is expected to widen the coexistence region. This can be appreciated [5, 6] from Fig. 6.

With respect to the order of such nonequilibrium transitions, there exist debates. Even if there exists a critical point, the universality related to such a second order transition remains highly debated [10].

Vicsek interaction has also been incorporated into single component systems [7] having passive inter-particle potential. In recent works, certain truncated, shifted, and force corrected Lennard–Jones (LJ) [41] potential was used as passive interaction. Note that the standard LJ potential has the form

$$V = 4\varepsilon\left[\left(\frac{\sigma}{r}\right)^{12} - \left(\frac{\sigma}{r}\right)^{6}\right], \tag{31}$$

where, like above, ε and σ are, respectively, the interaction strength and diameter. A popular modified version is [41, 64]

$$u(r) = V(r) - V(r_c) - (r - r_c)\frac{dV}{dr}\bigg|_{r=r_c}, \tag{32}$$

with truncation radius $r_c = 2.5\sigma$. This model provides temperature driven phase transition. Like in the case of the modified AO model, the truncation helps faster computation. Given that the interactions are already of short-range nature, such cuts do not alter the critical universality in the passive case. For these models, MD simulations have been performed in the canonical ensemble. To keep the temperature

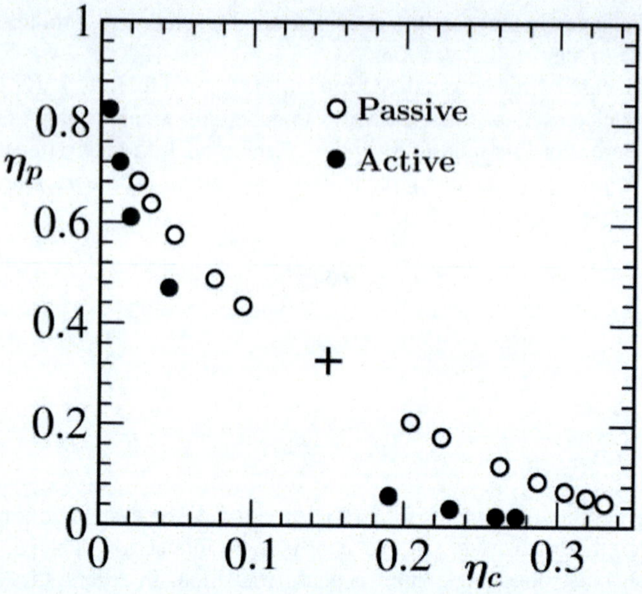

Fig. 6 Phase behavior of the (modified) Asakura–Oosawa model in η_C versus η_P plane, η_C and η_P being the packing fractions of colloids and polymers, respectively. We have shown results for pure passive case as well as for that obtained by making the colloids active via a Vicsek-like rule. See text for more details. The plus (+) symbol marks the location of the passive critical point. This figure is reproduced with due permission from S.K. Das et al., Phys. Rev. Lett. **112**, 198301 (2014) in a modified form

constant, if a Langevin thermostat is used, one needs to solve the dynamical equation (for the ith particle) [5–7]

$$m\ddot{\vec{r}}_i = -\vec{\nabla} u_i - \gamma m \dot{\vec{r}}_i + \sqrt{\ell m \gamma k_B T}\vec{R}(t). \tag{33}$$

Here, m is the mass of a particle, γ is a drag coefficient, and $\vec{R}$ is a noise delta-correlated in space and time as [5–7, 62]

$$\langle R_{i\mu} R_{j\nu} \rangle = \delta_{ij}\delta_{\mu\nu}\delta(t - t'), \tag{34}$$

where μ and ν are related to Cartesian components, while t and t' represent two different times. The value of the constant ℓ in Eq. (33) depends on uniform or Gaussian noise. In recent works [7, 15, 16], at the end of each such MD step the velocities of the particles were further updated by adding it with the average direction of their neighbors [7]

$$\vec{D}_n = \frac{\sum_j \vec{v}_j}{\left|\sum_j \vec{v}_j\right|}, \tag{35}$$

with a proportionality factor f_A. This was done in such a way that this Vicsek rule provides only the directional change. The value of f_A provides the strength of activity, influencing the quantitative change in direction of motion of a particle. For this model also, to be referred to as LJ-VM, it is observed that phase separation occurs much above the known critical temperature in the passive limit. In fact, the influence of Vicsek alignment is so robust in inducing phase separation that even for pure repulsive interaction in the passive limit, a coexistence region can be observed.

6 Kinetics of Flocking Transition: Structure Formation

Dynamics of evolution to steady states have been studied in the original VM as well as in variants of the latter. Like in the case of phase behavior, here also the objective is to find if scaling properties are similar to the passive matter case. Related results we present for the above mentioned LJ-VM, for $f_A = 1$, with reference to the $f_A = 0$ case, the pure LJ (passive) case, in $d = 2$. The state points are chosen in such a way that in each of the cases, there exists short-range crystalline order in the particle arrangement in the high density region.

In Fig. 7 we show [15] evolution snapshots, including the starting random initial configuration, that mimic quenches from high temperature phase, in $d = 2$. The final temperature lies inside the coexistence region of the model in the passive limit. Growth is clearly visible in the system. However, the structure appears different from that of the corresponding passive version of the model. Representative snapshots from the latter [65] are shown in Fig. 8 which were obtained via MD simulations with a hydrodynamic preserving Nosé-Hoover thermostat (NHT). Note that the overall density in the passive (0.35) and active (0.37) cases differ marginally. From the comparison between Figs. 7 and 8, it also transpires that the growth is much faster in the active case, despite the absence of hydrodynamics in the latter case.

The self-similarity in growth, for $f_A = 1$, has been demonstrated in Fig. 9. In Fig. 9a, we have shown the plots of two-point equal time correlation, versus r, from a few different times. The slower decay with the increase of time implies growth. In Fig. 9b, we present the same results, but here the distance axis has been scaled by the characteristic length scale, that was obtained from the decay of $C(r, t)$ as

$$C(r = \ell, t) = h, \tag{36}$$

by fixing h at a constant number (= 0.25). Note that $C(r, t)$ has been normalized in such a way that $C(0, t) = 1$. Clearly, data from different times nicely overlap with each other, implying self-similarity, despite the presence of fractality of some degree. Note that such scaling is well known in the literature of passive matter [44].

In Fig. 10, we present scaling plots for the structure factor, for $f_A = 1$. Here also nice collapse of data is visible. The large wave number data are reasonably consistent with the Porod law [44]. Note that here $d = 2$ and $n = 1$. Thus, for the tail part, one expects the behavior $S(k, t) \sim k^{-3}$. In the small k regime, the data are consistent

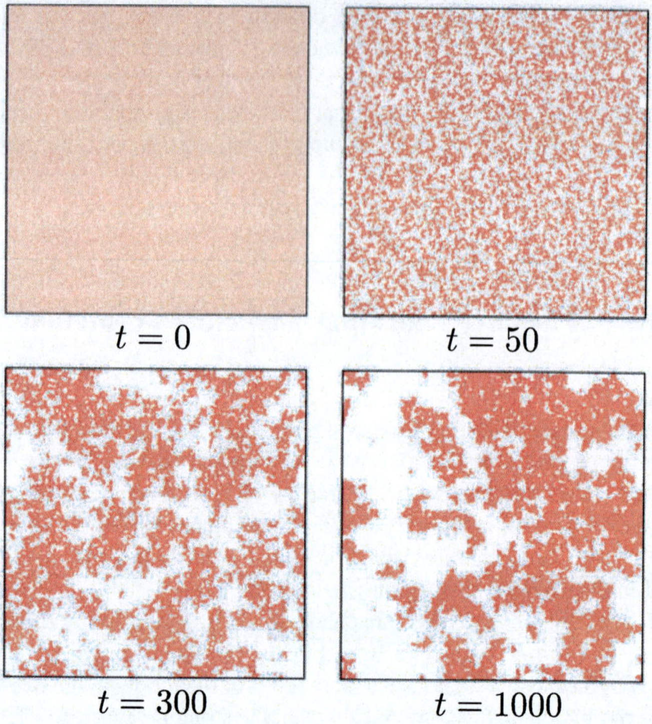

Fig. 7 Snapshots recorded following quench of a homogeneous system ($t = 0$) containing active particles with random initial velocities. These results are from the MD simulations of LJ-VM in $d = 2$. The quench temperature is set at $T = 0.25$ in units of ε/k_B. The overall density of particles in the system of linear dimension $L = 1024$ is 0.37. This figure is reproduced with permission from S. Chakraborty and S.K. Das, J. Chem. Phys. **153**, 044905 (2020) in a modified form

with the behavior

$$S(k \to 0, t) \sim k^x, \tag{37}$$

x being approximately 1.2. This number will be useful in future discussion.

For the above discussed results, the quench temperature was fixed at $T = 0.25$ and ρ was rather high, approximately the critical value for the passive transition. It appears that one gets bicontinuous structure during the evolution with this density. It is expected that the morphology will consist of disconnected clusters when the density is low. Below we discuss pictures for such quenches.

For the off-critical quench, we present the snapshots [16] for $f_A = 1$ in Fig. 11a. As expected, the morphology is made of disconnected clusters. Once again, for the purpose of comparison, in Fig. 11b, we have shown [16] the snapshots from the passive limit of the model, i.e., for $f_A = 0$. Here also the pattern consists of well-separated clusters, but in this case, the clusters are more filamental. Such difference must have its origin in dynamics that becomes different when the active interaction is

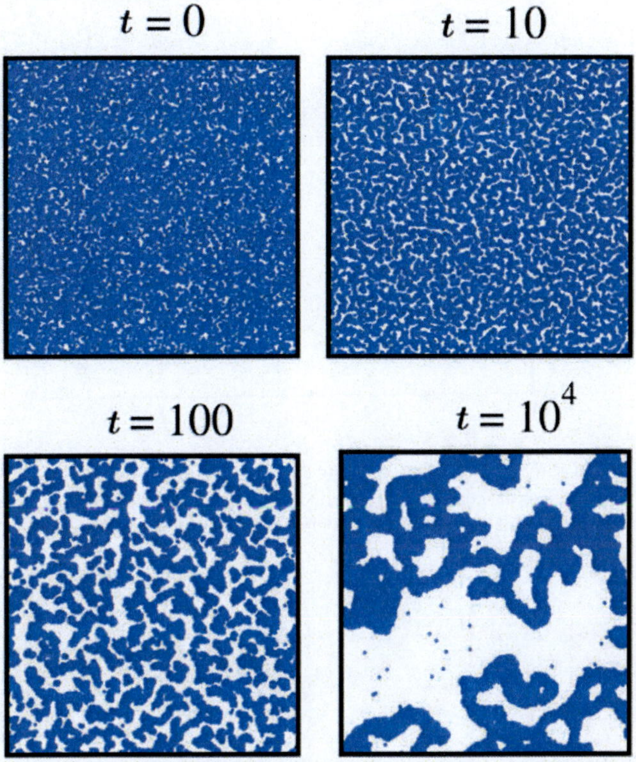

Fig. 8 Similar to Fig. 7 but here the activity was turned off and hydrodynamics was turned on. These snapshots correspond to $\rho = 0.35$. This figure is reproduced with due permission from J. Midya and S.K. Das, Phys. Rev. E **102**, 062119 (2020)

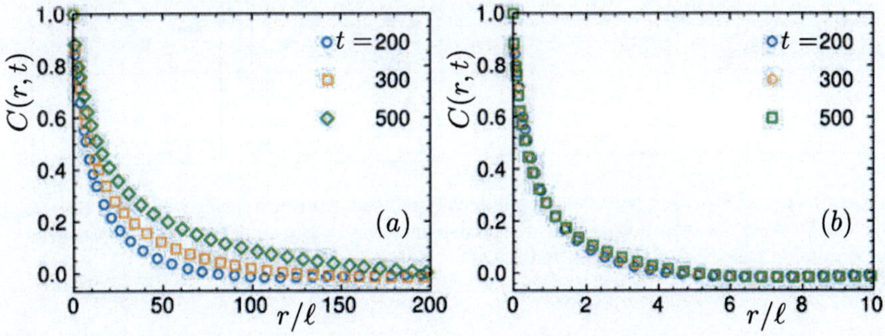

Fig. 9 **a** Two-point equal time correlation functions, $C(r, t)$, are plotted versus r. Data sets, corresponding to the evolution in Fig. 7, from a few different times are shown. **b** Same as **a** but here the distance axis is scaled by the average size of the domains. This figure is reproduced with permission from S. Chakraborty and S.K. Das, J. Chem. Phys. **153**, 044905 (2020)

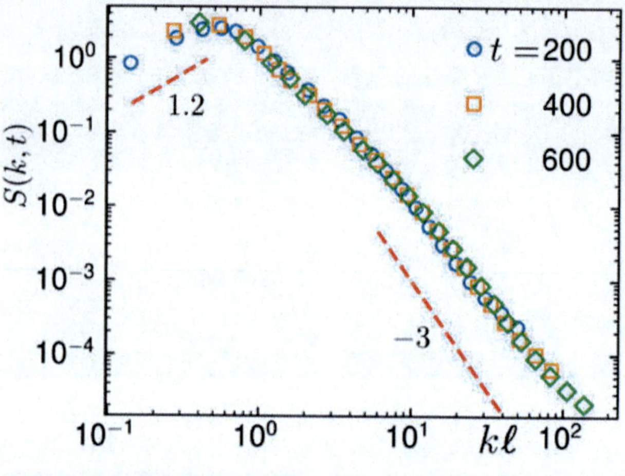

Fig. 10 Scaling plots of the structure factor, taking data from various different times, for the active matter model of Fig. 7. The dashed lines represent various power laws, quantifying the small and large k behavior. This figure is reproduced with permission from S. Chakraborty and S.K. Das, J. Chem. Phys. **153**, 044905 (2020)

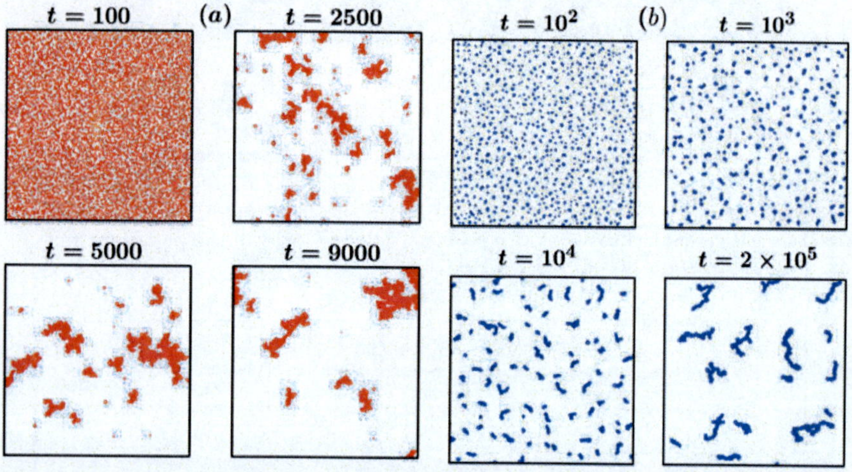

Fig. 11 **a** Same as Fig. 7 but here the overall density is fixed at a much lower value, viz. 0.05. **b** Same as **a** but here the snapshots are for the passive case, viz., $f_A = 0$. This figure is reproduced with permission from S. Paul, A. Bera and S.K. Das, Soft Matter **17**, 645 (2021)

implemented. In both the parts, the results are for $\rho = 0.05$ and $T = 0.1$, in absence of hydrodynamics.

7 Growth During Flocking Evolution

Kinetics of phase transition in passive matter systems are broadly divided into two categories [44], based on the dependence of the total value of the order parameter on time. Typically, for phase separation in mixtures, during vapor–liquid transition, etc., the order parameter does not change with time and belongs to the category of conserved order-parameter dynamics. The cases for which this quantity does not remain preserved over time, e.g., during ordering in a ferromagnet, belong to the category of non-conserved dynamics. Here we are focusing on the density field order parameter that remains preserved for both active and passive cases. Before discussing recent results from active systems, we first provide a theoretical background in the passive context.

7.1 Theoretical Background

In the passive conserved category, the growth rate is significantly influenced by the underlying phase, e.g., whether we are studying phase separation in fluids or in solid mixtures. In the case of solid mixtures, irrespective of the compositions, the growth occurs due to diffusive transport of particles via chemical potential (μ) gradient [51]. Thus, one writes the interface velocity v $(= d\ell/dt)$ as [66]

$$\frac{d\ell}{dt} \sim |\vec{\nabla}\mu|. \tag{38}$$

Given that $\mu \sim \gamma_s/\ell$, γ_s being the interfacial tension, and assuming that the gradient exists over the length scale of the domain, one obtains

$$\frac{d\ell}{dt} \sim \frac{\gamma_s}{\ell^2}. \tag{39}$$

Solution of Eq. (39) provides $\delta = 1/3$, referred to as the Lifshitz–Slyozov (LS) growth exponent. In kinetics of phase separation in solid mixtures, this remains true irrespective of the type of morphology, connected or disconnected.

During phase separation in fluids, it is expected that the solid-like diffusive picture and, thus, $\delta = 1/3$, will remain true at an early time. Beyond a critical length scale, depending upon various thermodynamic and transport properties, hydrodynamics becomes important in fluids and growth becomes much faster, due to advective transport of materials through tube-like elongated domains in the case of bicontinuous structure. A balance between the viscous stress and interfacial free energy density in this case,

$$\frac{6\pi\eta v}{\ell} = \frac{\gamma_s}{\ell}, \tag{40}$$

provides [44]

$$\ell \sim t. \tag{41}$$

However, this growth law does not describe the entire hydrodynamic regime. A further crossover occurs at much later time. In this case, one expects [44]

$$\ell \sim t^{2/3}, \tag{42}$$

which can be obtained via a balance between interfacial free energy density and kinetic energy density. Even though in the considered examples, particles in the high density regions exhibit crystalline ordering, this discussion will be useful. However, note that the hydrodynamic picture described above for $d = 3$ is questionable in $d = 2$. The common consensus is though that hydrodynamics enhances the rate of growth.

In the passive case, for state points close to a branch of the coexistence curve, clusters of the minority phase can move in a hydrodynamic environment, as opposed to the non-hydrodynamic situations. It can be anticipated then that the growth in such a situation will occur via a coalescence mechanism, assuming that the collisions among the clusters are sticky in nature. For diffusive motion of the clusters, a proposal was put forward by Binder and Stauffer [67, 68]. In this case, the droplet density n follows the equation

$$\frac{dn}{dt} = -Cn^2, \tag{43}$$

where C is a constant. Mass conservation demands

$$n \propto \frac{1}{\ell^d}. \tag{44}$$

From these latter equations, one obtains the dimension dependent growth exponent

$$\delta = \frac{1}{d}. \tag{45}$$

Even if one has a fluid-like continuum system, if somehow hydrodynamic conservations are destroyed the growth should follow the LS value. This should, e.g., be the case for MD simulations with Anderson or Langevin thermostats [62].

In an hydrodynamic environment where the clusters move in a very low density background phase, it is possible that the displacements are faster than diffusive. For ballistic motion of clusters, n may follow the dynamical equation [53, 69]

$$\frac{dn}{dt} = -\text{"collision-cross-section"} \times v_{rms} \times n^2, \tag{46}$$

where v_{rms} is the root-mean-squared velocity of the clusters. By inserting an appropriate expression for the collision-cross-section (Ξ) in d space dimension, in terms of cluster mass M, for fractal clusters of dimension d_f, e.g.

$$\Xi \sim M^{\frac{d-1}{d_f}}, \tag{47}$$

and taking

$$v_{rms} \sim M^z, \tag{48}$$

along with

$$n \sim \frac{1}{M}, \tag{49}$$

one arrives at the expression for growth exponent y, in $M \sim t^y$, as [34, 53]

$$y = \frac{d_f}{1 - d + d_f(1 - z)}. \tag{50}$$

Again, this picture will be valid in a hydrodynamic environment. Here we will show numerical results only for the disconnected morphologies. Due to certain complexity in scaling arising due to fractal nature of the clusters, instead of presenting results for ℓ, we will focus on M for most of the remaining part of the section.

7.2 *Computational Results on Growth of Fractal Clusters*

In Fig. 11b, we had shown typical evolution snapshots from the MD simulations of the passive LJ model with Langevin thermostat in $d = 2$. It is checked that these clusters are practically static. The plot of average domain mass for this case is presented [16] in Fig. 12. The growth is reasonably consistent with the expected LS exponent $1/3$.

In the presence of hydrodynamics, the situation is different from what we discussed now. In Fig. 13, we present the snapshots for this case that again show interesting fractal clusters [53]. Note that, in this case, the MD simulations were performed by using the NHT [62] that is known to preserve hydrodynamics well.

In the main frame of Fig. 14, we show [53] the plot of average mass as a function of time. The growth is much faster, i.e., $y \simeq 1.15$. In inset-I, we show a plot of mass versus R_g, the radius of gyration of the fractal clusters. Consistency of the data set with the power law line provides $d_f = 1.6$. In inset-II, the mean squared displacement (MSD) [70] of a typical cluster as a function of time is shown. This indicates ballistic motion. Note that at a higher temperature, when the vapor phase is reasonably dense, the motion may be closer to diffusive. In this case, the fractality may also be different. Cluster growth for such disconnected morphologies in a hydrodynamic environment should happen via a coalescence mechanism. Ballistic aggregation is expected for very low temperature quenches. In inset-III, we show a plot of v_{rms} versus M. This provides $z = -1/2$. Thus, the growth exponent quoted in the figure is consistent with the expected theoretical value [see Eq. (50)]. Note that we are working in $d = 2$.

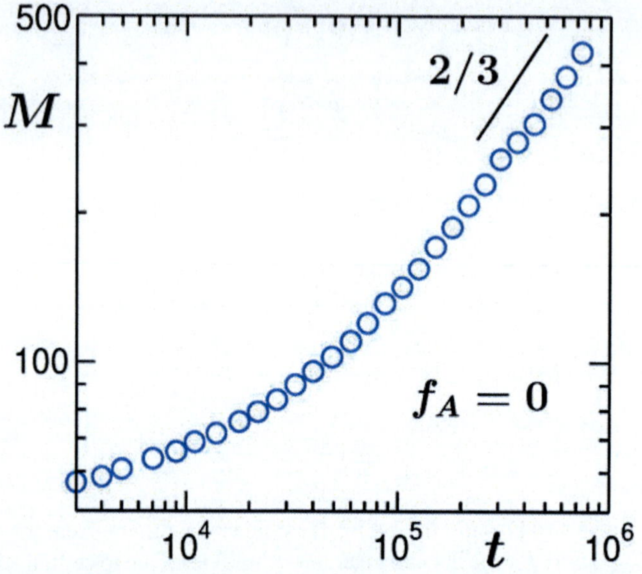

Fig. 12 Plot of average mass (M) of clusters as a function of time for the $f_A = 0$ case of the LJ-VM. The solid line there is a power law. Corresponding exponent is mentioned inside the figure. This figure is reproduced with permission from S. Paul, A. Bera, and S.K. Das, Soft Matter **17**, 645 (2021)

Next we return to the off-critical LJ-VM with Langevin thermostat and $f_A = 1$. Snapshots for this case are already discussed above. From Fig. 15a, where we show [16] the MSD of a cluster as a function of time, it is clear that these clusters are not only mobile, but they also exhibit ballistic motion, even though hydrodynamics is not applied. Interestingly, the growth is even faster than [16] the passive hydrodynamic case, with $y \simeq 1.8$—see Fig. 15b. For this case, the value of d_f and z were estimated to be $\simeq 1.7$ and $\simeq 0$, respectively. Thus, we expect $y \simeq 2.4$. The theory largely explains the rapid growth. There, of course, exist discrepancy. But we do not discuss it here. The self propulsion of Vicsek type brings hydrodynamics-like cluster motion. In this case, we also observe interesting topological defects in the velocity field. Instead of these, we briefly discuss some results on the aging phenomena below.

8 Aging in Evolving Active Matter System

We switch to the bicontinuous structure for the discussion of aging phenomena. Snapshots for this case were presented in Fig. 7. In this case, the growth is estimated [15] to occur with $\delta = 1$, much faster than the passive case.

In Fig. 16a, we show [15] the plots of the order-parameter autocorrelation function, versus $t - t_w$. As expected, the time translation invariance is violated. In Fig. 16b,

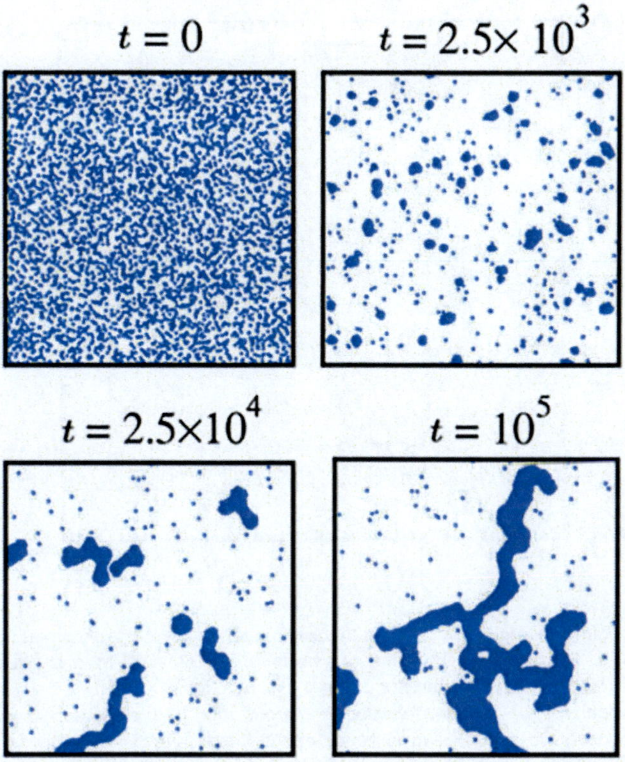

Fig. 13 Snapshots are shown to depict the evolution in the LJ-VM for $f_A = 0$, i.e., for the pure LJ model, with $\rho = 0.03$. In this case, a hydrodynamics preserving Nosé-Hoover thermostat was used for the molecular dynamics simulations. Only parts of the snapshots are shown. This figure is reproduced with permission from J. Midya and S.K. Das, Phys. Rev. Lett. **118**, 165701 (2017)

we present the same plots versus ℓ/ℓ_w. Nice collapse of data, as typically seen in the passive matter case, is observed. The decay of $C_{\text{ag}}(t, t_w)$ in the latter plot appears linear on the log–log scale, in the large ℓ/ℓ_w limit. This implies a power law behavior. The exponent is $\lambda \simeq 2.2$. See Eq. (18) for definition of λ. Note that rapid falls are related to the finite-size effects [7, 15].

In the literature of aging, there exists bounds on λ. From a general consideration, Yeung, Rao, and Desai arrived at [48]

$$\lambda \geqslant \frac{d + x}{2}. \tag{51}$$

We have previously observed in Fig. 10 that $x \simeq 1.2$ in this case. Note that x has been defined in Eq. (37). Thus, this bound is obeyed in the present case.

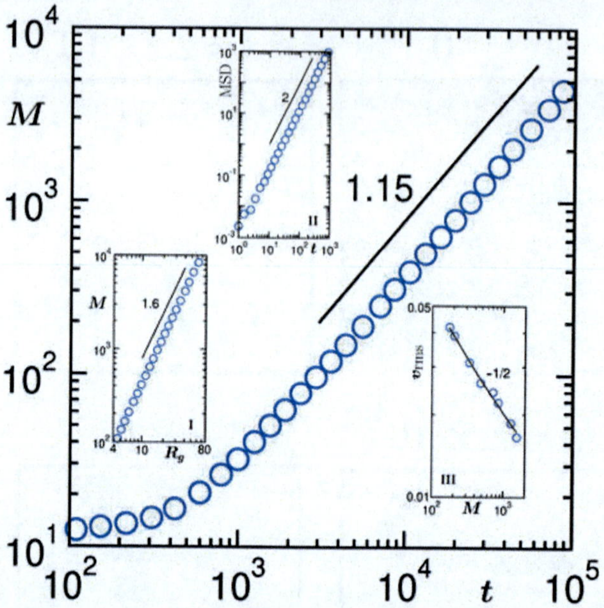

Fig. 14 Average cluster mass, corresponding to the evolution in Fig. 13, is shown as a function of t. Inset-I: A plot of M versus R_g, the radius of gyration of the clusters. Inset-II: Mean-squared-displacements (MSD) for a typical cluster are shown as a function of shifted time. Inset-III: Root-mean-squared velocity, v_{rms}, is plotted versus M. Various solid lines represent power laws, and values of the exponents are mentioned in appropriate places. This figure is reproduced with permission from J. Midya and S.K. Das, Phys. Rev. Lett. **118**, 165701 (2017)

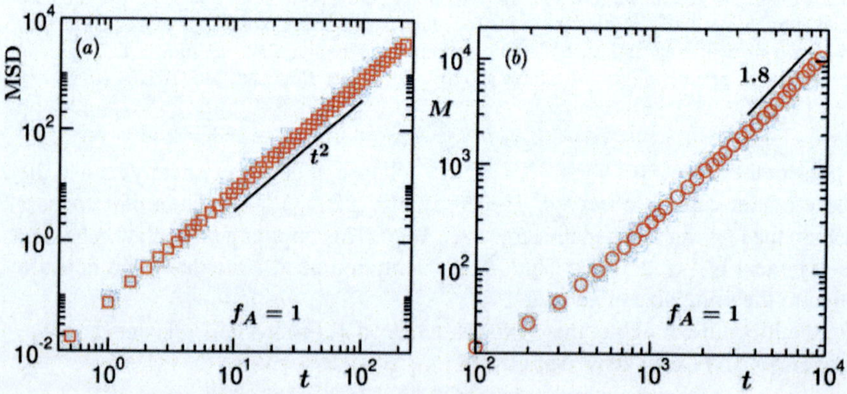

Fig. 15 **a** Mean-squared-displacements (MSD) are shown versus shifted time for the LJ-VM with $f_A = 1$ and Langevin thermostat. The solid line represents a power law corresponding to ballistic motion. **b** Average cluster mass for the LJ-VM with $f_A = 1$ and Langevin thermostat is plotted versus time, on a log–log scale. The solid line represents a power law with exponent 1.8. This figure is reproduced with permission from S. Paul, A. Bera and S.K. Das, Soft Matter **17**, 645 (2021)

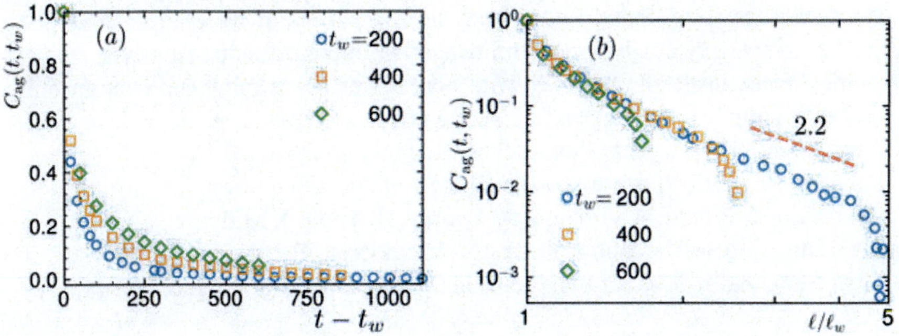

Fig. 16 **a** Plots of the order-parameter autocorrelation function for the high density LJ-VM are shown versus $t - t_w$. **b** Same data sets in **a** are shown versus ℓ/ℓ_w, on a log–log scale. The dashed line here is a power law having exponent 2.2. This figure is reproduced with permission from S. Chakraborty and S.K. Das, J. Chem. Phys. **153**, 044905 (2020)

In the active case, even though in many ways, the character of transition is different from the passive situation, it appears that the basic scaling properties remain valid. However, there exist quantitative differences.

9 Conclusion

We have provided an overview of active matter systems. These systems exhibit nonequilibrium phase transitions. The focus of this chapter was on the structure and dynamics associated with such transitions.

We have discussed phase behavior in the steady state situation. Critical phenomena were introduced in this context. The topic of evolution toward the steady states was covered by considering quenches to state points that provide different kinds of structure. In each of the cases, the background was set up by sketching the picture in the passive matter context.

It is seen that in the active matter case, qualitative picture remains the same with respect to various scaling laws. However, there exists disagreement with the passive case at the quantitative level. The universality is in general weaker in systems containing self-propelling particles. In this chapter, we have discussed only systems with alignment interactions [9]. There exists other interesting cases [10, 17, 71]. These should be explored appropriately to establish an accurate understanding with respect to universality in phase transition concerning active matters.

In the context of dynamics, the role of hydrodynamics needs to be studied [72]. There exist methods to implement hydrodynamics in such systems. However, large-scale simulations for studies of phase transitions have not been performed by using such techniques.

Another interesting domain is to study active matters under confinement [73–77]. These systems typically contain finite number of constituents. However, studies by putting these inside finite boxes with boundaries are gaining momentum only recently. Existing reports suggest fascinating effects of boundaries on both structure and dynamics. These aspects should, thus, be explored more.

Some of the results that are presented here were obtained through fruitful collaborations with K. Binder, P. Virnau, S.A. Egorov, B. Trefz, J. Midya, S. Chakraborty, S. Paul, and S. Roy. The author thanks N. Vadakkayil, P. Pathak, A.P. Tripathi, T. Paul, A. Bera, and K. Das for their helps in the preparation of the manuscript.

References

1. M.C. Marchetti, F.J. Joanny, S. Ramaswamy et al., Hydrodynamics of soft active matter. Rev. Mod. Phys. **85**, 1143–1189 (2013)
2. S. Ramaswamy, The mechanics and statistics of active matter. Ann. Rev. Cond. Mat. Phys. **1**, 323–345 (2010)
3. M.E. Cates, J. Tailleur, Motility-induced phase separation. Ann. Rev. Cond. Mat. Phys. **6**, 219–244 (2015)
4. S. Wang, P.G. Wolynes, Effective temperature and glassy dynamics of active matter. J. Chem. Phys. **135**, 051101 (2011)
5. S.K. Das, S.A. Egorov, B. Trefz et al., Phase behavior of active swimmers in depletants: molecular dynamics and integrated equation theory. Phys. Rev. Lett. **112**, 198301 (2014)
6. B. Trefz, S.K. Das, S.A. Egorov et al., Activity mediated phase separation: can we understand phase behaviour of all the nonequilibrium problem from an equilibrium approach? J. Chem. Phys. **144**, 144902 (2016)
7. S.K. Das, Pattern, growth and aging in aggregation kinetics of a Vicsek-like active matter model. J. Chem. Phys. **146**, 044902 (2017)
8. D. Loi, S. Mossa, L.F. Cugliandolo, Effective temperature of active complex matter. Soft Matter **7**, 3726–3729 (2011)
9. T. Vicsek, A. Czirók, E. Ben-Jacob et al., Novel type of phase transition in a system of self-driven particles. Phys. Rev. Lett. **75**, 1226–1229 (1995)
10. G. Baglietto, E.V. Albano, J. Candia, Criticality and the onset of ordering in the standard Vicsek model. Interface Focus **2**, 708–714 (2012)
11. A. Czirók, T. Vicsek, Collective behaviour of interacting self-propelled particles. Phys. A **281**, 17–29 (2000)
12. H. Chaté, F. Ginelli, G. Grégoire et al., Modeling collective motion: variations on the Vicsek model. Eur. Phys. J. B **64**, 451–456 (2008)
13. J. Schwarz-Linek, C. Valeriani, A. Cacciuto et al., Phase separation and rotor self-assembly in active particle suspensions. Proc. Natl. Acad. Sci. USA **109**, 4052–4057 (2012)
14. J. Palacci, S. Sacanna, A.P. Steinberg et al., Living crystals of light-activated colloidal surfers. Science **339**, 936–940 (2013)
15. S. Chakraborty, S.K. Das, Relaxation in a phase-separating two-dimensional active matter system with alignment interaction. J. Chem. Phys. **153**, 044905 (2020)
16. S. Paul, A. Bera, S.K. Das, How do clusters in a phase-separating active matter systems grow? a study for Vicsek activity in systems undergoing vapor-solid transition. Soft Matter **17**, 645–654 (2021)
17. J.T. Siebert, F. Dittrich, F. Schmid et al., Critical behavior of active Brownian particles. Phys. Rev. E **98**, 030601 (2018)
18. N. Kumar, H. Soni, S. Ramaswamy et al., Flocking at a distance in active granular matter. Nat. Comm. **5**, 4688 (2014)

19. Y. Fily, M.C. Marchetti, Athermal phase separation of self-propelled particles with no alignment. Phys. Rev. Lett. **108**, 235702 (2012)
20. A. Wysocki, R.G. Winkler, G. Gompper, Cooperative motion of active Brownian spheres in a in a three-dimensional dense suspensions. EPL **105**, 480004 (2014)
21. J. Bialké, T. Speck, H. Löwen, Active colloidal suspensions: clustering and phase behavior. J. Non-Cryst. Solids **407**, 367–375 (2015)
22. J.T. Siebert, J. Letz, T. Speck et al., Phase behavior of active Brownian disks, sphere and dimers. Soft Matter **13**, 1020–1026 (2017)
23. P. Digregorio, D. Levis, A. Suma et al., Full phase diagram of active Brownian disks: from melting to motility-induced phase separation. Phys. Rev. Lett. **121**, 098003 (2018)
24. R. Wittkouski, A. Tiribocchi, J. Stenhammer et al., Scalar ϕ^4 field theory for active-particle phase separation. Nat. Comm. **5**, 4351 (2014)
25. N.V. Brilliantov, H. Abutuqayqah, I.Y. Tyukin et al., Swirlonic state of active matter. Sci. Rep. **10**, 16783 (2020)
26. K. Binder, P. Virnau, Phase transitions and phase coexistence: equilibrium systems versus externally driven or active systems - Some perspectives. Soft. Mater. **19**, 267–285 (2021)
27. Photo source: Wikipedia, the free encyclopedia, *Red-Billed Queleas, the Most Numerous Species of Bird, Form Enormous Flocks—Sometimes Tens of Thousands Strong*. Reproduced following Creative Commons Attribution-Share Alike 2.0 Generic license. Get link via: https://en.wikipedia.org/wikiBird#media (2006)
28. B. de Giusti, An underwater picture taken in Moofushi Kandu, Maldives, showing predator bluefin trevally sizing up schooling anchovies, in *Wikipedia, the Free Encyclopedia*. Reproduced following Creative Commons Attribution-Share Alike 2.5 Italy license. See link: https://en.wikipedia.org/wiki/File:Moofushi_Kandu_fish.jpg (2006)
29. R. Zwart (2005) Sheep on the Meseta, Spain, in *Wikipedia, the Free Encyclopedia*. Reproduced following Creative Commons Attribution-Share Alike 2.5 Netherlands license. See link: https://commons.wikimedia.org/wiki/File:Meseta_herd.jpg
30. C. Dombrowski, L. Cisneros, S. Chatkaew et al., Self-concentration and large-scale coherence in bacterial dynamics. Phys. Rev. Lett. **93**, 098103 (2004)
31. I. Goldhirsch, G. Zanetti, Clustering instability in dissipative gases. Phys. Rev. Lett. **70**, 1619–1622 (1993)
32. I.S. Aranson, L.S. Tsimring, Patterns and collective behavior in granular media: theoretical concepts. Rev. Mod. Phys. **78**, 641–692 (2006)
33. N.V. Brilliantov, T. Poeschel, *Kinetic Theory of Granular Gases* (Oxford University Press, UK, 2004)
34. S. Paul, S.K. Das, Dimension dependence of clustering dynamics in models of ballistic aggregation and freely cooling granular gas. Phys. Rev. E **97**, 032902 (2018)
35. M.E. Fisher, The theory of equilibrium critical phenomena. Rep. Prog. Phys. **30**, 615–730 (1967)
36. K.G. Wilson, The renormalization group and critical phenomena. Rev. Mod. Phys. **55**, 583–600 (1983)
37. M.E. Fisher, Renormalization group theory: its basics and formulations in Statistical Physics. Rev. Mod. Phys. **70**, 653–681 (1998)
38. P.C. Hohenberg, B.I. Halperin, Theory of dynamic critical phenomena. Rev. Mod. Phys. **49**, 435–479 (1977)
39. J.V. Sengers, P.H. Keyes, Scaling of the thermal conductivity near the gas-liquid critical point. Phys. Rev. Lett. **26**, 70–73 (1971)
40. D.P. Landau, K. Binder, *A Guide to Monte Carlo Simulations in Statistical Physics* (Cambridge University Press, UK, 2009)
41. S. Roy, S.K. Das, Transport Phenomena in fluids: finite size scaling for critical behavior. EPL **94**, 36001 (2011)
42. A. Onuki, *Phase Transition Dynamics* (Cambridge University Press, UK, 2002)
43. A. Cavagna, D. Conti, C. Creato et al., Dynamical scaling in natural swarms. Nat. Phys. **13**, 914 (2017)

44. A.J. Bray, Theory of phase ordering kinetics **51**, 481–587 (2002)
45. S. Puri, V. Wadhawan (eds.), *Kinetics of Phase Transitions* (CRC Press, Boca Raton, 2009)
46. K. Binder, in *Phase Transform. Mater.*, vol. 5, ed. by R.W. Cahn, P. Haasen, E.J. Kramer (VCH, Weinheim, 1991), p. 405
47. D.S. Fisher, D.A. Huse, Nonequilibrium dynamics of spin glasses. Phys. Rev. B **38**, 373–385 (1988)
48. C. Yeung, M. Rao, R.C. Desai, Bounds on the decay of the autocorrelation in phase ordering dynamics. Phys. Rev. E **53**, 3073–3077 (1996)
49. G.F. Mazenko, Response functions in phase ordering kinetics. Phys. Rev. E **69**, 016114 (2004)
50. F. Corberi, E. Lippiello, A. Mukherjee et al., Crossover in growth law and violation of superuniversality in the random-field Ising model. Phys. Rev. E **85**, 021141 (2012)
51. I.M. Lifshitz, V.V. Slyozov, The kinetics of precipitation from supersaturated solid solutions. J. Phys. Chem. Solids **19**, 35–50 (1961)
52. A. Furukawa, H. Tanaka, Key role of hydrodynamic interactions in colloidal gelation. Phys. Rev. Lett. **104**, 245702 (2010)
53. J. Midya, S.K. Das, Kinetics of vapor-solid phase transitions: structure, growth, and mechanism. Phys. Rev. Lett. **118**, 165701 (2017)
54. J. Midya, S. Majumder, S.K. Das, Dimensionality dependence of aging in kinetics of diffusive phase separation: behaviour of order-parameter autocorrelation. Phys. Rev. E **92**, 022124 (2015)
55. A. Henkel, A. Picone, M. Pleimling, Two-time autocorrelation function in phase-ordering kinetics from local scale invariance. Europhys. Lett. **68**, 191–197 (2004)
56. E. Lorenz, W. Janke, Numerical tests of local scale invariance in ageing q-state Potts models. EPL **77**, 10003 (2007)
57. A.J. Bray, S. Puri, Asymptotic structure factor and power-law tails for phase ordering in systems with continuous symmetry. Phys. Rev. Lett. **67**, 2670–2673 (1991)
58. G. Gonnella, D. Marenduzzo, A. Suma et al., Motility-induced phase separation and coarsening in active matter. Comptes Rendus Physique **16**, 316–331 (2015)
59. N. Goldenfeld, *Lectures on Phase Transitions and the Renormalization Group* (CRC Press, Boca Raton, 2018)
60. S. Asakura, F. Oosawa, On interaction between two bodies immersed in a solution of macromolecules. J. Chem. Phys. **22**, 1255–1256 (1954)
61. R.L.C. Vink, J. Horbach, K. Binder, Critical phenomena in colloid-polymer mixtures: interfacial tension, order parameter, susceptibility, and coexistence diameter. Phys. Rev. E **71**, 011401 (2005)
62. D. Frenkel, B. Smit, *Understanding Molecular Simulations: From Algorithms to Applications* (Academic Press, San Diego California, 2002)
63. J. Zausch, P. Virnau, K. Binder et al., Statics and dynamics of colloid-polymer mixtures near their critical point of phase separation: a computer simulation study of a continuous Asakura-Oosawa model. J. Chem. Phys. **130**, 064906 (2009)
64. M.P. Allen, D.J. Tildesley, *Computer Simulation of Liquids* (Clarendon Press, Oxford, 1987)
65. J. Midya, S.K. Das, Kinetics of domain growth and aging in two-dimensional off-lattice system. Phys. Rev .E **102**, 062119 (2020)
66. D.A. Huse, Corrections to late-stage behavior in Spinodal decomposition: Lifshitz-Slyozov scaling and Monte Carlo simulations. Phys. Rev. B **34**, 7845–7850 (1986)
67. K. Binder, D. Stauffer, Theory for the slowing domain of the relaxation and Spinodal decomposition of binary mixtures. Phys. Rev. Lett. **33**, 1006–1009 (1974)
68. E.D. Siggia, Late stages of Spinodal decomposition in binary mixtures. Phys. Rev. A **20**, 595–605 (1979)
69. G.F. Carnevale, Y. Pomeau, W.R. Young, Statistics of ballistic agglomeration. Phys. Rev. Lett. **64**, 2913–2916 (1990)
70. H.P. Hansen, I.R. McDonald, *Theory of Simple Liquids* (Academic Press, London, 1986)
71. F. Dittrich, T. Speck, P. Virnau, Critical behavior in active lattice models of motility-induced phase separation. Eur. Phys. J. E **44**, 53 (2021)
72. A. Bera, S. Sahu, S. Thakur et al., Active particles in explicit solvent: dynamics of clustering for alignment interaction. Phys. Rev. E, **105**, 014606 (2022)

73. T.N. Shendruk, A. Doostmohammadi, K. Thijssen et al., Dancing disclinations in confined active nematics. Soft Matter **13**, 3853–3862 (2017)
74. G. Duclos, C. Blanch-Mercader, V. Yashunsky et al., Spontaneous shear flow in confined cellular nematics. Nat. Phys. **14**, 728–732 (2018)
75. H. Wioland, F.G. Woodhouse, J. Dunkel et al., Confinement stabilizes a bacterial suspension into a Spiral Vortex. Phys. Rev. Lett. **110**, 268102 (2013)
76. T. Gao, M.D. Betterton, A.S. Jhang et al., Analytical structure, dynamics, and coarse graining of a kinetic model of an active fluid. Phys. Rev. Fluids **2**, 093302 (2017)
77. J. Elgeti, G. Gompper, Wall accumulation of self-propelled spheres. Europhys. Lett. **101**, 48003 (2013)

Kinetic Theory of Binary Granular Suspensions at Low Density. Thermal Diffusion Segregation

Rubén Gómez González and Vicente Garzó

Abstract Transport properties of granular mixtures surrounded by an interstitial gas are determined by solving the Boltzmann kinetic equation by means of the Chapman–Enskog method. As usual, the influence of the viscous gas on solid particles is accounted for by an effective external force composed of two terms: a drag force proportional to the particle velocity plus a stochastic Langevin-like term. Before considering inhomogeneous situations, we study first the homogeneous steady state where collisional cooling and viscous friction are compensated for by the energy gained by grains due to their interaction with the interstitial gas. Then, the Chapman–Enskog method is used to solve the Boltzmann equation and express the Navier–Stokes transport coefficients in terms of the solutions of a set of coupled linear integral equations. Explicit forms are obtained here in the tracer limit for the diffusion transport coefficients which are explicitly determined by considering the so-called first Sonine approximation. As an application of the previous results, thermal diffusion segregation of an intruder immersed in a granular suspension is analyzed and compared with previous theoretical attempts where the effect of the interstitial gas was neglected.

1 Introduction

Granular matter in nature is generally surrounded by an interstitial fluid, like water or air. Although in many situations the effect of the surrounding fluid on the dynamic properties of grains can be neglected, there are also other situations (for instance, when the stress exerted by the fluid phase on grains is significant) where the influence of the interstitial fluid must necessarily be taken into account. A typical example of it refers to the species segregation in granular mixtures [1]. Since a granular suspension is a multiphase process, in the context of kinetic theory, one could start from a set of coupled kinetic equations for each one of the velocity distribution functions of the

R. Gómez González · V. Garzó (✉)
Departamento de Física and Instituto de Computación Científica Avanzada (ICCAEx), Universidad de Extremadura, 06006 Badajoz, Spain
e-mail: vicenteg@unex.es

L. Brenig et al. (eds.), *Nonequilibrium Thermodynamics and Fluctuation Kinetics*, Fundamental Theories of Physics 208,
https://doi.org/10.1007/978-3-031-04458-8_9

phases. However, this approach involves many technical intricacies, especially in the case of granular mixtures. Thus, to avoid this problem, the effect of the interstitial fluid on grains is usually taken into account by means of an effective external force [2]. This fluid–solid force is composed of two terms: (i) a viscous drag force (Stokes' law) proportional to the particles velocity and (ii) a stochastic force modeled as a Gaussian white noise. While the first term mimics the friction of grains with the surrounding gas, the second term accounts for the energy gained by grains due to their interaction with the particles of the gas phase (thermal reservoir).

An interesting and challenging problem is to assess the impact of gas phase on the Navier–Stokes transport coefficients of a binary granular mixture modeled as an ensemble of smooth inelastic hard spheres. This problem is not only relevant from a fundamental point of view but also from a realistic point of view since granular suspensions are present in nature formed by grains of different masses, sizes, densities, and coefficients of restitution. However, the determination of the transport coefficients of bidisperse gas–solid flows is a quite ambitious target due essentially to the large number of integro-differential equations involved as well as the wide parameter space of the system. For this reason and in order to offer a complete description, we consider here binary granular suspensions at low-density where the Boltzmann kinetic equation turns out to be a reliable starting point [3, 4].

As in previous papers [6–8], the Boltzmann equation (BE) is solved by means of the Chapman–Enskog (CE) method [9] adapted to dissipative dynamics. A subtle and important point of the expansion method is the choice of the reference distribution in the perturbation scheme. Although we are interested here in obtaining the transport coefficients under steady conditions, the presence of the surrounding fluid gives rise to a local energy unbalance in such a way the zeroth-order distributions $f_i^{(0)}$ of each species (reference states) are not in general stationary distributions. Thus, in order to determine the Navier–Stokes transport coefficients, one has to obtain first the unsteady integral equations defining the above transport coefficients and solve (approximately) then these equations in steady-state conditions. An important consequence of this procedure is that the transport properties depend not only on the steady temperature but also on quantities such as the derivatives of the temperature ratio on the temperature.

The plan of the paper is as follows. The granular suspension model as well as the balance equations for the densities of mass, momentum, and energy are derived in Sect. 2. Then, the steady homogeneous state is studied in Sect. 3 where the temperature ratio T_1/T_2 of both species is calculated and compared against the Monte Carlo simulations. Section 4 addresses the application of the CE method up to first order in the spatial gradients. As expected, transport coefficients are given in terms of the solutions of a set of coupled linear integral equations. These integral equations are approximately solved by considering the leading Sonine approximation; this procedure is explicitly displayed here for the diffusion transport coefficients in the special limit case where one of the components of the mixture is present in tracer concentration. As an application of the previous results, thermal diffusion segregation of an intruder or tracer particle is analyzed in Sect. 5. The paper is closed in Sect. 6 with a brief discussion of the results obtained in this work.

2 Granular Suspension Model

We consider a granular binary mixture of inelastic hard disks ($d = 2$) or spheres ($d = 3$) of masses m_i and diameters σ_i ($i = 1, 2$). The spheres are assumed to be completely smooth and so, the inelasticity of collisions is characterized by three constant (positive) coefficients of normal restitution $\alpha_{ij} \leq 1$. The solid particles are surrounded by a molecular gas of viscosity η_g and temperature T_{ex}. As said before, the influence of the interstitial gas on grains is modeled via a fluid–solid force constituted by two terms: a deterministic drag force plus a stochastic force. In the low-density limit and taking into account the above terms, the one-particle velocity distribution function of each species verifies the Boltzmann kinetic equation [8]

$$\frac{\partial f_i}{\partial t} + \mathbf{v} \cdot \nabla f_i - \gamma_i \Delta \mathbf{U} \cdot \frac{\partial f_i}{\partial \mathbf{v}} - \gamma_i \frac{\partial}{\partial \mathbf{v}} \cdot \mathbf{V} f_i - \frac{\gamma_i T_{ex}}{m_i} \frac{\partial^2 f_i}{\partial v^2} = \sum_{j=1}^{2} J_{ij}[f_i, f_j], \quad (1)$$

where $J_{ij}[f_i, f_j]$ is the Boltzmann collision operator [4]. In addition, $\Delta \mathbf{U} = \mathbf{U} - \mathbf{U}_g$, $\mathbf{V} = \mathbf{v} - \mathbf{U}$ is the peculiar velocity,

$$\mathbf{U} = \rho^{-1} \sum_{i=1}^{2} \int d\mathbf{v} \; m_i \mathbf{v} f_i(\mathbf{v}) \quad (2)$$

is the mean flow velocity of the solid particles, and $\mathbf{U}_g$ is the known mean flow velocity of the interstitial gas. The friction coefficients γ_i are proportional to the gas viscosity η_g and are functions of the partial volume fractions $\phi_i = (\pi^{d/2}/(2^{d-1} d\Gamma(d/2)) n_i \sigma_i^d$, where

$$n_i = \int d\mathbf{v} \; f_i(\mathbf{v}) \quad (3)$$

is the number density of species i. In the dilute limit, every particle is only subjected to its respective Stokes' drag [10] so that for hard spheres ($d = 3$) γ_i is

$$\gamma_i = \gamma_0 R_i, \quad \gamma_0 = \frac{18\eta_g}{\rho \sigma_{12}^2}, \quad R_i = \frac{\rho \sigma_{12}^2}{\rho_i \sigma_i^2} \phi_i. \quad (4)$$

Here, $\rho_i = m_i n_i$, $\rho = \rho_1 + \rho_2$ is the total mass density, and $\sigma_{12} = (\sigma_1 + \sigma_2)/2$. Apart from the partial densities n_i and the mean flow velocity $\mathbf{U}$, the other relevant hydrodynamic field is the granular temperature T, defined as

$$T = \frac{1}{n} \sum_{i=1}^{2} \int d\mathbf{v} \frac{m_i}{d} V^2 f_i(\mathbf{v}), \quad (5)$$

where $n = n_1 + n_2$ is the total number density.

The Boltzmann collision operators $J_{ij}[f_i, f_j]$ conserve the number densities of each species and the total momentum but the total energy is not conserved:

$$\int d\mathbf{v}\left\{1, \sum_{i,j} m_i \mathbf{v}, \sum_{i,j} m_i V^2\right\} J_{ij}[\mathbf{v}|f_i, f_j] = \left\{0, \mathbf{0}, -dnT\zeta\right\}, \tag{6}$$

where ζ is the total cooling rate due to inelastic collisions among all species. The macroscopic balance equations for the densities of mass, momentum, and energy can be easily obtained by multiplying both sides of the BE (1) by 1, $m_i\mathbf{v}$, and $m_i V^2$; integrating over $\mathbf{v}$; and taking into account the properties (6). After some algebra, one gets

$$D_t n_i + n_i \nabla \cdot \mathbf{U} + \frac{\nabla \cdot \mathbf{j}_i}{m_i} = 0, \tag{7}$$

$$D_t \mathbf{U} + \rho^{-1} \nabla \cdot \mathsf{P} = -\rho^{-1} \Delta \mathbf{U} \sum_{i=1}^{2} \rho_i \gamma_i - \rho^{-1} (\gamma_1 - \gamma_2) \mathbf{j}_1, \tag{8}$$

$$D_t T - \frac{T}{n} \sum_{i=1}^{2} \frac{\nabla \cdot \mathbf{j}_i}{m_i} + \frac{2}{dn} (\nabla \cdot \mathbf{q} + \mathsf{P} : \nabla \mathbf{U}) = -\frac{2}{dn} \Delta \mathbf{U} \cdot \sum_{i=1}^{2} \gamma_i \, \mathbf{j}_i$$
$$+2 \sum_{i=1}^{2} x_i \gamma_i \left(T_{\text{ex}} - T_i\right) - \zeta T. \tag{9}$$

In the above equations, $D_t = \partial_t + \mathbf{U} \cdot \nabla$ is the material derivative,

$$\mathbf{j}_i = m_i \int d\mathbf{v} \, \mathbf{V} f_i(\mathbf{v}) \quad (\mathbf{j}_1 = -\mathbf{j}_2) \tag{10}$$

is the mass flux for the component i relative to the local flow $\mathbf{U}$,

$$\mathsf{P} = \sum_{i=1}^{2} \int d\mathbf{v} \, m_i \mathbf{V}\mathbf{V} f_i(\mathbf{v}) \tag{11}$$

is the pressure tensor, and

$$\mathbf{q} = \sum_{i=1}^{2} \int d\mathbf{v} \, \frac{m_i}{2} V^2 \mathbf{V} f_i(\mathbf{v}) \tag{12}$$

is the heat flux. In addition, the partial kinetic temperature T_i is

$$T_i = \frac{m_i}{dn_i} \int d\mathbf{v}\, V^2 f_i(\mathbf{v}). \tag{13}$$

The partial temperature T_i measures the mean kinetic energy of particles of species i. The relationship between the granular temperature T and the partial temperatures T_i is simply given by $T = \sum_i x_i T_i$, where $x_i = n_i/n$ is the concentration of species i. The breakdown of energy equipartition in granular systems ($T_i \neq T$) predicted by kinetic theory [4] has been confirmed in computer simulations [11] as well as in real experiments [12].

It is quite apparent that the balance equations (7)–(9) become a closed set of differential equations for n_1, n_2, $\mathbf{U}$, and T when the fluxes, the cooling rate, and the partial temperatures are expressed in terms of the hydrodynamic fields and their gradients. These constitutive equations for $\mathbf{j}_i$, P, $\mathbf{q}$, ζ, and T_i may be derived by solving the BE (1) by the CE expansion up to first order in spatial gradients. This will be analyzed in Sect. 4.

3 Homogeneous Steady States

As a first step and before studying inhomogeneous situations, we consider homogeneous states. In this case, n_i and T are spatially uniform, and with a convenient choice of the reference frame, the mean velocities vanish ($\mathbf{U} = \mathbf{U}_g = \mathbf{0}$). For times longer than the mean free time, it is expected that the suspension achieves a steady state ($\partial_t f_i = 0$) where the BE (1) reads

$$-\gamma_i \frac{\partial}{\partial \mathbf{v}} \cdot \mathbf{v} f_i - \frac{\gamma_i T_{\text{ex}}}{m_i} \frac{\partial^2 f_i}{\partial v^2} = \sum_{j=1}^{2} J_{ij}[f_i, f_j]. \tag{14}$$

The balance equation for the partial temperature T_i can be easily derived by multiplying both sides of Eq. (14) by $m_i v^2$ and integrating over velocity:

$$2\gamma_i (T_{\text{ex}} - T_i) = \zeta_i T_i, \tag{15}$$

where the partial cooling rates ζ_i for the partial temperatures T_i are defined as

$$\zeta_i = -\frac{m_i}{dn_i T_i} \sum_{j=1}^{2} \int d\mathbf{v}\, v^2 J_{ij}[f_i, f_j], \quad (i = 1, 2). \tag{16}$$

The relationship between ζ and ζ_i is

$$\zeta = \sum_{i=1}^{2} x_i \tau_i \zeta_i, \tag{17}$$

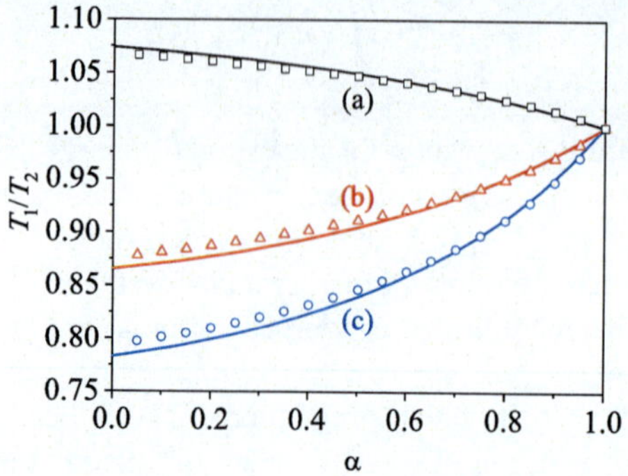

Fig. 1 Temperature ratio T_1/T_2 versus the (common) coefficient of restitution α for $d = 3$, $x_1 = 0.5$, $\sigma_1/\sigma_2 = 1$, $T_{\text{ex}}^* = 1$, and three different values of the mass ratio: $m_1/m_2 = 0.5$ (**a**), $m_1/m_2 = 4$ (**b**), and $m_1/m_2 = 10$ (**c**). Lines are the theoretical results while symbols refer to the Monte Carlo simulations

where $\tau_i = T_i/T$ is the temperature ratio of the species i. Upon deriving Eq. (17), use has been made of the relation $T = \sum_i x_i T_i$ and Eq. (6).

For elastic collisions ($\alpha_{ij} = 1$), $\zeta = \zeta_i = 0$, Eq. (15) yields $T_i = T_{\text{ex}} = T$ so that the Maxwellian distribution with a common temperature is a solution of the BE (14). On the other hand, for inelastic collisions ($\alpha_{ij} \neq 1$), ζ_i and ζ are different from zero and to date the solution to Eq. (14) is unknown. Thus, one has to consider an approximate form for the distributions f_i to estimate ζ_i. Here, we take the simplest approximation for both distributions, namely the Maxwellian distributions $f_{i,\text{M}}$ defined with the partial temperatures T_i:

$$f_i(\mathbf{v}) \to f_{i,\text{M}}(\mathbf{v}) = n_i \left(\frac{m_i}{2\pi T_i}\right)^{d/2} \exp\left(-\frac{m_i v^2}{2T_i}\right). \tag{18}$$

The partial cooling rates can be computed from Eq. (16) by replacing f_i by $f_{i,\text{M}}$. The result is [4]

$$\begin{aligned}\zeta_1 = &\frac{\sqrt{2}\pi^{(d-1)/2}}{d\Gamma\left(\frac{d}{2}\right)} n_1 \sigma_1^{d-1} \left(\frac{2T_1}{m_1}\right)^{1/2} (1-\alpha_{11}^2) + \frac{4\pi^{(d-1)/2}}{d\Gamma\left(\frac{d}{2}\right)} n_2 \sigma_{12}^{d-1} \mu_{21} \\ &\times \left(\frac{2T_1}{m_1} + \frac{2T_2}{m_2}\right)^{1/2} (1+\alpha_{12}) \left[1 - \frac{\mu_{21}}{2}(1+\alpha_{12})\left(1 + \frac{m_1 T_2}{m_2 T_1}\right)\right],\end{aligned} \tag{19}$$

where $\mu_{ij} = m_i/(m_i + m_j)$. The expression for ζ_2 can be easily obtained from Eq. (19) by making the change $1 \leftrightarrow 2$.

The partial temperatures T_i can be obtained from Eq. (15) (for $i = 1, 2$) when the expressions (19) for ζ_1 and ζ_2 are considered. Figure 1 plots the temperature ratio T_1/T_2 as a function of the (common) coefficient of restitution $\alpha \equiv \alpha_{ij}$ for $x_1 = 0.5$, $\sigma_1 = \sigma_2$, $T_{\text{ex}}^* = 1$, and three different values of the mass ratio. Here,

$T_{ex}^* = T_{ex}/(\overline{m}\sigma_{12}^2\gamma_0^2)$ and $\overline{m} = (m_1 + m_2)/2$. Theory is compared against the Monte Carlo simulations [13]. As expected, energy equipartition is broken for inelastic collisions; the extent of the energy violation is greater when the mass disparity is large. The excellent agreement found between theory and computer simulations is also quite apparent, except for quite small values of α (extreme inelasticity) where small discrepancies appear.

4 Chapman–Enskog Method. First-Order Solution

We assume now that the homogeneous steady state is perturbed by small spatial gradients. The existence of these gradients gives rise to nonzero contributions to the fluxes of mass, momentum, and energy. To first order in spatial gradients, the knowledge of the above fluxes allows one to identify the relevant Navier–Stokes transport coefficients of the granular suspension. As usual in the CE scheme [9], for times longer than the mean free time and distances larger than the mean free path, we suppose that the system achieves a hydrodynamic regime. This means that (i) the system has completely "forgotten" its initial preparation (initial conditions) and (ii) only the bulk domain of the system (namely, far away from the boundaries) is considered. Under these conditions, the BE (1) admits an special solution: the so-called *normal* or hydrodynamic solution where all space and time dependence of the distributions $f_i(\mathbf{r}, \mathbf{v}; t)$ is through a functional dependence on the hydrodynamic fields. This means that in the hydrodynamic regime, $f_i(\mathbf{r}, \mathbf{v}; t)$ adopts the normal form

$$f_i(\mathbf{r}, \mathbf{v}; t) = f_i\big[\mathbf{v}|n_1(t), n_2(t), T(t), \mathbf{U}(t)\big]. \tag{20}$$

The notation on the right-hand side of Eq. (20) indicates a functional dependence on the partial densities, temperature, and flow velocity. For small Knudsen numbers, the functional dependence (20) can be made local in space by expanding f_i in powers of the spatial gradients

$$f_i = f_i^{(0)} + \epsilon f_i^{(1)} + \epsilon^2 f_i^{(2)} + \cdots, \tag{21}$$

where ϵ is a bookkeeping parameter that denotes an implicit spatial gradient (for instance, a term of order ϵ is of first order in gradients). This parameter is taken to be equal to 1 at the end of the calculations.

An important point in the CE expansion is to characterize the magnitude of the friction coefficients γ_i and the term $\Delta\mathbf{U}$ with respect to the spatial gradients. On the one hand, since γ_i does not create any flux, then it is assumed to be to zeroth order in ϵ. On the other hand, because $\Delta\mathbf{U} = \mathbf{0}$ in the absence of gradients, it should be considered to be at least of first order in spatial gradients (first order in ϵ).

The implementation of the CE method to solve the BE (1) to first order in spatial gradients is very large and beyond the scope of the present contribution. We refer the interested reader to Ref. [8] for specific details. Since we want here to analyze thermal diffusion segregation in a granular suspension, in order to show more clearly the different competing mechanisms appearing in this phenomenon, we consider a binary mixture where the concentration of one of the species (let's say species 1) is much smaller than that of the other species 2 (tracer limit, $x_1 \to 0$). The consideration of this simple situation allows us to offer a simplified theory where a segregation criterion can be explicitly obtained.

In the tracer limit, the pressure tensor P_{ij}, the heat flux $\mathbf{q}$, and the cooling rate ζ of the binary mixture are the same as that of the excess species. While the fluxes P_{ij} and $\mathbf{q}$ are of first order in the spatial gradients in the Navier–Stokes description, the expression of ζ must retain terms up to second order in gradients. Part of these second-order contributions to ζ have been computed by Brey et al. [5] for dry (dilute) granular gases, while the complete set of these contributions has been determined by Brilliantov and Pöschel [14] for granular gases of viscoelastic particles. Nevertheless, it has been shown [5] that these second-order contributions to ζ are negligible as compared with its zeroth-order counterparts. We expect that the same occurs for the case of binary granular suspensions and hence, they can be ignored.

4.1 Tracer Limit. Diffusion Transport Coefficients

In the tracer limit, the first-order contribution $\mathbf{j}_1^{(1)}$ to the mass flux is [8]

$$\mathbf{j}_1^{(1)} = -\frac{m_1^2}{\rho} D_{11} \nabla n_1 - \frac{m_1 m_2}{\rho} D_{12} \nabla n_2 - \frac{\rho}{T} D_1^T \nabla T - D_1^U \Delta \mathbf{U}, \tag{22}$$

where the diffusion transport coefficients are defined as

$$D_{11} = -\frac{\rho}{\rho_1 d} \int d\mathbf{v} \mathbf{V} \cdot \boldsymbol{\mathcal{B}}_{11}(\mathbf{V}), \quad D_{12} = -\frac{1}{d} \int d\mathbf{v} \mathbf{V} \cdot \boldsymbol{\mathcal{B}}_{12}(\mathbf{V}), \tag{23}$$

$$D_1^T = -\frac{m_1}{\rho d} \int d\mathbf{v} \mathbf{V} \cdot \boldsymbol{\mathcal{A}}_1(\mathbf{V}), \quad D_1^U = -\frac{m_1}{d} \int d\mathbf{v} \mathbf{V} \cdot \boldsymbol{\mathcal{E}}_1(\mathbf{V}). \tag{24}$$

The unknowns $\boldsymbol{\mathcal{A}}_1(\mathbf{V})$, $\boldsymbol{\mathcal{B}}_{11}(\mathbf{V})$, $\boldsymbol{\mathcal{B}}_{12}(\mathbf{V})$, and $\boldsymbol{\mathcal{E}}_1(\mathbf{V})$ are the solutions of a set of coupled linear integral equations [8]. In the tracer limit, this set reads

$$-\left(2\gamma_2\theta^{-1} + \frac{1}{2}\zeta^{(0)}\right)\boldsymbol{\mathcal{A}}_1 - \gamma_1 \frac{\partial}{\partial \mathbf{V}} \cdot (\mathbf{V}\boldsymbol{\mathcal{A}}_1) - \gamma_1 \frac{T_{\text{ex}}}{m_1} \frac{\partial^2 \boldsymbol{\mathcal{A}}_1}{\partial v^2} - J_{12}[\boldsymbol{\mathcal{A}}_1, f_2^{(0)}] = \mathbf{A}_1 + J_{12}[f_1^{(0)}, \boldsymbol{\mathcal{A}}_2], \tag{25}$$

$$-\gamma_1 \frac{\partial}{\partial \mathbf{V}} \cdot (\mathbf{V}\boldsymbol{\mathcal{B}}_{11}) - \gamma_1 \frac{T_{\text{ex}}}{m_1} \frac{\partial^2 \boldsymbol{\mathcal{B}}_{11}}{\partial v^2} - J_{12}[\boldsymbol{\mathcal{B}}_{11}, f_1^{(0)}] = \mathbf{B}_{11}, \tag{26}$$

$$-\gamma_1\frac{\partial}{\partial \mathbf{V}}\cdot(\mathbf{V}\boldsymbol{\mathcal{B}}_{12})\ -\gamma_1\frac{T_{\rm ex}}{m_1}\frac{\partial^2\boldsymbol{\mathcal{B}}_{12}}{\partial v^2}-J_{12}[\boldsymbol{\mathcal{B}}_{12},f_2^{(0)}]=\mathbf{B}_{12}+J_{12}[f_1^{(0)},\boldsymbol{\mathcal{B}}_{21}]$$
$$+\left[\zeta^{(0)}-2\gamma_2\left(\theta^{-1}-1\right)+2\lambda_1\frac{\partial\tau_1}{\partial\lambda_1}\right]\boldsymbol{\mathcal{A}}_1, \tag{27}$$

$$-\gamma_1\frac{\partial}{\partial \mathbf{V}}\cdot(\mathbf{V}\boldsymbol{\mathcal{E}}_1)-\gamma_1\frac{T_{\rm ex}}{m_1}\frac{\partial^2\boldsymbol{\mathcal{E}}_1}{\partial v^2}-J_{12}[\boldsymbol{\mathcal{E}}_1,f_2^{(0)}]=\mathbf{E}_1. \tag{28}$$

In the integral equations (25)–(28), $\zeta^{(0)}$ is the zeroth-order approximation to the cooling rate, $\theta=T/T_{\rm ex}$, $\lambda_1=(2T_{\rm ex}^*)^{-1/2}(R_1/n\sigma_{12}^d)$, and $\boldsymbol{\mathcal{A}}_2$ and $\boldsymbol{\mathcal{B}}_{21}$ refer to quantities of the excess species 2. These quantities obey certain integral equations; their explicit forms are not needed for evaluating the diffusion transport coefficients in the so-called first Sonine approximation. In addition, the expression of the derivative $\partial_{\lambda_1}\tau_1$ can be found in Appendix A of Ref. [8] while the inhomogeneous terms $\mathbf{A}_1$, $\mathbf{B}_{11}$, $\mathbf{B}_{12}$, and $\mathbf{E}_1$ are given, respectively, by

$$\mathbf{A}_1(\mathbf{V})=-\mathbf{V}\frac{\partial f_1^{(0)}}{\partial T}-\frac{p}{\rho}\frac{\partial f_1^{(0)}}{\partial \mathbf{V}},\quad \mathbf{B}_{11}(\mathbf{V})=-\mathbf{V}n_1\frac{\partial f_1^{(0)}}{\partial n_1}, \tag{29}$$

$$\mathbf{B}_{12}(\mathbf{V})=-\mathbf{V}n_2\frac{\partial f_1^{(0)}}{\partial n_2}-\frac{T}{m_2}\frac{\partial f_1^{(0)}}{\partial \mathbf{V}},\quad \mathbf{E}_1(\mathbf{V})=(\gamma_1-\gamma_2)\frac{\partial f_1^{(0)}}{\partial \mathbf{V}}. \tag{30}$$

Note that Eqs. (25)–(28) have been obtained under steady-state conditions, namely when the conditions (15) apply. Furthermore, in order to obtain the above set of coupled integral equations, we have taken into account that while in the tracer limit D_{11} is independent of x_1, the coefficients D_{12}, D_1^T, and D_1^U are proportional to x_1. This dependence on x_1 will be then self-consistently confirmed. Accordingly, $\boldsymbol{\mathcal{A}}_1\propto x_1$, $\boldsymbol{\mathcal{B}}_{12}\propto x_1$, and $\boldsymbol{\mathcal{E}}_1\propto x_1$.

Although the exact form of the zeroth-order distributions $f_i^{(0)}$ is not known, dimensional analysis requires that they have the scaled form $f_i^{(0)}(\mathbf{V})=n_i v_{\rm th}^{-d}\varphi_i(\mathbf{c},\gamma_i^*,\theta)$. Here, $\mathbf{c}=\mathbf{V}/v_{\rm th}$ and $\gamma_i^*=\gamma_i/\nu_0$, where $\nu_0=n\sigma_{12}^{d-1}v_{\rm th}$ is an effective collision frequency, and $v_{\rm th}=\sqrt{2T/\overline{m}}$ is the thermal velocity. Thus, one has the property

$$T\frac{\partial f_i^{(0)}}{\partial T}=-\frac{1}{2}\frac{\partial}{\partial \mathbf{V}}\cdot\mathbf{V}f_i^{(0)}+n_i v_{\rm th}^{-d}\theta\frac{\partial\varphi_i}{\partial\theta}. \tag{31}$$

4.2 Leading Sonine Approximation

Equations (25)–(28) are still exact. However, the determination of the diffusion transport coefficients requires to solve the above integral equations as well as to know the zeroth-order distributions $f_i^{(0)}$. The results derived for driven granular mixtures [15] have shown that non-Gaussian corrections to $f_i^{(0)}$ (which are measured through the fourth cumulants c_i) are in general very small. Thus, $f_i^{(0)}$ is well represented by its Maxwellian form (18) and so, a theory incorporating the cumulants c_i seems to be unnecessary in practice for computing the diffusion transport coefficients. Regarding the functions $\boldsymbol{\mathcal{A}}_i$, $\boldsymbol{\mathcal{B}}_{ij}$, and $\boldsymbol{\mathcal{E}}_i$, as usual we consider the leading terms in a series expansion of these quantities in Sonine polynomials. In this case, $\boldsymbol{\mathcal{A}}_2 \to 0$, $\boldsymbol{\mathcal{B}}_{21} \to 0$, and the quantities $\boldsymbol{\mathcal{A}}_1, \boldsymbol{\mathcal{B}}_{11}, \boldsymbol{\mathcal{B}}_{12}$, and $\boldsymbol{\mathcal{E}}_1$ corresponding to the tracer species are approximated by

$$\boldsymbol{\mathcal{A}}_1(\mathbf{V}) \to -f_{1,\mathrm{M}}\mathbf{V}\frac{\rho}{n_1T_1}D_1^T, \quad \boldsymbol{\mathcal{B}}_{11}(\mathbf{V}) \to -f_{1,\mathrm{M}}\mathbf{V}\frac{m_1^2}{\rho T_1}D_{11}, \tag{32}$$

$$\boldsymbol{\mathcal{B}}_{12}(\mathbf{V}) \to -f_{1,\mathrm{M}}\mathbf{V}\frac{m_1}{n_1T_1}D_{12}, \quad \boldsymbol{\mathcal{E}}_1(\mathbf{V}) \to -f_{1,\mathrm{M}}\mathbf{V}\frac{1}{n_1T_1}D_1^U. \tag{33}$$

Now, we substitute first Eqs. (32) and (33) into the integral equations (25)–(28), multiply them by $m_1\mathbf{V}$, and integrate over $\mathbf{v}$. After some algebra, D_{11}, D_1^T, D_{12}, and D_1^U can be written, respectively, as

$$D_{11} = \frac{\rho T}{m_1^2\nu_0}\frac{\tau_1}{\nu_D^* + \gamma_1^*}, \quad D_1^T = \frac{nT}{\rho\nu_0}x_1\frac{\theta\frac{\partial\tau_1}{\partial\theta} - (\mu - \tau_1)}{\nu_D^* + \gamma_1^* - 2\gamma_2^*\theta^{-1} - \frac{1}{2}\zeta_0^*}, \tag{34}$$

$$D_{12} = \frac{x_1\rho T}{m_1m_2\nu_0}\frac{\left[\zeta_0^* - 2\gamma_2^*\left(\theta^{-1} - 1\right)\right]x_1^{-1}D_1^{T*} - \mu - 2\lambda_1\frac{\partial\tau_1}{\partial\lambda_1}}{\nu_D^* + \gamma_1^*}, \tag{35}$$

$$D_1^U = \rho_1\frac{\gamma_1^* - \gamma_2^*}{\gamma_1^* - \nu_D^*}. \tag{36}$$

Here, $\mu = m_1/m_2$ is the mass ratio, the derivative $\partial_\theta\tau_1$ is given in Appendix A of Ref. [8],

$$\zeta_0^* = \frac{\pi^{(d-1)/2}}{d\Gamma\left(\frac{d}{2}\right)}\left(\frac{\sigma_2}{\sigma_{12}}\right)^{d-1}\mu_{21}^{-1/2}\left(1 - \alpha_{22}^2\right), \tag{37}$$

and the reduced collision frequency ν_D^* is

$$\nu_D^* = \frac{2\sqrt{2}\pi^{(d-1)/2}}{d\Gamma\left(\frac{d}{2}\right)}\mu_{21}^{3/2}\left(\frac{1+\beta}{\beta}\right)^{1/2}(1+\alpha_{12}), \tag{38}$$

where $\beta = \beta_1/\beta_2 = \mu/\tau_1$.

Figure 2 shows the dependence of the reduced diffusion transport coefficients $D_{ij}(\alpha)/D_{ij}(1)$, $D_1^T(\alpha)/D_1^T(1)$, and $D_1^U(\alpha)/D_1^U(1)$ for $\sigma_1/\sigma_2 = 1$, $m_1/m_2 = 10$, and $T_{\mathrm{ex}}^* = 0.1$. Here, $D_{ij}(1)$, $D_1^T(1)$, and $D_1^U(1)$ are the values of these coefficients

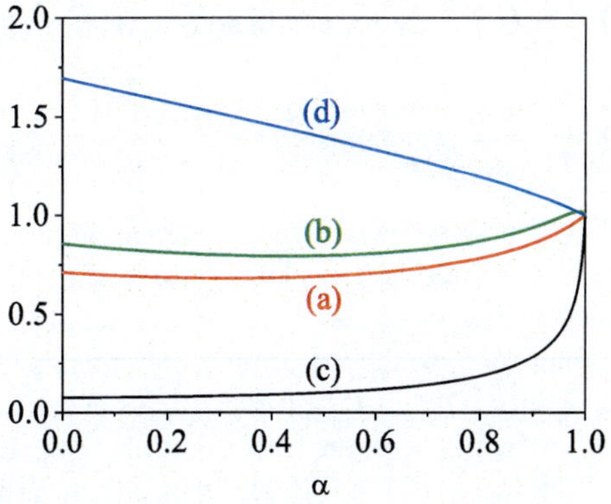

Fig. 2 Plot of the (reduced) transport coefficients $D_{11}(\alpha)/D_{11}(1)$ (**a**), $D_{12}(\alpha)/D_{12}(1)$ (**b**), $D_1^T(\alpha)/D_1^T(1)$ (**c**), and $D_1^U(\alpha)/D_1^U(1)$ (**d**) as a function of the common coefficient of restitution for a binary mixture of hard spheres ($d=3$) in the tracer limit ($x_1 \to 0$) with $\sigma_1/\sigma_2 = 1$, $m_1/m_2 = 10$, and $T_{\text{ex}}^* = 0.1$

for elastic collisions. We observe that the impact of inelasticity on those coefficients is in general quite important since they differ clearly from their elastic forms, especially in the case of the thermal diffusion coefficient D_1^T. Moreover, a comparison with the results obtained in the dry granular limit (no gas phase) shows important qualitative differences between both theories (see, for instance, Fig. 6.3 of Ref. [4] for the diffusion coefficient D_{11}).

5 Thermal Diffusion Segregation of an Intruder in a Granular Suspension

A nice application of the previous results is the study of thermal diffusion segregation of an intruder or tracer particle in a granular suspension. Needless to say, segregation and mixing of dissimilar grains are one of the most interesting problems in granular mixtures, not only from a fundamental point of view but also from a more practical perspective. This problem has been widely studied in the past few years for dry granular mixtures. The objective here is to assess the influence of the interstitial gas phase on the segregation criterion.

Thermal diffusion is originated by the relative motion of the components of a mixture due to the presence of a temperature gradient. Due to this motion, concentration gradients appear in the mixture producing ordinary diffusion. A steady state is finally achieved in which the separating effect emerging from thermal diffusion is offset by the remixing effect arising from ordinary diffusion [16]. The partial separation between both components of the mixture is then observed; this effect is usually referred to as the Soret effect.

The amount of segregation parallel to the thermal gradient may be characterized by the so-called thermal diffusion factor Λ. This quantity is defined in an inhomogeneous non-convecting ($\mathbf{U} = \mathbf{U}_g = \mathbf{0}$) steady state with zero mass flux ($\mathbf{j}_1^{(1)} = \mathbf{0}$) as

$$-\Lambda \frac{\partial \ln T}{\partial z} = \frac{\partial}{\partial z} \ln \left(\frac{n_1}{n_2}\right), \tag{39}$$

where only gradients along the z-axis have been assumed for simplicity. Let us assume that the bottom plate is hotter than the top plate ($\partial_z \ln T < 0$). If Λ is supposed to be constant over the relevant ranges of composition and temperature of the system, according to Eq. (39), when $\Lambda > 0$, the tracer particle tends to rise with respect to the gas particles 2, i.e., $\partial_z \ln(n_1/n_2) > 0$ (tracer particles accumulate near the cold plate). On the other hand, when $\Lambda < 0$, the tracer particle tends to fall with respect to the gas particles 2, i.e., $\partial_z \ln(n_1/n_2) < 0$ (tracer particles accumulate near the hot plate).

Let us determine the thermal diffusion factor. The mass flux $j_{1,z}^{(1)}$ is given by Eq. (22) with $\Delta\mathbf{U} = \mathbf{0}$. Since $j_{1,z}^{(1)} = 0$ in the steady state and $\mathbf{U} = \mathbf{U}_g = \mathbf{0}$, then Eq. (8) yields $\partial_z(nT) = 0$ and so,

$$\partial_z \ln T = -\partial_z \ln n_2. \tag{40}$$

Here, we have taken into account that $n \simeq n_2$ in the tracer limit. The factor Λ can be written in terms of the diffusion coefficients when one takes into account Eq. (40) and that $j_{1,z}^{(1)} = 0$. Its expression is finally given by

$$\Lambda = \frac{x_1^{-1} D_1^{T^*} - D_{11}^* - x_1^{-1} D_{12}^*}{D_{11}^*}, \tag{41}$$

where we have introduced the dimensionless transport coefficients

$$D_{11}^* = \frac{m_1^2 \nu_0}{\rho T} D_{11}, \quad D_{12}^* = \frac{m_1 m_2 \nu_0}{\rho T} D_{11}, \quad D_1^{T*} = \frac{\rho \nu_0}{n T} D_1^T. \tag{42}$$

The explicit dependence of Λ on the parameters of the granular suspension (mass and size ratios, the coefficients of restitution α_{12} and α_{22}, and the dimensionless external temperature T_{ex}^*) can be obtained when one substitutes Eqs. (34) and (35) of D_{11}, D_1^T, and D_{12}, respectively, into Eq. (41). Since $D_{11}^* > 0$, the condition $\Lambda = 0$ is

$$x_1^{-1} D_1^{T^*} = D_{11}^* + x_1^{-1} D_{12}^*. \tag{43}$$

Equation (43) gives the marginal segregation curve separating intruder segregation toward the cold wall ($\Lambda > 0$) from intruder segregation toward the hot wall ($\Lambda < 0$). On the other hand, since the number of parameters involved in the segregation problem is still large, it is not easy to disentangle the influence of each mechanism (mass and size ratios, inelasticity in collisions, external temperature, ...) on the intruder segregation problem. Thus, it is convenient first to consider some simple situations.

5.1 Mechanically Equivalent Particles

This is quite trivial case since the system is in fact monodisperse ($m_1 = m_2, \sigma_1 = \sigma_2$, $\alpha_{11} = \alpha_{22} \equiv \alpha$, $\gamma_2 = \gamma_1 \equiv \gamma$). In this limit case, $\tau_1 = 1$ and according to Eqs. (34) and (35), $x_1^{-1} D_1^{T*} = 0$ and $D_{11}^* = -x_1^{-1} D_{12}^* = (\nu_D^* + \gamma^*)^{-1}$. Therefore, Eq. (41) yields $\Lambda = 0$ for any value of the coefficients of restitution α_{12} and α_{22} and the bath temperature T_{ex}^*. This implies that no segregation is possible, as expected.

5.2 Elastic Collisions

For elastic collisions ($\alpha_{22} = \alpha_{12} = 1$), $\zeta_0^* = 0, \tau_1 = \theta = 1, \beta = \mu$, and Eq. (43) leads to the condition

$$\left(\nu_D^* + \gamma_1^*\right) \frac{\partial \tau_1}{\partial \theta} = 2\gamma_2^* \left(\mu - 1\right). \tag{44}$$

Upon deriving Eq. (44), we have considered the region of parameter space where $\nu_D^* + \gamma_1^* - 2\gamma_2^* \neq 0$. It is quite apparent that even for elastic collisions, the segregation criterion ($\Lambda = 0$) is not simple and differs from the simple segregation criterion obtained in the dry case ($\mu = 1$). Figure 3 shows a phase diagram in the $\{m_2/m_1, \sigma_2/\sigma_1\}$-plane at $T_{\text{ex}}^* = 1$. For a given value of the mass ratio m_2/m_1, it is quite apparent that the region $\Lambda > 0$ (tracer particle falls with respect to excess granular gas) is dominant when the size of gas particles is much larger than that of the tracer particle. This tendency increases with increasing the mass ratio m_2/m_1.

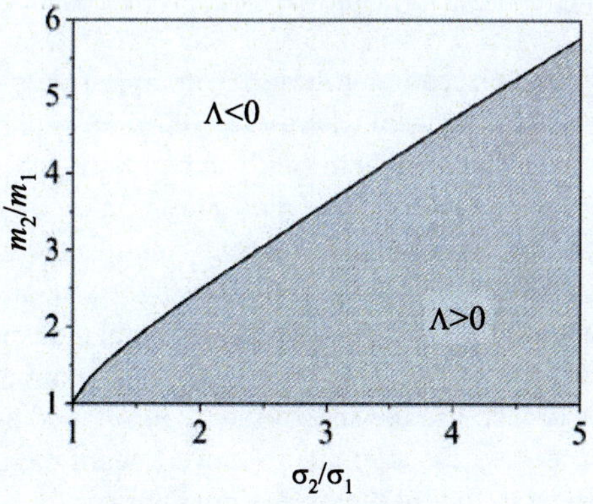

Fig. 3 Plot of the marginal segregation curve ($\Lambda = 0$) for $d = 3$, $\alpha_{22} = \alpha_{12} = 1$, and $T_{\text{ex}}^* = 1$. Points below (above) the curve correspond to $\Lambda > 0$ ($\Lambda < 0$)

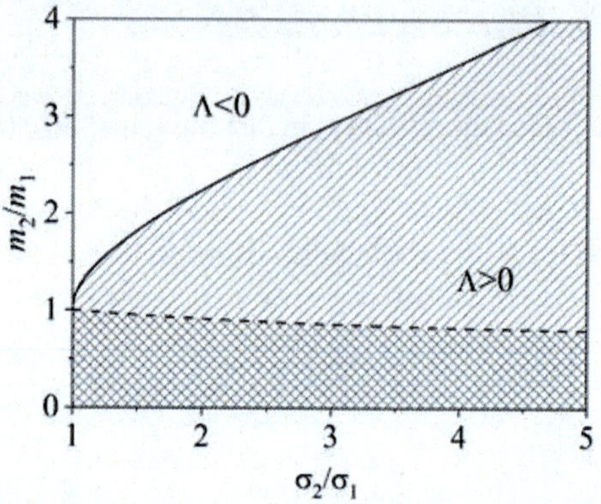

Fig. 4 The same as in Fig. 3 but for $\alpha_{22} = \alpha_{12} = 0.9$. The solid line corresponds to the segregation criterion for granular suspensions while the dashed line refers to the one derived for dry granular mixtures. Points below (above) each curve correspond to $\Lambda > 0$ ($\Lambda < 0$)

5.3 *Inelastic Collisions*

We consider now the general case where α_{22} and α_{12} are different from 1. In this case, considering the region of parameter space where $\nu_D^* + \gamma_1^* - 2\gamma_2^*\theta^{-1} - \frac{1}{2}\zeta_0^* \neq 0$, Eq. (43) yields

$$\left[\nu_D^* + \gamma_1^* - \zeta_0^* + 2\gamma_2^*\left(\theta^{-1} - 1\right)\right]\left(\theta\frac{\partial\tau_1}{\partial\theta} - \mu + \tau_1\right) = \left(\tau_1 - \mu - 2\lambda_1\frac{\partial\tau_1}{\partial\lambda_1}\right) \times\left(\nu_D^* + \gamma_1^* - 2\gamma_2^*\theta^{-1} - \frac{1}{2}\zeta_0^*\right). \tag{45}$$

This is quite a complex segregation criterion in comparison with the one derived in the dry granular case (no gas phase) where $\Lambda = 0$ if $\mu = \tau_1$ [17, 18]. To illustrate more clearly the differences between both (with and without gas phase) segregation criteria, Fig. 4 shows the marginal segregation curve ($\Lambda = 0$) for the (common) coefficient of restitution 0.9. Figure highlights that the impact of gas phase on tracer segregation is quite significant since, at a given value of the size ratio, the value of the mass ratio m_2/m_1 at which $\Lambda = 0$ is greater in the granular suspension than in the dry granular system. In addition, we also observe that the main effect of gas phase on tracer segregation is to increase the size of the region $\Lambda > 0$ as σ_2/σ_1 increases. This means that the tracer particle attempts to move toward the cold regions as its size decreases with respect to that of the excess granular gas.

6 Concluding Remarks

The primary objective of this short review has been to derive the Navier–Stokes hydrodynamic equations of a binary granular suspension within the context of the inelastic version of the Boltzmann kinetic equation. As usual, the effect of the interstitial surrounding gas on grains has been modeled through an effective external force constituted by a deterministic drag force plus an stochastic Langevin-like force. This way of modeling gas–solid flows is essentially based on the following assumptions and/or simplifications. First, assuming that the granular mixture is rarefied, one supposes that the state of the surrounding gas is not perturbed by the presence of grains and so, it can be treated as a thermostat. Second, the impact of gas phase on collision dynamics is very weak, and consequently, the Boltzmann collision operator is not affected by the presence of the interstitial gas. As a third simplification, one considers the friction coefficients appearing in the thermal-drag forces to be scalar quantities. Finally, as a fourth simplification, one assumes low Reynolds numbers so that only laminar flows are considered.

The road map for obtaining the Navier–Stokes hydrodynamic equations needs to characterize first the homogeneous state. This is important because the Navier–Stokes transport coefficients are obtained from the CE expansion around the above state. Given that the transport coefficients are given in terms of the solutions to a set of coupled linear integral equations, these equations are approximately solved by considering the leading terms in a series expansion of Sonine polynomials. This road map is large and involves many technical steps. Here, for the sake of simplicity, we have obtained the mass flux of a binary granular suspension where the concentration of one of the species is negligible (tracer limit). The tracer limit allows us to provide expressions that are easy to handle for potential applications. In particular, we have briefly analyzed here the thermal diffusion segregation of an intruder or tracer particle in a granular suspension. The segregation criterion obtained here shows significant discrepancies with respect to the one previously derived for dry granular mixtures [17, 18]. These differences between both situations (with and without interstitial gas) are clearly illustrated in Figs. 3 and 4 for elastic and inelastic collisions, respectively.

Multicomponent granular suspensions exhibit a wide range of interesting phenomena for which kinetic theory and hydrodynamics (in the broader sense) may be considered as useful tools for understanding the behavior of such complex materials. However, due to their complexity, many of their features are not still completely understood. For this reason, from the theoretical side, one has to introduce new ingredients in the model for approaching more realistic situations. In this context, the extension of the results presented in this review to inertial suspensions of inelastic rough hard spheres could be an interesting and challenging problem in the near future.

Acknowledgements The present work has been supported by the Spanish Government through Grant No. PID2020-112936GB-I00 funded by MCIN/AEI/10.13039/501100011033 and by the Junta de Extremadura (Spain) Grant Nos. IB20079 and GR18079, partially financed by "Fondo

Europeo de Desarrollo Regional" funds. The research of Rubén Gómez González has been also supported by the predoctoral fellowship BES-2017-079725 from the Spanish Government.

References

1. See for instance, M.E. Möbius, B.E. Lauderdale, S.R. Nagel, H.M. Jaeger, Brazil-nut effect: size separation of granular particles. Nature **414**, 270 (2001); P. Sánchez, M.R. Swift, P.J. King, Stripe formation in granular mixtures due to the differential influence of drag. Phys. Rev. Lett. **93**, 184302 (2004); J.C. Pastenes, J.C. Géminard, F. Melo, Interstitial gas effect on vibrated granular columns. Phys. Rev. E **89**, 062205 (2014)
2. D.L. Koch, R.J. Hill, Inertial effects in suspensions and porous-media flows. Annu. Rev. Fluid Mech. **33**, 619 (2001)
3. N.V. Brilliantov, T. Pöschel, *Kinetic Theory of Granular Gases* (Oxford University Press, Oxford, 2004)
4. V. Garzó, *Granular Gaseous Flows* (Springer Nature, Cham, 2019)
5. J.J. Brey, J.W. Dufty, C.S. Kim, A. Santos, Hydrodynamics for granular flow at low-density. Phys. Rev. E **58**, 4638 (1998)
6. V. Garzó, J.W. Dufty, Dense fluid transport for inelastic hard spheres. Phys. Rev. E **59**, 5895 (1999); V. Garzó, J.W. Dufty, Hydrodynamics for a granular binary mixture at low density. Phys. Fluids **14**, 1476 (2002); V. Garzó, J.W. Dufty, C.M. Hrenya, Enskog kinetic theory for polydisperse granular mixtures. I. Navier–Stokes order transport. Phys. Rev. E **76**, 031303 (2007); V. Garzó, C.M. Hrenya, J.W. Dufty, Enskog kinetic theory for polydisperse granular mixtures. II. Sonine polynomial approximation. Phys. Rev. E **76**, 031304 (2007)
7. N. Khalil, V. Garzó, Transport coefficients for driven granular mixtures at low-density. Phys. Rev. E **88**, 052201 (2013)
8. R. Gómez González, N. Khalil, V. Garzó, Enskog kinetic theory for multicomponent granular suspensions. Phys. Rev. E **101**, 012904 (2020)
9. S. Chapman, T.G. Cowling, *The Mathematical Theory of Nonuniform Gases* (Cambridge University Press, Cambridge, 1970)
10. X. Yin, S. Sundaresan, Fluid-particle drag in low-Reynolds-number polydisperse gas-solid suspensions. AIChE J. **55**, 1352 (2009)
11. See for instance, J.M. Montanero, V. Garzó, Monte Carlo simulation of the homogeneous cooling state for a granular mixture. Granular Matter **4**, 17 (2002); S.R. Dahl, C.M. Hrenya, V. Garzó, J.W. Dufty, Kinetic temperatures for a granular mixture. Phys. Rev. E 66, 041301 (2002); A. Barrat, E. Trizac. Molecular dynamics simulations of vibrated granular gases. Phys. Rev. **66**, 051303 (2002); P. Krouskop, J. Talbot, Mass and size effects in three-dimensional vibrofluidized granular mixtures. Phys. Rev. E **68**, 021304 (2003); M. Schröter, S. Ulrich, J. Kreft, J.B. Swift, H.L. Swinney, Mechanisms in the size segregation of a binary granular mixture. Phys. Rev. E **74**, 011307 (2006)
12. R.D. Wildman, D.J. Parker, Coexistence of two granular temperatures in binary vibrofluidized beds. Phys. Rev. Lett. **88**, 064301 (2002); K. Feitosa, N. Menon, Breakdown of energy equipartition in a 2D binary vibrated granular gas. Phys. Rev. Lett. **88**, 198301 (2002)
13. G.A. Bird, *Molecular Gas Dynamics and the Direct Simulation of Gas Flows* (Oxford University Press, Oxford, 1994)
14. N. Brilliantov, T. Pöschel, Hydrodynamics and transport coefficients for dilute granular gases. Phys. Rev. E **67**, 061304 (2003)
15. N. Khalil, V. Garzó, Homogeneous states in driven granular mixtures: Enskog kinetic theory versus molecular dynamics simulations. J. Chem. Phys. **140**, 164901 (2014)
16. J. Kincaid, E.G.D. Cohen, M. López de Haro, The Enskog theory for multicomponent mixtures. IV. Thermal diffusion. J. Chem. Phys. **86**, 963 (1987)

17. J.J. Brey, M.J. Ruiz-Montero, F. Moreno, Energy partition and segregation for an intruder in a vibrated granular system under gravity. Phys. Rev. Lett. **95**, 098001 (2005)
18. V. Garzó, Segregation in granular binary mixtures: thermal diffusion. Europhys. Lett. **75**, 521–527 (2006)

Boltzmann Equation in Aggregation Kinetics

Nikolai V. Brilliantov, Alexander I. Osinsky, and Thorsten Pöschel

Abstract We consider the application of the Boltzmann equation to aggregation kinetics, where the transport mechanism is the ballistic motion of particles. This refers to molecular gases, granular gases, and, hypothetically, dark matter. Two aggregation models are analyzed—random and impact energy-dependent aggregation. The latter is associated with different interparticle forces responsible for agglomeration. We start from the Boltzmann equation governing the evolution of the mass–velocity distribution functions of different species—the agglomerates of different sizes and derive generalized Smoluchowski equations. These describe the time dependence of the agglomerates densities and their mean kinetic energy (partial temperatures). We obtain exact solutions to these equations for simplified cases and develop a scaling theory for the asymptotic behavior of the system. We explore numerically, the agglomeration kinetics and observe a very rich behavior of the system. We reveal new surprising regimes and construct the according kinetic phase diagram. The scaling theory is in excellent agreement with the simulation results.

1 Introduction

The celebrated Boltzmann equation [1] derived by Ludwig Boltzmann in 1872 remains one of the main pillars of non-equilibrium statistical mechanics and thermodynamics. It is used in many areas of kinetic theory, ranging from classical gas dynamics and dynamics of granular gases [2–5], to aggregation and fragmentation

N. V. Brilliantov (✉) · A. I. Osinsky
Skolkovo Institute of Science and Technology, 30 Bolshoi Boulevard, Moscow 121205, Russia
e-mail: n.brilliantov@skoltech.ru; nb144@leicester.ac.uk

A. I. Osinsky
e-mail: A.Osinskiy@skoltech.ru

T. Pöschel
Institute for Multiscale Simulation, Friedrich-Alexander-Universität Erlangen-Nürnberg, Cauerstraße 3, 91058 Erlangen, Germany
e-mail: thorsten.poeschel@fau.de

L. Brenig et al. (eds.), *Nonequilibrium Thermodynamics and Fluctuation Kinetics*, Fundamental Theories of Physics 208,
https://doi.org/10.1007/978-3-031-04458-8_10

phenomena [6–13], traffic and active matter modeling, e.g., [14, 15]. The aggregation kinetics described by the Boltzmann equation refers to atmospheric phenomena, such as coagulation of dust or airborne particles, e.g., [16–19], the behavior of astrophysical systems—planetary rings and interstellar dust clouds, e.g., [6, 7, 9, 20–24]; modern theories of the evolution of dark matter also exploit the Boltzmann equation [10–13]. A complete description of aggregation is very complicated. Therefore, several simplifying assumptions are applied. First, it is assumed that each aggregate may be completely characterized by its mass, which is determined by the number of elementary units (monomers) comprising the cluster. Second, in the realm of the Boltzmann approach, it is assumed that it is also characterized by velocity. Hence, the shape of an aggregate and its angular motion is ignored; under these assumptions, one can formulate Boltzmann-like equations for aggregating particles. Nevertheless, the approach based on the Boltzmann equation is rather flexible and, in principle, allows further generalizations.

The transport mechanism is crucial for aggregation. The first theory of aggregation kinetics was developed by Smoluchowski for Brownian coagulation, where the diffusional transport was assumed [25]. Such aggregation processes underlie polymerization and many chemical reactions in solutions [14, 26]. The main quantities of interests there are $n_k(t)$—the densities of clusters comprised of k monomers (clusters of size k), at time t. The rates of the aggregation processes $[i] + [j] \to [i + j]$ are determined by the clusters size i and j and properties of the solvent; they do not alter in the course of time. Another transport mechanism underlying numerous aggregation phenomena is ballistic transport. Despite of intensive studies [27–38], ballistic agglomeration (BA) is not well understood yet.

The main difference between BA and diffusion-driven aggregation (DDA) is the role of the kinetic energy of particles. For DDA, energy is an external parameter, as the reaction rates are determined by the temperature of the surrounding solvent. In contrast, the kinetic energy is lost due to agglomeration events of BA. The decay of the kinetic energy of particles results in the slowing down of the aggregation process; that is, it determines the agglomeration rates. Hence, in addition to the mass of clusters, as for DDA, one needs to account for cluster velocities and consider a joint mass-velocity distribution function; it obeys Boltzmann–Smoluchowski equations discussed below [8, 9, 33, 34].

2 Boltzmann–Smoluchowski Equations for Ballistic Agglomeration

2.1 Boltzmann Equation

To illustrate the main concepts, we start with a simpler example—the Boltzmann equation (BE) for equal, non-aggregating particles. Here, we will address space-uniform systems described by the velocity distribution function $f(\mathbf{v}, t)$. It gives the density of particles with velocity $\mathbf{v}$ in the small interval $(\mathbf{v}, \mathbf{v} + d\mathbf{v})$ at time t. The BE describes the evolution of this function [2]:

$$\frac{\partial}{\partial t} f(\mathbf{v}, t) = I^{\text{res}} = I^{\text{gain}} - I^{\text{loss}}. \tag{1}$$

Here, I^{res} is the collision integral for restitutive impacts, that is, for impacts without agglomeration. It may be written as a sum of the loss and gain terms. The loss term I^{loss} quantifies the rate at which the number of particles (in a unit volume) with the velocity $\mathbf{v} \in (\mathbf{v}, \mathbf{v} + d\mathbf{v})$ decreases due to collisions with other particles with the velocity $\mathbf{v}'$; such collisions are called direct collisions. I^{loss} reads [2]:

$$I^{\text{loss}} = \sigma^2 \int d\mathbf{v}' \int d\mathbf{e} \Theta(-\Delta\mathbf{v} \cdot \mathbf{e}) \, |\Delta\mathbf{v} \cdot \mathbf{e}| f(\mathbf{v}, t) f(\mathbf{v}', t). \tag{2}$$

Here, σ is the particle diameter, which determines the collision cross section. $\Delta\mathbf{v} = \mathbf{v} - \mathbf{v}'$ gives the relative velocity of the colliding pair, and $\mathbf{e}$ is the unit vector directed along the vector joining the particles' centers at the collision instant. Hence $(\sigma^2 d\mathbf{e}|\Delta\mathbf{v} \cdot \mathbf{e}|)$ is the volume of the collision cylinder for the particles moving with the velocities $\mathbf{v}$ and $\mathbf{v}'$, whose collision is specified by the unit vector $\mathbf{e}$. Note that the factor Δt in the length of the collision cylinder, $(\Delta v \Delta t)$ is skipped, as Eq. (2) defines the rate. Correspondingly, $f(\mathbf{v}', t)(\sigma^2 d\mathbf{e}|\Delta\mathbf{v} \cdot \mathbf{e}|)$, is the number of particles with velocity $\mathbf{v}'$ in the collision cylinder, see e.g., [2]. To obtain the total number of collisions with the set $(\mathbf{v}, \mathbf{v}', \mathbf{e})$ in a unit volume, one needs to multiply $f(\mathbf{v}', t)(\sigma^2 d\mathbf{e}|\Delta\mathbf{v} \cdot \mathbf{e}|)$ with $f(\mathbf{v}, t)$. The integration in Eq. (2) is to be performed over all velocities $\mathbf{v}'$ and directions of the unit vector $\mathbf{e}$. The factor $\Theta(-\Delta\mathbf{v} \cdot \mathbf{e})$, with $\Theta(x)$ being the unit step-function [$\Theta(x) = 1$ for $x \geq 0$ and $\Theta(x) = 0$ for $x < 0$], selects approaching particles.

The gain term quantifies, correspondingly, the rate of collisions, which results in velocity $\mathbf{v}$ of one of the colliding particles. That is, if the pre-collision velocities were $\mathbf{v}_*$ and $\mathbf{v}'_*$, the post-collision velocities would be $\mathbf{v}$ and $\mathbf{v}'$. Such collisions are called *inverse* collisions. The velocities $\mathbf{v}_*$ and $\mathbf{v}'_*$ of an inverse collision may be expressed in terms of the precollisional velocities $\mathbf{v}$ and $\mathbf{v}'$ and the unit vector $\mathbf{e}$ specifying the impact. Hence, the gain term may be written as [2],

$$I^{\text{gain}} = \sigma^2 \int d\mathbf{v}' \int d\mathbf{e} \Theta(-\Delta\mathbf{v} \cdot \mathbf{e}) \, |\Delta\mathbf{v} \cdot \mathbf{e}| \chi f(\mathbf{v}_*, t) f(\mathbf{v}'_*, t), \tag{3}$$

the meaning of all terms in the above equation is the same as in Eq. (2). The factor χ accounts for the difference of the volume of the collision cylinders in the direct and inverse collisions and for the transformation Jacobian from the velocities $(\mathbf{v}_*, \mathbf{v}'_*)$ to $(\mathbf{v}, \mathbf{v}')$. For elastic collisions, $\chi = 1$. For inelastic collisions, it depends on the restitution coefficient ε defined as the ratio of the post-collisional normal component of the impact velocity, $(\Delta\mathbf{v}' \cdot \mathbf{e})$ and the pre-collisional one, $(\Delta\mathbf{v} \cdot \mathbf{e})$:

$$\varepsilon = \left| \frac{(\Delta \mathbf{v}' \cdot \mathbf{e})}{(\Delta \mathbf{v} \cdot \mathbf{e})} \right| . \tag{4}$$

Here, the prime denotes post-collisional quantities. If ε does not depend on the impact velocity, $\chi = 1/\varepsilon^2$, otherwise a more complicated expression is to be applied [2].

2.2 *Boltzmann–Smoluchowski Equations*

Now we consider Boltzmann–Smoluchowski equations (BSE). There are two main differences between BE and BSE. First, the BSE deals with mass–velocity distribution functions $f_k(m_k, \mathbf{v}_k, t)$, which give the densities of aggregates of size k with the velocity $\mathbf{v}_k$ from the small interval $(\mathbf{v}_k, \mathbf{v}_k + d\mathbf{v}_k)$ at time t. These aggregates have the diameter $\sigma_k = \sigma_1 k^{1/3}$ and mass $m_k = m_1 k$.[1] Second, apart from the restitutive collisions, the aggregative collisions play an important role in the evolution of the system. That is, the BSE read [8, 9]

$$\frac{\partial}{\partial t} f\left(m_k, \mathbf{v}_k, t\right) = I_k^{\text{agg}} + I_k^{\text{res}}, \tag{5}$$

where I_k^{res} and I_k^{agg} describe, respectively, the restitutive and aggregative collisions. The first collision integral reads

$$I_k^{\text{res}} = \sum_i \sigma_{ki}^2 \int d\mathbf{v}_i d\mathbf{e}\, \Theta\left(-\mathbf{v}_{ki} \cdot \mathbf{e}\right) \left|\mathbf{v}_{ki} \cdot \mathbf{e}\right| \left(\chi f_k'' \cdot f_i'' - f_k f_i\right) \Theta\left(E_{ki} - W_{ki}\right) , \tag{6}$$

where we abbreviate $f_{k/i} = f_{k/i}\left(m_{k/i}, \mathbf{v}_{k/i}, t\right)$ and $f_{k/i}'' = f_{k/i}\left(m_{k/i}, \mathbf{v}_{k/i}'', t\right)$, with $\mathbf{v}_k$, $\mathbf{v}_i$ and $\mathbf{v}_k''$, $\mathbf{v}_i''$ being the velocities in the direct and inverse collision. $\sigma_{ij} = (\sigma_i + \sigma_j)/2$ is the cross section for the colliding aggregates of size i and j and $\mathbf{v}_{ij} = \mathbf{v}_i - \mathbf{v}_j$ is their relative velocity. The last factor $\Theta\left(E_{ki} - W_{ki}\right) = \Theta_{ki}^{\text{res}}$ in the integrand in Eq. (6) discriminates aggregative and restitutive collisions: If the kinetic energy of the relative motion of the two aggregates, $E_{ij} = m_{ij} v_{ij}^2/2$ exceeds the binding energy, W_{ij}, the aggregates rebound, otherwise they stick. Finally, $m_{ij} = m_i m_j/(m_i + m_j)$ is the reduced mass of the pair of colliding particles.

The collision integral for aggregation has a similar form [8, 9]:

$$\begin{aligned} I_k^{\text{agg}} = {} & \frac{1}{2} \sum_{i+j=k} \sigma_{ij}^2 \int d\mathbf{v}_i d\mathbf{v}_j d\mathbf{e} \Theta\left(-\mathbf{v}_{ij} \cdot \mathbf{e}\right) \left|\mathbf{v}_{ij} \cdot \mathbf{e}\right| f_i f_j \delta\left(m_k \mathbf{v}_k - m_i \mathbf{v}_i - m_j \mathbf{v}_j\right) \Theta_{ij}^{\text{agg}} \\ & - \sum_j \sigma_{kj}^2 \int d\mathbf{v}_j d\mathbf{e} \Theta\left(-\mathbf{v}_{kj} \cdot \mathbf{e}\right) \left|\mathbf{v}_{kj} \cdot \mathbf{e}\right| f_k f_j \Theta_{kj}^{\text{agg}} . \end{aligned} \tag{7}$$

[1] We assume that the aggregates are spherical and compact; the generalization for fractal aggregates is straightforward.

Here, $\Theta_{ij}^{\mathrm{agg}} \equiv \Theta(W_{ij} - E_{ij})$ selects agglomerating collisions for which the condition of aggregation, $E_{ij} < W_{ij}$, is fulfilled. The first sum in the right-hand side of Eq. (7) describes the aggregation of clusters of size i and j with the velocities $\mathbf{v}_i$ and $\mathbf{v}_j$ resulting in a cluster of size k with the velocity $\mathbf{v}_k$. The sum is performed over all i and j, with $i + j = k$ and the factor $1/2$ prevents double counting. Here, $m_k = m_i + m_j$ and $m_k\mathbf{v}_k = \mathbf{v}_i m_i + m_j\mathbf{v}_j$, due to the mass and momentum conservation, which is reflected by the corresponding factor in the integrand, $\delta\left(m_k\mathbf{v}_k - m_i\mathbf{v}_i - m_j\mathbf{v}_j\right)$. All other factors have the same meaning as before. The second sum accounts for the aggregative collisions of clusters of size k and velocity $\mathbf{v}_k$, with all other aggregates. Note that in some aggregation models for atomic gases the opposite agglomeration condition, $\Theta_{ij}^{\mathrm{agg}} = \Theta(E_{ij} - W_{ij})$, is used.

3 Smoluchowski Equations for Space-Uniform Systems

3.1 Velocity Distribution Function and Its Moments

To illustrate the main concepts, we consider again the Boltzmann equation (1) for a space-uniform non-aggregating gas of identical particles. The first three moments of the velocity distribution function define the number density, n, the flux velocity, $\mathbf{u}$, and the temperature of the gas, T:

$$n = \int d\mathbf{v} f(\mathbf{v}, t); \qquad n\mathbf{u} = \int d\mathbf{v} m\mathbf{v} f(\mathbf{v}, t); \qquad 3nT = \int d\mathbf{v} m\mathbf{v}^2 f(\mathbf{v}, t), \quad (8)$$

where m is the particles mass. Here, we consider flux-free systems, $\mathbf{u} = 0$. For molecular gases, the temperature is measured in units of the Boltzmann constant k_B; for granular gases, it is measured in units of energy [2]. According a theorem, the collision integral vanishes, $I^{\mathrm{res}} = 0$, for the stationary velocity distribution function. It depends in this case only on three first moments and has the Gaussian form [4],

$$f(\mathbf{v}, t) = \frac{n(t)}{\pi^{3/2} v_0^3(t)} e^{-v^2/v_0^2}, \qquad (9)$$

where $v_0(t) = \sqrt{2T(t)/m}$ is thermal velocity and we take into account that the average flux velocity vanishes.

In agglomerating systems, there are many different species characterized by the size of the aggregates. The abundance of aggregates of the same size, i, establishes a sub-gas, characterized by their partial number densities, $n_i(t)$, and partial temperatures, $T_i(t)$; additionally, it is convenient to consider the total number density $N(t)$ and average temperature $T(t)$ [39]. These quantities are defined by

$$n_i = \int d\mathbf{v}_i \, f(m_i, \mathbf{v}_i)\,; \qquad N = \sum_i n_i; \tag{10}$$
$$3n_i T_i = \int d\mathbf{v}_i \, m_i \mathbf{v}_i^2 f(m_i, \mathbf{v}_i)\,; \qquad NT = \sum_i n_i T_i\,.$$

Correspondingly, the distribution function of the sub-gas, i, may be approximated by a Gaussian function:

$$f_i \equiv f(m_i, \mathbf{v}_i, t) = \frac{n_i(t)}{\pi^{3/2} v_{0i}^3(t)} e^{-v_i^2/v_{0i}^2}\,, \tag{11}$$

where $v_{0i}(t) = \sqrt{2T_i(t)/m_i}$ is the thermal velocity of the particles of size i.

3.2 Smoluchowski Equations. Randomly Aggregative Collisions

The condition of aggregative collisions used in Eqs. (6) and (7) is based on the comparison of the binding energy W_{ij} and the kinetic energy of the relative motion of the pair E_{ij}. The microscopical expression for the binding energy is addressed in the next section. In the current section, we consider a more simple model of random agglomeration. Namely, we assume that instead of the deterministic factors $\Theta_{ij}^{\rm agg} \equiv \Theta(W_{ij} - E_{ij})$ and $\Theta_{ij}^{\rm res} \equiv \Theta(E_{ij} - W_{ij})$, which determine whether a collision is aggregative or restitutive, we consider an aggregation as a random event, which occurs with the probability q. This probability represents the influence of degrees of freedom that are not explicitly considered by the Boltzmann equation. For instance, orientation degrees of freedom may determine the collision outcome when the aggregation can happen only for a certain mutual orientation of colliding particles. Additionally, we assume that the restitutive collisions are elastic, that is, $\varepsilon = 1$.

Randomly aggregating systems are described by Eqs. (5)–(7) with

$$\Theta_{ij}^{\rm agg} = q, \qquad \Theta_{ij}^{\rm res} = 1 - q.$$

Furthermore, we consider first the so-called reaction-controlled limit, when only a tiny fraction of collisions leads to merging; this corresponds to small $q \ll 1$. In this case, the system rapidly attains a quasi-stationary state, with $\partial f_k/\partial t \approx 0$, where all distribution functions, f_k, are close to the Gaussian (11). Since, typically, the particles collide many times before they merge, the energy equipartition holds, that is, $T_i = T$ for all i; therefore, we apply here the approximation (11) with equal temperatures.

To obtain the equations for n_k, defined in Eq. (10), we integrate Eq. (7) over $\mathbf{v}_k$. On the left-hand side of the equation, we then obtain dn_k/dt [see Eq. (10)]. On the right-hand side, we obtain $\int d\mathbf{v}_k I_k^{\rm res} = 0$, since the restitutive collisions do not alter the partial number densities (see [9, 34] for the formal proof). For aggregative collision, we encounter the integrals:

$$\mathcal{B}_{kj} = q\sigma_{kj}^2 \int d\mathbf{v}_k d\mathbf{v}_j d\mathbf{e}\, \Theta(-\mathbf{v}_{kj}\cdot\mathbf{e})\, |\mathbf{v}_{kj}\cdot\mathbf{e}|\, f_k f_j, \tag{12}$$

with f_k, f_j given by (11) with $T_i = T$. Such integrals appear in the second sum in Eq. (7). The integrals in the first sum of this equations attain essentially the same form, after integration out the δ-function. Integration over the unit vector $\mathbf{e}$ in (12) may be performed in the spherical coordinates with the OZ-axis directed along the vector $\mathbf{v}_{kj}$:

$$\int d\mathbf{e}\, \Theta(-\mathbf{v}_{kj}\cdot\mathbf{e})\, |\mathbf{v}_{kj}\cdot\mathbf{e}| = \int_0^{2\pi} d\phi \int_{\pi/2}^{\pi} v_{kj} |\cos\theta|\, \sin\theta d\theta = \pi v_{kj}. \tag{13}$$

To proceed with the computation of $\mathcal{B}_{kj}$, we make thc transformation, $(\mathbf{v}_k, \mathbf{v}_j) \to (\mathbf{v}_{kj}, \mathbf{V})$, to the center of mass velocity $\mathbf{V} = (m_k\mathbf{v}_k + m_j\mathbf{v}_j)/(m_k + m_j)$ and the relative velocity $\mathbf{v}_{kj} = \mathbf{v}_k - \mathbf{v}_j$. The product of the velocity distribution functions becomes

$$f_k(v_k)\, f_j(v_j) = \frac{n_k n_j}{\pi^3 v_{0,k}^3 v_{0,j}^3} \exp\left[-\frac{\mu_{kj} v_{kj}^2 + M_{kj} V^2}{2T} \right], \tag{14}$$

where $\mu_{kj} = m_k m_j/(m_k + m_j)$ is the reduced mass and $M_{kj} = m_k + m_j$. Substituting (14) and (13) into Eq. (12), and using the identity $d\mathbf{v}_k d\mathbf{v}_j = d\mathbf{V} d\mathbf{v}_{kj}$, we obtain

$$\begin{aligned}\mathcal{B}_{kj} &= q\pi\sigma_{kj}^2 \int d\mathbf{v}_k d\mathbf{v}_j v_{kj} f_k f_j \\ &= q\pi\sigma_{kj}^2 \frac{n_k n_j}{\pi^3 v_{0,k}^3 v_{0,j}^3} \int d\mathbf{v}_{kj} v_{kj} e^{-\mu_{kj} v_{kj}^2/2T} \int d\mathbf{V}\, e^{-M_{kj} V^2/2T},\end{aligned} \tag{15}$$

which is the product of two Gaussian integrals, that may be straightforwardly computed:

$$\mathcal{B}_{ij} = \sqrt{T}\, q\, K_{ij} n_i n_j\,. \tag{16}$$

That is, the agglomeration rates are proportional to $\sqrt{T}$. Here, the mass-dependent factors of the rates read

$$K_{ij} = K_0 (i^{1/3} + j^{1/3})^2 \sqrt{i^{-1} + j^{-1}}, \tag{17}$$

where $K_0 = \sigma_1^2 \sqrt{\pi/(2m_1)}$; see [8, 9, 34] for details of the calculation. Hence, we obtain the equations for densities $n_k(t)$, which are essentially Smoluchowski equations

$$\frac{dn_k}{dt} = T^{1/2} q \left[\frac{1}{2} \sum_{i+j=k} K_{ij} n_i n_j - n_k \sum_{i\geq 1} K_{ki} n_i \right]. \tag{18}$$

Still, in contrast to the standard Smoluchowski equations, which form a closed set for the densities $n_k(t)$, one needs to supplement Eqs. (18) with the equation for temperature, which evolution depends on the evolution of the densities.

Now, we will obtain the equation for the total kinetic energy, $(3/2)NT$, as defined in Eq. (10). In an impact of aggregates of size k and j, which results in merging, the total energy of the pair is reduced by the energy of their relative motion, $\mu_{kj}v_{kj}^2/2$. To obtain the rate equation for the decay of the energy $(3/2)NT$, we multiply the integrand in Eq. (12) by $\mu_{kj}v_{kj}^2/2$ and integrate over all possible velocities $\mathbf{v}_k$ and $\mathbf{v}_j$. Here, we again obtain Gaussian integrals, which can be evaluated. Summing up over k and j, we arrive at the energy equation

$$\frac{d}{dt}NT = -\frac{2}{3}T^{3/2}q\sum_{i\geq 1}\sum_{j\geq 1}K_{ij}n_in_j\,. \tag{19}$$

Here, we ignore the energy loss in the restitutive collisions, which are elastic ($\varepsilon = 1$). The generalization for inelastic collisions, as in granular gases [2], is straightforward. Equations (18) and (19) form a complete set of temperature-dependent Smoluchowski equations for the simplest case of random collisional aggregation.

3.3 *Smoluchowski Equations. Impact Energy-Dependent Aggregation*

Now we consider a more general case when aggregation depends on the kinetic energy of the relative motion of a colliding pair. When two particles come in close contact at the beginning of a collision, they experience an attraction. The attraction alters to repulsion, as the centers of the particles move closer. To get separated at the end of the impact, the moving away particles need to overcome a potential barrier W_{ij}. For a pair (i, j), it depends on the size of the particles and the nature of the attractive forces between them. For instance, the potential barrier of adhesive forces reads, $W_{ij} = A(\sigma_i\sigma_j)^{4/3}(\sigma_i+\sigma_j)^{-4/3}$, where the constant A depends on the material parameters and $\sigma_{i/j}$ are the particles' diameters [40]. Generally, it may be put into the form,

$$W_{ij} = a\frac{\left(i^{1/3}j^{1/3}\right)^{\lambda_1}}{\left(i^{1/3}+j^{1/3}\right)^{\lambda_2}}\,, \tag{20}$$

where the constant a has the dimension of energy. Different exponents λ_1 and λ_2 allow to describe various interactions forces. Namely, $\lambda_1 = \lambda_2 = 4/3$ correspond to the adhesive forces, as has been mentioned above. $\lambda_1 = 3$, $\lambda_2 = 1$ characterize gravitational or Coulomb interactions when the particles' charges scale as their masses. Similarly, $\lambda_1 = \lambda_2 = 3$ stand for dipole–dipole interactions [41], etc.

We will use W_{ij} from Eq. (20) in the factors $\Theta_{ij}^{\rm agg}$ and $\Theta_{ij}^{\rm res} \equiv \Theta(E_{ij} - W_{ij})$ which define the conditions of aggregative or restitutive collisions, respectively, see Eqs.

(6) and (7). To account for inelastic collisions with the restitution coefficient ε, we re-scale E_{ij}, entering the factors $\Theta_{ij}^{\rm agg}$ and $\Theta_{ij}^{\rm res}$, as follows, $E_{ij} \to \varepsilon^2 E_{ij}$; this reflects the decrease of the relative kinetic energy at the end of a collision due to dissipation. Moreover, since we do not consider explicitly particles' rotation, we assume that its impact may be effectively accounted by the factor a in Eq. (20).

To obtain the set of equations for the cluster densities, we again integrate Eq. (5), with the collision integrals (6) and (7) over $\mathbf{v}_k$. As previously, the restitutive collision integral does not contribute, and we obtain, instead of Eq. (12),

$$\mathcal{B}_{kj} = \sigma_{kj}^2 \int d\mathbf{v}_k d\mathbf{v}_j d\mathbf{e}\, \Theta(-\mathbf{v}_{kj} \cdot \mathbf{e})\, |\mathbf{v}_{kj} \cdot \mathbf{e}|\, f_k f_j\, \Theta\left(W_{kj} - \frac{1}{2}\varepsilon^2 \mu_{kj} v_{kj}^2\right), \tag{21}$$

with W_{ij} given by Eq. (20). To evaluate the above expression for $\mathcal{B}_{kj}$ we again apply the Gaussian approximation (9) for the velocity distribution functions f_k and f_j. Under this approximation, the integrals in Eq. (21) are Gaussian and hence may be evaluated. Referring for computation details to [34], we provide here the final result:

$$\frac{dn_k}{dt} = \frac{1}{2}\sum_{i+j=k} C_{ij} n_i n_j - n_k \sum_j C_{kj} n_j, \tag{22}$$

where

$$\begin{aligned} C_{ij} &= 2\sqrt{2\pi}\, \sigma_{ij}^2 \sqrt{\theta_i + \theta_j}\,\left(1 - F_{ij}\right) \\ F_{ij} &= \left(1 + Q_{ij}\right) e^{-Q_{ij}} \\ Q_{ij} &= \frac{W_{ij}}{\varepsilon^2 \mu_{ij} (\theta_i + \theta_j)} \\ \theta_i &= T_i / m_i \,. \end{aligned} \tag{23}$$

Equation (22) have the form of standard Smoluchowski equations; the rate coefficients, C_{ij}, depend, however, on the partial temperatures, $T_i = \theta_i m_i$. To find the equations for T_i, we multiply the BE by $m_k v_k^2/2$ and integrate over $\mathbf{v}_k$. Using the definitions (10), we observe that after the integration, the left-hand side of the equation turns into $3m_k d(n_k\theta_k)/dt$. In the right-hand side of this equation, we obtain the Gaussian integrals, similar to these in Eq. (21). Computation of these integrals yields the set of equations for $\theta_k = T_k/m_k$ (see [34] for the detail),

$$\frac{d}{dt} n_k \theta_k = \frac{1}{2} \sum_{i+j=k} B_{ij} \frac{n_i n_j \theta_i \theta_j}{\theta_i + \theta_j} - \sum_j D_{kj} \frac{n_k n_j \theta_k \theta_j}{\theta_k + \theta_j}. \tag{24}$$

Here, the new kinetic rates read,

$$
\begin{aligned}
B_{ij} =& 2\sqrt{2\pi}\sigma_{ij}^2\sqrt{\theta_i+\theta_j} \times \left[1 - F_{ij} + \frac{4}{3}\frac{(\theta_i\Delta_{ij} - \theta_j\Delta_{ji})^2}{\theta_i\theta_j}(1-G_{ij})\right] \\
D_{ij} =& 2\sqrt{2\pi}\sigma_{ij}^2\sqrt{\theta_i+\theta_j} \times \left[1 - F_{ij} + \frac{4}{3}\frac{\theta_i}{\theta_j}(1-G_{ij}) + \frac{4}{3}\frac{\mu_{ij}}{m_i\theta_j}(\theta_i+\theta_j)\right. \\
& \left.\times(1+\varepsilon)\left(1 - \frac{1}{2}(1+\varepsilon)\frac{\mu_{ij}}{m_i\theta_i}(\theta_i+\theta_j)\right)G_{ij}\right] \\
G_{ij} =& e^{-Q_{ij}}\left(1 + Q_{ij} + \frac{1}{2}Q_{ij}^2\right),
\end{aligned}
\tag{25}
$$

where Q_{ij} has been defined above and $\Delta_{ij} = m_i/(m_i + m_j)$.

If we assume the energy equipartition, $T_i = T$ for all i, the kinetic rates C_{ij} will take the form,

$$
C_{ij} = \sqrt{T}K_{ij}\left[1 - \left(1 + \tilde{W}_{ij}/T\right)e^{-\tilde{W}_{ij}/T}\right], \tag{26}
$$

where $\tilde{W}_{ij} = W_{ij}/\varepsilon^2$ and K_{ij} are defined in Eq. (17). Correspondingly, the temperature equation reads,

$$
\frac{d}{dt}NT = -\sum_{i,j} P_{ij}n_in_j, \tag{27}
$$

with

$$
\begin{aligned}
P_{ij} &= \frac{2}{3}T^{3/2}K_{ij}\left(1 - G_{ij} + \frac{1}{2}(1-\varepsilon^2)G_{ij}\right) \\
G_{ij} &= \left(1 + \tilde{W}_{ij}/T + \frac{1}{2}\tilde{W}_{ij}^2/T^2\right)e^{-\tilde{W}_{ij}/T}.
\end{aligned}
\tag{28}
$$

In the limit $\tilde{W}_{ij} \to \infty$ (all collisions are aggregative) Eq. (27) coincide, up to the factor q, with Eq. (19). Equations (22)–(25) comprise a complete set of temperature-dependent Smoluchowski equations. Correspondingly, Eqs. (22), (26), (27) and (28) comprise the respective set for the case of temperature equipartition.

3.4 *Application to the Hypothetical Evolution of Dark Matter*

In recent theories of the evolution of dark matter, it was hypothesized that it could be described in terms of ballistic agglomeration of dark nuclei from dark nucleons [11–13]. This leads to the formation of dark nuclei with very wide spectrum of masses, that is, the BSE may be applied [11–13]. In the framework of our approach, the governing equations are (18) and (19), with the according replacement $\frac{d}{dt} \to \frac{d}{dt} + 3H$, where $H = H(t)$ is the Hubble parameter accounting for the expansion of the Universe. The transformation

$$n_k = hm_k, \qquad N = hm, \qquad h(t) = \exp\left[-3\int_{t_0}^{t} dt'\, H(t')\right] \tag{29}$$

recasts these equations into the form

$$\frac{1}{hT^{1/2}}\frac{dm_k}{dt} = \frac{1}{2}\sum_{i+j=k} K_{ij}m_i m_j - m_k \sum_{i\geq 1} K_{ki}m_i \tag{30}$$

$$\frac{1}{hT^{3/2}}\frac{d\,mT}{dt} = -\frac{2}{3}\sum_{i\geq 1}\sum_{j\geq 1} K_{ij}m_i m_j \tag{31}$$

that differ from (18)–(19) only by an extra factor $h(t)$. The authors of Ref. [11] assumed that the dark nuclei were in a contact with a bath of lighter particles, which determined their temperature. In our model, we do not need the hypothesis of additional bath of lighter particle; instead, the temperature is determined by the agglomeration and Hubble expansion only. The time dependence for the factor $h(t)$ may be obtained from the basic cosmological equations (see [10]):

$$h(t) = [1 + 2H_0(t - t_0)]^{-3/2}, \tag{32}$$

where $H_0 = H(t_0) = (8\pi G\rho_0/3)^{1/2}$, G is the gravitational constant and ρ is the density. The initial time t_0 is the time, when the agglomeration in the early Universe becomes a dominating process [10]. The analysis of the Eqs. (29)–(31) with the rates defined by Eq. (17), along with the implications to the asymptotic mass distribution of dark nuclei, is given in Ref. [10].

4 Exact and Scaling Solutions

4.1 Exact Solutions

The exact solutions play an important role as the reference result for equations which may be generally solved only numerically. The accuracy of the numerical schemes may be tested on the exact solutions. For the standard Smoluchowski equations, of the form (22), with the time-independent rate coefficients C_{ij}, the exact solutions are known only for a limited class of kernels. Namely, for $C_{ij} = a + b(i + j) + c(ij)$, where a, b, c are constants [14, 26]. Here, we demonstrate how one can obtain the exact solutions for the temperature-dependent Smoluchowski equations, using the available solutions for the standard Smoluchowski equations.

4.1.1 Systems with Energy Equipartition

We start from the simplest case of random agglomeration, which implies the equipartition $T_i = T$ and consider $qK_{ij} = \text{const.} = 1$.[2] Then Eqs. (18) and (19) attain the form

$$\frac{dn_k}{d\tau} = \frac{1}{2}\sum_{i+j=k} n_i n_j - \sum_{i\geq 1} n_k n_i \tag{33}$$

$$\frac{dNT}{d\tau} = -\frac{2}{3}T\sum_{i,j\geq 1} n_i n_j = -\frac{2}{3}TN^2, \tag{34}$$

where we introduce the new time variable,

$$\tau = \int_0^t \sqrt{T(t')}dt'. \tag{35}$$

Equation (33) is the standard Smoluchowski equation written for the time variable τ. For the mono-disperse initial conditions, $n_k(0) = n_0\delta_{k,1}$ the solution reads [26],

$$n_k(\tau) = \frac{4n_0\tau^{k-1}}{(2+\tau)^{k+1}}. \tag{36}$$

Summing up Eq. (33), we obtain, $dN/d\tau = -N^2/2$, which may be easily solved for $N(\tau)$. Substituting this into Eq. (34), we find $T(\tau) = T_0/(1+\tau/2)^{1/3}$, where T_0 is the initial temperature T_0. Then $\tau(t)$ follows from Eq. (35):

$$\tau = 2(1+t/\tau_0)^{6/7} - 2, \tag{37}$$

where $\tau_0 = (7/12)n_0\sqrt{T_0}$, in the time units, where $qK_{ij} = qK_{11} = 1$. Hence, the time dependence of temperature and total cluster density in the laboratory time reads,

$$T(t) = \frac{T_0}{(1+t/\tau_0)^{2/7}}, \qquad N(t) = \frac{n_0}{(1+t/\tau_0)^{6/7}}. \tag{38}$$

For $t \gg 1$ and $k \gg 1$, the solution (36) attains a *scaling* form. Using Eq. (37), we write $n_k(t)$ in the laboratory time:

$$n_k(t) = \frac{1}{s^2(t)}\Phi\left(\frac{k}{s(t)}\right); \qquad s(t) \sim t^{6/7}; \qquad \Phi(x) = e^{-x}. \tag{39}$$

Here, $s(t)$ is the typical cluster size and $\Phi(x)$ is the scaling distribution function. The above results, Eqs. (36)–(39), also describe the case of energy-dependent agglom-

[2] Note that the constant may be always set to one using the appropriate time units.

eration for $W_{ij} \to \infty$, and $K_{ij} = K_{11} = 1$. Similar analysis may be performed for other kernel, e.g., for $qK_{ij} = (i + j)$.

4.1.2 Systems with Different Partial Temperatures

As it has been mentioned above, for the case of small sticking probability (when only a tiny fraction of all collisions is aggregative), the temperature equipartition, $T_i = T_j$, holds true. Generally, this is not the case, and the partial temperatures T_j differ. The dependence of the rate coefficients on the partial temperatures is determined by the nature of agglomeration process and system parameters; it may not be chosen arbitrarily. Still, the exact solutions are very important for testing the accuracy of numerical schemes. Hence, it is worth to find exact solutions for specially tailored rate kernels, even if the physical background for such kernels is lacking. By carefully choosing the kernels B_{ij} and D_{ij} and the form of temperature dependence of the kernel C_{ij}, one can obtain infinitely many exact solutions, based on the exact solutions of the standard (temperature-independent) Smoluchowski equations. Here is one such example:

$$C_{ij} = T_i + T_j; \qquad B_{ij} = (T_i + T_j)^2; \qquad D_{ij} = (T_i + T_j + 1)T_i, \tag{40}$$

for the full system

$$\begin{aligned} \frac{d}{dt}n_k &= \frac{1}{2}\sum_{i+j=k} C_{ij}\left(T_i, T_j\right) n_i n_j - \sum_{i\geq 1} C_{kj}\left(T_k, T_j\right) n_k n_j. \\ \frac{d}{dt}\left(n_k T_k\right) &= \frac{1}{2}\sum_{i+j=k} B_{ij}\left(T_i, T_j\right) n_i n_j - \sum_{j\geq 1} D_{kj}\left(T_k, T_j\right) n_k n_j. \end{aligned} \tag{41}$$

After multiplying the first set of equations by T_k, subtracting it from the second one, and substituting the kernels (40), we arrive at

$$n_k \frac{d}{dt} T_k = \frac{1}{2}\sum_{i+j=k} \left(T_i + T_j\right)\left(T_i + T_j - T_k\right) n_i n_j - \sum_{j\geq 1} T_k n_k n_j.$$

Let us search for the solution in the form $T_k(t) = kf(t)$. Then the first sum in the right-hand side vanishes and after substitution we are left with

$$\frac{d}{dt} f(t) = -f(t)\sum_{j=1}^{\infty} n_j = -fN. \tag{42}$$

With this substitution, we also recast the first equation for n_k in Eq. (41) into the form of standard Smoluchowski equation with the linear kernel, $C_{ij} = i + j$:

$$\frac{dn_k}{d\tau} = \frac{1}{2} \sum_{i+j=k} (i+j)\, n_i n_j - \sum_{j \geq 1} (k+j)\, n_k n_j ,$$

where we again introduce the new time variable, $\tau = \int_0^t f(t)dt$. For the linear kernel, the exact solution is known [14]; for mono-disperse initial condition, it reads

$$n_k(\tau) = \frac{k^{k-1}}{k!} e^{-\tau} \left(1 - e^{-\tau}\right)^{k-1} \exp\left(-k\left(1 - e^{-\tau}\right)\right).$$

Also, $N = e^{-\tau}$. Substituting N into (42) yields, for $N(0) = 1$ and $T_1(0) = f(0) = 1$, the solution

$$T_k(t) = \frac{k}{t+1}, \qquad \tau = \ln(t+1).$$

As a result, the number densities for the above rate coefficients are

$$n_k(t) = \frac{k^{k-1}}{k!\,(t+1)} \left(\frac{t}{t+1}\right)^{k-1} e^{-kt/(t+1)}. \tag{43}$$

Other examples of kernels, which allow an exact solution for temperature-dependent Smoluchowski equations may be found in Ref. [42].

4.2 Scaling Analysis

4.2.1 Scaling Functions and Exponents

Scaling analysis is applicable for the asymptotic behavior of the aggregating system for $t \gg 1$ and $k \gg 1$. Here, we present it for the case of energy equipartition, which is described by the coupled Eqs. (22) and (27):

$$\frac{dn_k}{dt} = \frac{1}{2} \sum_{i+j=k} C_{ij} n_i n_j - n_k \sum_j C_{kj} n_j \tag{44}$$

$$\frac{d}{dt} NT = - \sum_{i,j} P_{ij} n_i n_j . \tag{45}$$

Scaling is applicable when the rate coefficient may be represented in the form:

$$C_{ij} = cT^{\nu_c} \tilde{C}_{ij}; \qquad P_{ij} = pT^{\nu_p} \tilde{P}_{ij}, \tag{46}$$

where c, p are the dimensional constants, the exponents ν_c, ν_p quantify the dependence of the rates on temperature. $\tilde{C}_{ij}$ and $\tilde{P}_{ij}$ are dimensionless homogeneous functions of i and j, satisfying for any $b > 0$ the relation,

$$\tilde{C}_{bi\,bj} = b^{\mu_c}\tilde{C}_{i\,j}; \qquad \tilde{P}_{bi\,bj} = b^{\mu_p}\tilde{P}_{i\,j}. \tag{47}$$

Here, μ_c and μ_p are, correspondingly, the homogeneity exponents for $\tilde{C}_{ij}$ and $\tilde{P}_{ij}$.

Now we assume that the densities n_k obey the scaling form, while temperature obeys a power law:

$$n_k(t) = \frac{1}{t^{2z}}\Phi\left(\frac{k}{t^z}\right); \qquad T \sim t^{-\beta} \tag{48}$$

Using (48), we can write the left-hand side of Eq. (44) as (see also [43]):

$$\frac{dn_k}{dt} = -\frac{2z}{t^{2z+1}}\Phi - \frac{1}{t^{2z}}\frac{zk}{t^{z+1}}\Phi' = -\frac{z}{t^{2z+1}}\left(2\Phi + x\frac{d\Phi}{dx}\right), \tag{49}$$

with $x = k/t^z$. Changing from the discrete variables i, j to the continuous ones, $C_{ij} \to C(i, j)$, one can write the right-hand side of the same Eq. (44) as

$$\frac{dn_k}{dt} = \frac{1}{2}cT^{\nu_c}\int_0^k \tilde{C}(k-j, j)\frac{1}{t^{2z}}\Phi\left(\frac{k-j}{t^z}\right)\frac{1}{t^{2z}}\Phi\left(\frac{j}{t^z}\right)dj \tag{50}$$

$$- cT^{\nu_c}\frac{1}{t^{2z}}\Phi\left(\frac{k}{t^z}\right)\int_0^\infty \tilde{C}(k, j)\frac{1}{t^{2z}}\Phi\left(\frac{j}{t^z}\right)dj$$

$$= -cT^{\nu_c}t^{\mu_c z-3z}\,\mathcal{K}(x), \tag{51}$$

where

$$\mathcal{K}(x) \equiv \left[\int_0^\infty \tilde{C}(x, y)\Phi(x)\Phi(y)dy - \frac{1}{2}\int_0^x \tilde{C}(x-y, y)\Phi(x-y)\Phi(y)dy\right], \tag{52}$$

and we take into account that $k = xt^z$, $j = yt^z$ and use

$$\tilde{C}(xt^z, yt^z) = (t^z)^{\mu_c}\tilde{C}(x, y) \tag{53}$$

and similar relations for homogeneous kernels. With $T \sim t^{-\beta}$, Eq. (44) transforms into

$$\frac{z}{t^{2z+1}}\left(2\Phi(x) + x\frac{d\Phi}{dx}\right) = ct^{\mu_c z-3z-\beta\nu_c}\,\mathcal{K}(x). \tag{54}$$

If we divide both sides of the above equation by $z\,t^{\mu_c z-3z-\beta\nu_c}(2\Phi + x\Phi')$, we observe that one side of the equation depends only on t, while the other one only on x. Hence, we conclude [43],

$$t^{\nu_c\beta+z(1-\mu_c)-1} = w = \text{const}, \tag{55}$$

where w is the separation constant. This condition yields the relation

$$\nu_c\beta + z(1-\mu_c) - 1 = 0\,. \tag{56}$$

and the equation for the scaling function:

$$w\left[x\Phi' + 2\Phi\right] = \Phi(x)\int_0^\infty \tilde{C}(x,y)\Phi(y)dy - \frac{1}{2}\int_0^x \tilde{C}(x-y)\Phi(x-y)\Phi(y)dy\,. \tag{57}$$

Using the scaling solution (48), we write for $N = \sum_i n_i$:

$$N = \int_0^\infty t^{-2z}\Phi(k/t^z)dk = Ct^{-z}, \qquad C = \int_0^\infty \Phi(x)dx\,. \tag{58}$$

Then one can recast Eq. (45) into the form,

$$\begin{aligned} \frac{d}{dt}NT \sim t^{-z-\beta-1} &= pT^{\nu_p}\int_0^\infty dx\int_0^\infty dy\tilde{P}(xt^z, yt^z)\frac{1}{t^{2z}}\Phi(x)\Phi(y) \\ &\sim t^{-\beta\nu_p + z(\mu_p - 2)}, \end{aligned} \tag{59}$$

which yields another scaling relation:

$$-z - \beta - 1 = -\beta\nu_p + z(\mu_p - 2)\,. \tag{60}$$

Thus, Eqs. (56) and (60) allow to find the scaling exponents z and β in terms of the homogeneity exponents μ_c and μ_p.

4.2.2 Scaling in Aggregation Regimes

Generally, the kernels C_{ij} and P_{ij}, given by Eqs. (26) and (28) are not homogeneous functions of i and j. However, they posses the scaling properties (46) and (47) in some limiting cases.

Aggregation with temperature growth. Consider the case when $(1-\varepsilon^2) \ll (W_{ij}/T)^2 \ll 1$, which corresponds to a molecular gas with $\varepsilon = 1$ or nearly elastic granular gas. The expansion of C_{ij} and P_{ij} in terms of small (W_{ij}/T) yields

$$C_{ij} = cT^{-3/2}(ij)^{2\lambda_1/3-1/2}(i+j)^{1/2}(i^{1/3}+j^{1/3})^{2-2\lambda_2} \tag{61}$$

$$P_{ij} = pT^{-3/2}(ij)^{\lambda_1-1/2}(i+j)^{1/2}(i^{1/3}+j^{1/3})^{2-3\lambda_2}, \tag{62}$$

with the dimensional constants c and p. Here, we use Eq. (20) for W_{ij}. The exponents follow from the above equations: $\nu_c = -3/2$, $\nu_p = -3/2$ and $\mu_c = 2/3\,(2\lambda_1 - \lambda_2) + 1/6$, $\mu_p = \Lambda + 1/6$, where $\Lambda = 2\lambda_1 - \lambda_2$. Solving Eqs. (56) and (60) for z and β with the above exponents, we obtain for this regime

$$z = \frac{6}{5 - \Lambda}; \qquad \beta = -\frac{2\Lambda}{5 - \Lambda}, \tag{63}$$

The above result indicates, that if $\Lambda > 0$, which is commonly the case, the exponent β is negative. That is, $T \sim t^{|\beta|}$, and the average temperature of an aggregating gas increases. This surprising effect has been confirmed in numerical simulations [34, 44], see also below. Note, however, that in seemingly close case of $(W_{ij}/T) \ll 1$ and $(1 - \varepsilon^2) \gg (W_{ij}/T)^2$ the temperature growth is not observed.

Our simplified scaling analysis presented here is, strictly speaking, valid only when $\mu_c < 2/3$ or, equivalently, $\Lambda < 3/4$ (which corresponds to the Case III kernels [26]). Moreover, the case of $\mu_c > 1$ (i.e., $\Lambda > 5/4$) may indicate a gelation [14, 26, 43], or run-away growth [41]; this requires a more subtle analysis. We will return to this problem in Sect. 5.4.

Cold aggregating gas. In the opposite case of $W_{ij}/T \gg 1$, which corresponds to a cold gas, all collisions are aggregative and the rate coefficients read,

$$C_{ij} = c_1 T^{1/2} K_{ij}; \qquad P_{ij} = p_1 T^{3/2} K_{ij}, \tag{64}$$

with K_{ij} from Eq. (17). Here $\mu_c = \mu_p = 1/6$ and $\nu_c = 1/2$, $\nu_p = 3/2$, yielding eventually the exponents,

$$z = 1; \qquad \beta = 1/3. \tag{65}$$

5 Numerical Methods in Aggregating Kinetics

The ballistic aggregation may be studied by various numerical methods, which differ by the level of microscopic detail. The molecular dynamic (MD) simulations are the most microscopic approach, as it provides the trajectories for all particles. The MD has been successfully applied to study the aggregation kinetics, see e.g., [35–38]. Still, however, it does not allow to simulate systems, that are large enough to obtain a reliable statistics for the velocity distribution of different species. An important alternative to the MD is the Direct Simulation Monte Carlo (DSMC), specially designed to directly solve the BE. This numerical technique was first developed for dilute molecular gas by Bird [45] and later generalized to model dense gases and dissipative granular gases, see e.g., [46–50]. DSMC allows to simulate much larger systems as it deals with the collision only and does not trace particles' trajectories. Still, the velocities of particles of all species are explicitly treated. The next level of coarse-graining is the solution of rate equations, which deal with the densities of the species. The rate equations, in their turn, may be simulated on the microscopic level, which takes into account fluctuations, or on the "mean-field" level, where fluctuations are ignored. In the former case, stochastic modeling is applied; in the latter one—the deterministic solvers for the systems of ordinary differential equations (ODE). Below, we briefly discuss each of the approaches.

5.1 *Direct Simulation Monte Carlo of the Boltzmann Equation*

Referring to the detailed description of the method in the introductory literature, e.g., [51, 52], we sketch here the main ideas of the method. We will also emphasize the detail, where our method [34] deviates from the conventional DSMC. In short, the DSMC does not model the actual motion of particles, but only probabilistic collisions. That is, this method operates with the masses and velocities of different particles which are stored. Each time step a particular pair of particles is randomly chosen with the probability proportional to the probability of their collision. Then a random impact vector **e**, specifying the inter-center direction at a collision instant is generated. For a given pre-collision set (particles mass, velocities and the impact vector), the condition of the collision type (merging or bouncing) is checked and the according after-collision set is computed. The result—mass(es) and velocity(es) are stored and the next pair of particles is randomly chosen. As one deals with only collisions, a question arises about how to model the temporal characteristics of the process. The DSMC algorithm exploits the number of collisions (or collision frequency) to measure the laboratory time.

Non-aggregating particles. Consider first non-aggregating particles. Let N_p particles in a volume V undergo $C_{\Delta t}$ collisions during the time interval Δt. The collision frequency ν is defined as a number of collisions per particle, per unit time. This quantity is also given by the kinetic theory in terms of particles number density $n_p = N_p/V$, thermal velocity $v_0 = \sqrt{2T/m}$ and packing fraction, $\sim\sigma^2 n_p$ [2]. That is,

$$\nu = \frac{\Delta C_{\Delta t}}{N_p \Delta t} = 2\sqrt{2\pi}\, v_0 \sigma^2 n_p, \qquad (66)$$

where we tacitly assume that T and N_p are constant. Hence, for given N_p and V, the collision frequency may be determined from the current temperature $T = 2E/(3N_p)$, where the kinetic energy of the system reads, $E = \sum_i m v_i^2/2$. Therefore, using the number of collisions $C_{\Delta t}$ and measuring the energy of the system E, one can easily compute the time interval Δt corresponding to this number of collisions, $\Delta t = C_{\Delta t}/(N_p \nu)$. In this way, one can determine the laboratory time from the current total number of collisions.

Aggregating particles. The number of different species—monomers, dimers, i-mers, as well as their energy rapidly changes with time in the aggregating systems. The total number of particles permanently decreases, and after a relatively short time, it becomes very difficult (if possible) to obtain accurate data due to the poor statistics. To keep the number of particles approximately constant, we regularly expand the system. The simulations start with the number of monomers equal N_p, but when the total number of particles decreases due to agglomeration below $N_p/2$, we duplicate all particles. Namely, each particle (monomer, dimer, etc.) is replaced by two identical particles with the same mass and velocity. This operation corresponds to the duplicating the size (volume V) of the system. We define a technical variable

$F = V/V_0$, where V_0 is the initial volume. After the first re-scaling of the system size $F = 2$, after the doubling is performed k times $F = 2^k$. All the extensive quantities are to be re-scaled accordingly, for instance, $N_p^{\rm MC} = FN_p$ and $C_{\Delta t}^{\rm MC} = FC_{\Delta t}$, where the variables with the superscript "MC" are the quantities which refer to the current time corresponding to a certain temperature. In this way, the total number of particles will always belong to the interval $N_p^{\rm MC} \in [N_p/2 + 1, N_p]$. The above re-scaling of the system size yields a reliable statistical data for all simulated processes.
Laboratory time for aggregative systems. In dilute gases, each species may be used to measure the laboratory time from the collision frequency; the collisions are to be between particles of the same kind. Still, the accuracy of the measurements sensitively depends on the abundance of such particles: The larger the amount of particles of a particular type, the more accurate is the laboratory time measurement. Therefore, in the DSMC simulations of agglomerating systems, we determine the time based on the most abundant at the current moment particles. Initially, the monomers are the most abundant and are used for time measurement. Then dimers, triplets, and so on are used for this purpose; again, we apply Eq. (66), see Ref. [34] for more detail.

5.2 *Solution of Temperature-Dependent Smoluchowski Equations by Monte Carlo Method*

Consider now another numerical method that also exploits the Monte Carlo technique. It, however, is applied to the Smoluchowski equations, resulting from the BE, and not to the BE itself. The detailed discussion is given in Refs. [53, 54]; here, we give a brief overview. To make our simulations equivalent to the temperature-dependent Smoluchowski equations, we assume that the velocity distribution of particles is Gaussian for all species [34]. This implies the application of the kernels C_{ij}, B_{ij}, and D_{ij}, as defined in Eqs. (23)–(25). We will use these kernels to determine the collision frequencies and the corresponding temperature variation.

The time step τ between collisions can be determined using Eq. (22). Note that the factor $1 - F_{ij}$ in the kernel C_{ij} defines the aggregation probability. Therefore, to determine the time between any (not only aggregative) collisions, we can use the kernel $\hat{C}_{ij} = C_{ij}/\left(1 - F_{ij}\right)$. Let us define the system volume as $V = N_p(t = 0)/N(t = 0)$, where $N_p(t)$ is the total number of agglomerates and $N(t)$ is the total agglomerate density. It may be shown that the time interval τ, corresponding to the decreases of the total total number of particles by one (due to aggregation), reads [53],

$$\tau = \frac{2V}{\sum\limits_{i,j} \hat{C}_{ij} N_i N_j} = \frac{2V\left(1 - F_{ij}\right)}{\sum\limits_{i,j} C_{ij} N_i N_j}, \tag{67}$$

where N_i is the number of particles of size i. Similarly, we can find $\tau_{i,j}^{agg}$—the average time interval between the events, when a particle of size i merges with a particle of size j:

$$\tau_{i,j}^{agg} = \frac{V}{C_{ij} N_i N_j}. \tag{68}$$

The average change of temperatures during the aggregative collision of particles of size i and j may be expressed in terms of the kernels B_{ij} and D_{ij} [53]:

$$\frac{\Delta\left(n_i T_i\right)}{\tau_{i,j}^{agg}} = \frac{\left(N_i - 1\right)\left(\theta_i + \Delta\theta_i\right) - N_i\theta_i}{V\tau_{i,j}^{agg}} \approx -D_{ij}^{agg} N_i N_j / V^2, \tag{69}$$

where $\Delta\left(n_j T_j\right)$ has the same form with the interchange $(i, j) \to (j, i)$ and where we use only the part D_{ij}^{agg} of the kernel D_{ij}, that determines the temperature variation in the aggregative collisions. Substitution $\tau_{i,j}^{agg}$ from Eq. (68) into Eq. (69) yields for the temperature increments of the species, involved in the aggregation $i + j \to k$:

$$\Delta\theta_i = \frac{\theta_i - D_{ij}^{agg}/C_{ij}}{N_i - 1} = -\frac{1}{N_i - 1}\left(\frac{4(1-G_{ij})}{3(1-F_{ij})} - 1\right)\frac{\theta_i^2}{\theta_i + \theta_j} \tag{70}$$

$$\Delta\theta_k = \frac{B_{ij}/C_{ij} - \theta_k}{N_k + 1} = \frac{1}{N_k + 1}\,\frac{\theta_i\theta_j + \frac{4(1-G_{ij})}{3(1-F_{ij})}\cdot\frac{\left(i\theta_i - j\theta_j\right)^2}{(i+j)^2}}{\theta_i + \theta_j} - \frac{\theta_k}{N_k + 1},$$

where $\Delta\theta_j$ may be obtained from the above equation for $\Delta\theta_i$, by changing $(i, j) \to (j, i)$. In the same way one obtains the equations for the temperature increments in the restitutive collisions $i + j \to i + j$; here we use $D_{ij}^{res} = D_{ij} - D_{ij}^{agg}$, that is, the other part of the kernel D_{ij}:

$$\Delta\theta_i = -\frac{\hat{D}_{ij}^{res}}{N_i}, \quad \Delta\theta_j = -\frac{\hat{D}_{ji}^{res}}{N_j}, \quad \hat{D}_{ij}^{res} = \frac{4G_{ij}}{3F_{ij}}\frac{j\,(1+\varepsilon)}{i+j}\left(\theta_i - \frac{(1+\varepsilon)\,j}{2\,(i+j)}\left(\theta_i + \theta_j\right)\right).$$

Exploiting the above procedure, we directly compute the temperature increments, which is significantly faster than randomly choosing an aggregate velocity from a Gaussian distribution. In this version of MC, we essentially use virtual particles which correspond to an ensemble of real particles. Therefore, it is straightforward to double the system size a desired number of times.

After specifying the main steps of the simulation approach, we consider the most efficient implementation of these steps. We apply the low-rank approximation of the coagulation kernel. It significantly outperforms the conventional techniques, like Gillespie [55] or inverse [56] method. Below, we briefly sketch the idea of the method.

To approximate the collision frequencies $\hat{C}_{ij}$, we use the following low-rank matrix A:

$$A_{ij} = \sqrt{\pi/2}\left(i^{1/3} + j^{1/3}\right)^2\sqrt{\theta_i}, \qquad \left(A_{ij} + A_{ji}\right)/\sqrt{2} \leqslant \hat{C}_{ij} \leqslant A_{ij} + A_{ji}. \tag{71}$$

Equation (71) shows that we can pick the probabilities up from the matrix $2A$ instead of $\hat{C}$ ($A_{ji} = A_{ij}$ due to the collision symmetry). We use the rejection sampling for the cases, where we have an overestimate, which happens with the probability $(\sqrt{\theta_i} + \theta_j)(\sqrt{\theta_i} + \sqrt{\theta_j})^{-1} \leqslant 1 - 1/\sqrt{2} < 0.3$. This idea is similar to the majorant kernels approach exploited in Ref. [57]. After the multiplication of the matrix A by the vectors composed of n_i and n_j, we still get the sum of rank-1 matrices. Indeed, let a matrix A be written as a sum of direct products of M-dimensional vectors (M is the largest aggregate size):

$$A = \sum_{k=1}^{r} u_{(k)} v_{(k)}^T, \quad u_{(k)}, v_{(k)} \in \mathbb{R}^M,$$

then $A_{ij} n_i n_j = \left(A \circ nn^T\right)_{ij}$, where $\circ$ denotes the element-wise product. It can be also written as

$$A \circ nn^T = \sum_{k=1}^{r} \left(nu_{(k)}\right) \left(nv_{(k)}\right)^T$$

i.e., also in terms of a direct product. This is essentially the basis for fast computations.

In case of ballistic agglomeration (71), we have $r = 3$:

$$u_{(1)}(i) = \sqrt{\pi/2} i^{2/3} \sqrt{\theta_i}, \qquad u_{(2)}(i) = 2\sqrt{\pi/2} i^{1/3} \sqrt{\theta_i}, \quad u_{(3)}(i) = \sqrt{\pi/2}\sqrt{\theta_i},$$
$$v_{(1)}(j) = 1, \qquad v_{(2)}(j) = j^{1/3}, \quad v_{(3)}(j) = j^{2/3}.$$

In order to keep track of the sums and quickly choose the sizes of the colliding particles, one can construct a segment tree on each of the vectors $u_{(k)}$ and $v_{(k)}$. Then the update of the full structure requires $O(r \log M)$ operations. In particular, segment trees contain the sums of all elements of $u_{(k)}$ and $v_{(k)}$ and thus we also know the sums of the $u_{(k)} v_{(k)}^T$ elements. To choose a pair of colliding particles, we firstly pick one of the three rank-1 matrices up with the probability, proportional to the sum of its elements ($O(r)$ operations), then the element of $u_{(k)}$ with the probability, proportional to its value ($O(\log M)$), and finally, a column of $v_{(k)}$ (another $O(\log M)$ operations). Therefore, the total complexity of our method of one collision computing is $O(r \log M)$. This is significantly smaller than the complexity, $O(M)$, (or even, $O\left(N_p\right)$) of conventional methods.

5.3 *Solution of Temperature-Dependent Smoluchowski ODE with Low-Rank Methods*

This method guarantees an efficient (fast and accurate) solution for very large systems of ODE corresponding to temperature-dependent Smoluchowski equations. Initially, this method was developed to solve conventional Smoluchowski equations

[58, 59], but later was generalized for its temperature-dependent counterpart [44]. The main idea of the method is again the usage of a smart representation of the quantities $(C_{ij}n_in_j)$, $(B_{ij}n_in_j)$ and $(D_{ij}n_in_j)$, comprising the main blocks of the Smoluchowski equations. These matrices may be represented through the sums of direct products of appropriately chosen vectors. Numerical operations with a few vectors are significantly less time-consuming than with a whole matrix. For instance, the application of the low-rank technique to the solution of a few thousand equations accelerates computations by about 50 times. Although the implementation of the method is straightforward, even its brief description is lengthy; therefore, we address an interested reader to the special literature [44, 58, 59].

5.4 *Phase Diagram for the Temperature-Dependent Smoluchowski Equations*

Using the advanced computational methods discussed in the previous section, we investigate the behavior of the temperature-dependent Smoluchowski equations in a wide range of parameters. It occurs to be very rich, with numerous surprising regimes. Here, we present the kinetic phase diagram in terms of the quantities specifying the aggregation energy W_{ij} [see Eq. (20)]: Λ and $q = (a/T_1(0))(\varepsilon^2 2^{\Lambda})^{-1}$ (please do not mix with the aggregation probability q in Sect. 3.2); in simulations we used $\lambda_1 = \lambda_2$. The parameter Λ characterizes the dependence of the rate kernels on the aggregates' size and q gives the ratio of the potential energy of monomers at contact and their initial kinetic energy. The phase diagram is depicted in Fig. 1.

All the regimes are described in detail in Ref. [53], where we mention the most unusual ones—the aggregation with temperature increase and with density separation. These are illustrated in Figs. 2 and 3, respectively. The increase in average temperature may occur when the number of particles decreases due to agglomeration faster than the total energy of the system. This results in the increasing energy per particle, that is, in the rise of the system temperature. The effect of density separation is more subtle: It is related to the quasi-gelation that may initially commence, as the gelation condition is satisfied. Later on, this condition becomes violated for very large clusters. Overall, this results in the depletion of densities of middle-size aggregates [53]. Among other regimes, one can mention the conventional aggregation with temperature decay, the regime of non-monotonous temperature growth, ending by temperature decay, and the non-aggregating regime of granular gas cooling, in accordance with Haff's law.

6 Discussion

We consider the application of the Boltzmann equation to aggregation kinetics. This includes the aggregation of particles in molecular gases, agglomeration of grains in

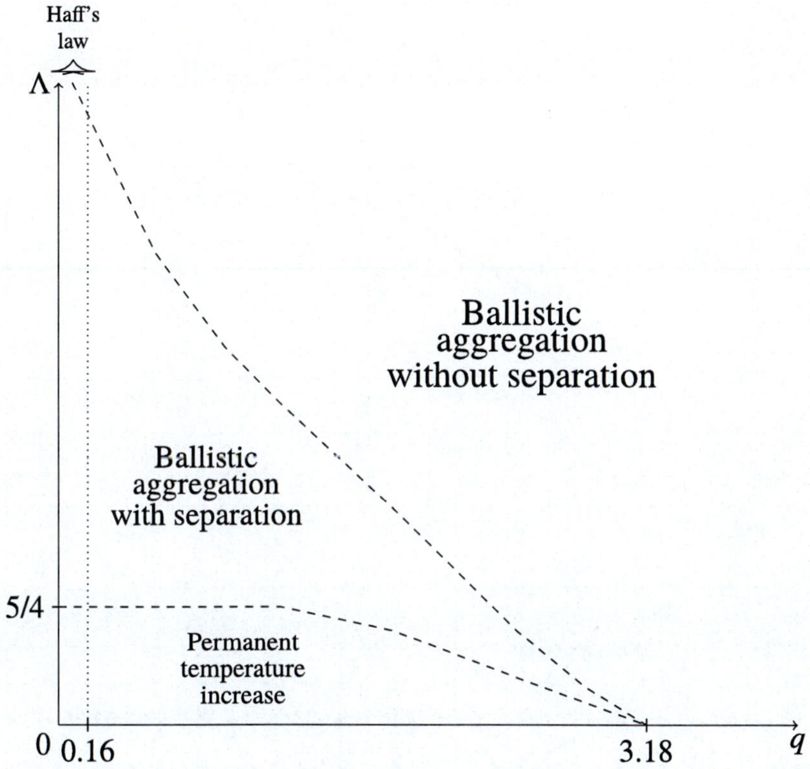

Fig. 1 Phase diagram for the temperature-dependent Smoluchowski equations, Eqs. (22) and (24). Different kinetic regimes are demarcated by dashed lines, see [54]. The simulations were performed for $\varepsilon = 0.99$, $m_1 = 1$, and $\sigma_1 = 1$, for mono-disperse initial conditions, initial dimensionless density $(\pi/6)n_1(0)\sigma_1^3 = 0.05$ and initial temperature $T_1(0) = 1$

granular gases, or even hypothetical agglomeration of nuclei of the dark matter. The main transport mechanism underlying all these processes is the ballistic motion of particles. We explore two agglomeration models—a random agglomeration model and the impact energy-dependent one. The former model accounts for collisional parameters which are not explicitly treated (e.g., orientational and rotational degrees of freedom of particles). The latter model accounts for the binding energy barrier for different interparticle interactions—adhesive, gravitational, or electrostatic.

We start from the Boltzmann–Smoluchowski equations for the mass–velocity distribution functions for aggregates of different sizes and derive temperature-dependent Smoluchowski equations. These describe the evolution of densities of the aggregates of different sizes and of their partial temperatures. For the above kinetic equations, we obtain microscopic expressions for the reaction rate coefficients. For some simplified models of the rate coefficients, we find exact solutions to the temperature-dependent Smoluchowski equations and develop a scaling theory which describes the asymptotic behavior of the system. We elaborate on a couple of effective numerical methods

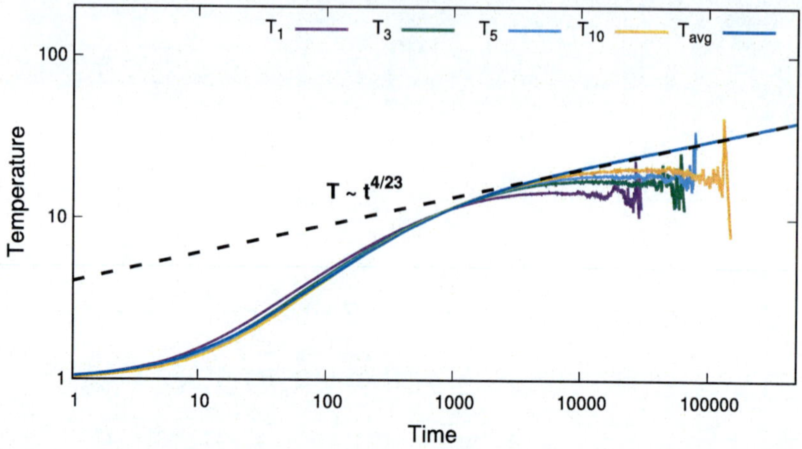

Fig. 2 Regime of aggregation with temperature growth, see Eq. (63), for $\Lambda = 0.4$, $q = 1.8$. The theoretical exponent $\beta = -2\Lambda/(5-\Lambda) = -4/23$ is in an excellent agreement with the simulation data. The system parameters are the same as for Fig. 1

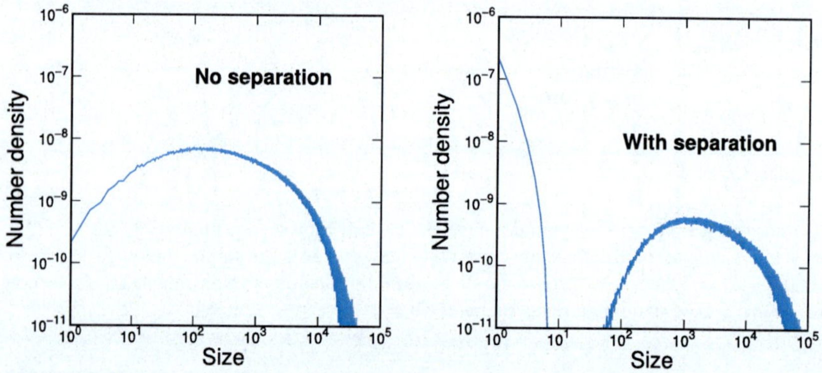

Fig. 3 The size distribution of the aggregates, n_k, as the function of their size k at $t = 10,000$. Left panel: Conventional size distribution ($\Lambda = 1.4$, $q = 3.8$). Right panel: Distribution of aggregate size with separation ($\Lambda = 1.4$, $q = 1.8$); middle-size densities are extremely small. The system parameters are the same as for Fig. 1

to model the ballistic-driven agglomeration. These include the modification of the Direct Simulation Monte Carlo for the Boltzmann equation and low-rank methods for the generalized Smoluchowski equations.

We explore a wide range of system parameters and construct a phase diagram illustrating the kinetic regimes. We observe that the behavior of the system is very rich, including such surprising regimes as aggregation with temperature growth and aggregation with density separation, where the densities of middle-size clusters practically vanish. A comparison of the scaling theory with the numerical results demonstrates an excellent agreement between the theory and simulations.

Acknowledgements The study was supported by a grant from the Russian Science Foundation No. 21-11-00363, https://rscf.ru/project/21-11-00363/. We also thank the German Science Foundation (DFG) for funding through Grant P0472/40-1 and the Interdisciplinary Center for Nanostructured Films (IZNP). Support from ZISC and IZ-FPS at FAU Erlangen–Nürnberg is gratefully acknowledged.

References

1. L. Boltzmann, *Leçons sur la théorie des gaz* (Gauthier-Villars, Paris, 1902–1905)
2. N.V. Brilliantov, T. Pöschel, *Kinetic Theory of Granular Gases* (Oxford University Press, Oxford, 2004)
3. V. Garzo, *Granular Gaseous Flows* (Springer Nature Switzerland AG, Cham, Switzerland, 2019)
4. S. Chapman, T.G. Cowling, *The Mathematical Theory of Non-uniform Gases* (Cambridge University Press, New York, 1970)
5. J. Ferziger, H. Kaper, *The Mathematical Theory of Transport Processes in Gases* (North-Holland, Amsterdam, 1972)
6. F. Spahn, N. Albers, M. Sremcevic, C. Thornton, Europhys. Lett. **67**, 545 (2004)
7. L. Esposito, *Planetary Rings* (Cambridge University Press, Cambridge, UK, 2006)
8. N.V. Brilliantov, A.S. Bodrova, P.L. Krapivsky, J. Stat. Mech. **P06011** (2009)
9. N.V. Brilliantov, P.L. Krapivsky, A. Bodrova, F. Spahn, H. Hayakawa, V. Stadnichuk, J. Schmidt, Proc. Natl. Acad. Sci. USA **112**, 9536 (2015)
10. N.V. Brilliantov, A.I. Osinsky, P.L. Krapivsky, Phys. Rev. E **102**, 042909 (2020)
11. E. Hardy, R. Lasenby, J. March-Russell, S.W. West, JHEP **06**, 011 (2015)
12. G. Krnjaic, K. Sigurdson, Phys. Lett. B **751**, 464 (2015)
13. M.I. Gresham, H.K. Lou, K.M. Zurek, Phys. Rev. D **97**, 036003 (2018)
14. P.L. Krapivsky, A. Redner, E. Ben-Naim, *A Kinetic View of Statistical Physics* (Cambridge University Press, Cambridge, UK, 2010)
15. C. Thuroff, F. Weber, E. Frey, Phys. Rev. X **4**, 041030 (2014)
16. G.R. Hidy, J.R. Brock, *The Dynamics of Aerocolloidal Systems, International Reviews in Aerosol Physics and Chemistry* (Pergamon Press, Oxford, 1970)
17. R.L. Drake, in G.M. Hidy, J.R. Brock (eds.), *Topics in Current Aerosol Research*, vol. 3, part 2 (Pergamon Press, New York, 1972)
18. R.C. Shrivastava, J. Atom. Sci. **39**, 1317 (1982)
19. S.K. Friedlander, *Smoke, Dust and Haze* (Oxford University Press, Oxford, 2000)
20. G.B. Field, W.C. Saslaw, Astrophys. J. **142**, 568 (1965)
21. J.J. Lissauer, Ann. Rev. Astron. Astrophys. **31**, 129 (1993)
22. A. Chokshi, A.G.G. Tielens, D. Hollenbach, Astrophys. J. **407**, 806 (1993)
23. C. Dominik, A.G.G. Tielens, Astrophys. J. **480**, 647 (1997)
24. V. Ossenkopf, Astron. Astrophys. **280** (1993)
25. M.V. Smoluchowski, Z. Phys. Chem. **92**, 129 (1917)
26. F. Leyvraz, Phys. Rep. **383**, 95 (2003)
27. G.F. Carnevale, Y. Pomeau, W.R. Young, Phys. Rev. Lett. **64**, 2913 (1990)
28. E. Trizac, J.P. Hansen, Phys. Rev. Lett. **74**, 4114 (1995)
29. L. Frachebourg, Phys. Rev. Lett. **82**, 1502 (1999)
30. L. Frachebourg, P.A. Martin, J. Piasecki, Phys. A **279** (2000)
31. L. Frachebourg, P. Martin, J. Piasecki, Phys. A **279**, 69 (2000)
32. E. Trizac, P.L. Krapivsky, Phys. Rev. Lett. **91**, 218302 (2003)
33. N.V. Brilliantov, F. Spahn, Math. Comput. Simul. **72**, 93 (2006)
34. N.V. Brilliantov, A. Formella, T. Pöschel, Nat. Commun. **9**, 797 (2018)
35. J. Midya, S.K. Das, Phys. Rev. Lett. **118**, 165701 (2017)

36. S. Paul, S.K. Das, Phys. Rev. E **97**, 032902 (2018)
37. C. Singh, M.G. Mazza, Sci. Rep. **9**, 9049 (2019)
38. C. Singh, M.G. Mazza, Phys. Rev. E **97**, 022904 (2018)
39. V. Garzo, J.W. Dufty, Phys. Rev. E **59**, 5895 (1999)
40. N.V. Brilliantov, N. Albers, F. Spahn, T. Pöschel, Phys. Rev. E **76**, 051302 (2007)
41. A. Ivelv, G. Morfill, U. Konopka, Phys. Rev. Lett. **89**, 195502 (2002)
42. A.I. Osinsky and N.V. Brilliantov, Exact solutions of temperature-dependent Smoluchowski equations, Unpublished (2022)
43. P.G.J. van Dongen, M.H. Ernst, Phys. Rev. Lett. **54**, 1396 (1985)
44. A.I. Osinsky, J. Comput. Phys. **422**, 109764 (2020)
45. G.A. Bird, *Molecular Gas Dynamics and the Direct Simulation of Gas Flows* (Oxford University Press, Oxford, 1994)
46. M. Hopkins, H. Shen, J. Fluid Mech. **244**, 477 (1992)
47. J.M. Montanero, A. Santos, Phys. Rev. E **54**, 438 (1996)
48. J.M. Montanero, A. Santos, Phys. Fluids **9**, 2057 (1997)
49. J.J. Brey, M.J. Ruiz-Montero, Comput. Phys. Commun. **121/122**, 278 (1999)
50. J.M. Montanero, V. Garzo, A. Santos, J.J. Brey, J. Fluid Mech. **389**, 391 (1999)
51. J.J. Brey, D. Cubero, *Lecture Notes in Physics*, vol. 564 (2001), p. 59
52. T. Pöschel, T. Schwager, *Computational Granular Dynamics: Models and Algorithms* (Springer, Berlin, 2005)
53. A.I. Osinsky, N.V. Brilliantov, Phys. Rev. E. **105**, 034119 (2022)
54. A. Kalinov, A.I. Osinsky, S.A. Matveev, W. Otieno, N.V. Brilliantov, arXiv:2103.09481 (2021)
55. D. Gillespie, J. Comput. Phys. **22**, 403 (1976)
56. A. Garcia, L. Alejandro, C. van den Broeck, M. Aertsens, R. Serneels, Phys. A **143**, 535 (1987)
57. A. Eibeck, W. Wagner, SIAM J. Sci. Comput. **22**, 802 (2000)
58. S. Matveev, P. Krapivsky, A. Smirnov, E. Tyrtyshnikov, N. Brilliantov, Phys. Rev. Lett. **119**, 260601 (2017)
59. S. Matveev, N. Ampilogova, V. Stadnichuk, E. Tyrtyshnikov, A. Smirnov, N. Brilliantov, Comp. Phys. Commun. **224**, 154 (2018)

Large-Scale Fluctuations in Collisional Dusty Plasmas with Regard to Grain Charging Processes

A. G. Zagorodny and A. I. Momot

Abstract The kinetic theory of large-scale electric fluctuations in a collisional weakly ionized dusty (complex) plasma is formulated with due regard to the grain charging dynamics. The dependencies of the charging frequencies and effective collision frequencies for electrons and ions on dusty plasma parameters are studied in detail. In order to describe resonant properties of collective fluctuations, the analysis of the ion-acoustic wave spectrum in dusty plasmas is presented for a wide range of ion collisions for both non-isothermal and isothermal plasmas. The electron density correlation spectra are calculated for various values of the grain density, grain size, and ion collisionality.

1 Introduction

Electromagnetic fluctuations are important and often provide the only source of information about the medium's parameters. They are closely related to electromagnetic, kinetic, and thermodynamic properties of macroscopic systems. Obviously, the more detailed is the description of fluctuations, the more information can be obtained from the experimental spectra of fluctuations. Thus, we may assume that the theory of fluctuations developed from the first principles (i.e., using the microscopic description) is the most general. Today, however, no general microscopic calculations are known that would give explicit results for the correlation functions of fluctuations in non-equilibrium systems with arbitrary correlation times. For example, existing theories that use assumptions about the Bogolyubov's hierarchy of characteristic evolution times [1, 2] make it possible to describe fluctuations with the correlation

A. G. Zagorodny (✉)
M.M. Bogolubov Institute for Theoretical Physics, The National Academy of Science of Ukraine, 14b, Metrologichna Str., Kyiv 03680, Ukraine
e-mail: azagorodny@bitp.kiev.ua

A. I. Momot
Faculty of Physics, Taras Shevchenko National University of Kyiv, 64/13, Volodymyrs'ka Str., Kyiv 01601, Ukraine

L. Brenig et al. (eds.), *Nonequilibrium Thermodynamics and Fluctuation Kinetics*, Fundamental Theories of Physics 208,
https://doi.org/10.1007/978-3-031-04458-8_11

time much shorter than the relaxation time (time needed to approach the local equilibrium distribution) [3–7]. In order to get rid of such restrictions (i.e., to describe fluctuations with the correlation times of the order of the relaxation time) one can employ either the kinetic theory of fluctuations based on the kinetic equations with appropriate Langevin sources (see, for example [3]), or the description in terms of microscopic quantities smoothed over the physically infinitesimal time, or spatial intervals ($\tau_{\mathrm{cor}} \ll \tau_{\mathrm{ph}} \ll \tau_{\mathrm{rel}}$ here τ_{cor} and τ_{rel} are the correlation and relaxation times, respectively) [4, 5, 7]. Such an approach has the advantage that it provides a possibility to calculate directly the correlation functions of fluctuation sources in the case of non-equilibrium systems and thus to develop the consistent theory of fluctuations without any additional assumptions. Namely such approach was used to take into account the contribution of large-scale fluctuations into the collision terms in the kinetic equations [5, 6], to calculate large-scale fluctuation spectra in weakly ionized plasmas [7–10], to perform semi-phenomenological description of fluctuations in turbulent plasmas [11–13], and to generalize the theory of fluctuations in homogeneous plasmas to the non-homogeneous case [14, 15].

It is of interest to note here that the application of the approach in terms of the microscopic equations smoothed over physically infinitesimal intervals leads to the fluctuation evolution equations with collision terms generated by microscopic fluctuations, that is, the Boltzmann-type collision term in the case of a weakly non-ideal gas, or the Balescu-Lenard collision term in the case of a fully ionized plasma. Since the solution of evolution equations with such collision integrals cannot be obtained in the general case, model collision terms are usually used. The Bhatnagar-Gross-Crook (BGK) collision term is a good example of such a model collision integral often used in the calculations of fluctuation spectra [3, 7, 8, 11–13]. However, the BGK collision term cannot be used to account for the difference between the bulk viscosity and shear viscosity of many-particle systems. For this, more sophisticated collision terms have been proposed [16, 17].

In the present contribution, we propose a theory of large-scale fluctuations in dusty plasmas, i.e., plasmas with fine dispersed solid objects (dust particles, or grains) that accumulate and carry large electric charge. Before considering the details of such theory it should be noted that dusty plasmas make an example of an open system since the dust particle charge appears due to the absorption of electrons and ions from the surrounding plasma whose stationary state is maintained by external plasma sources. The absolute value of the electron current on the surface of the initially uncharged grain is greater than the ionic one due to higher thermal velocity of electrons, and thus the grain is being charged negatively. This leads to the increase in the ion current and to the decrease in electron current. The negative grain charge grows until the total charging current becomes equal to zero (absolute values of electron and ion currents become equal) [18].

There are two reasons for grain charge fluctuations [19, 20]. The first one is the random and discrete nature of the charging process. Such a type of fluctuations is important for small grains [20] and for the description of electromagnetic fluctuations in dusty plasmas [21]. Spatial and temporal variations of plasma parameters that influence the charging processes are the second reason for grain charge fluctuations that are considered in the present contribution. In this case the charging processes are assumed to be continuous. The grain charge becomes a time-dependent quantity and one has to treat it as a dynamic variable that is coupled self-consistently with other dynamical variables in the plasma such as density and potential [22]. Obviously, the presence of dust particles can considerably influence the eigenfrequencies and damping rates of waves in dusty plasmas and thus to produce changes in fluctuation spectra. Besides that, due to charged grains new types of waves, particularly dust-acoustic waves, can appear. The importance of the charge fluctuations as the source of oscillation damping was suggested at first for astrophysical dusty plasmas [23, 24]. Later, it was shown that grain charge fluctuations can influence the propagation of waves in dusty plasmas [22, 25, 26].

The existence of ion-acoustic waves in an isothermal collisionless dusty plasmas was discussed for the first time in Ref. [27] that later was proved experimentally [28, 29]. Properties of ion-acoustic waves in dusty plasmas with regard to charging dynamics preferably were studied using fluid equations [30–34]. The kinetic approach for the description of ion acoustic waves was also employed in several papers [35–38], but only for the case of collisionless plasmas. Thus, the description of ion-acoustic waves taking into account self-consistent grain charging and particle collisions is an integral part of the theory of large-scale fluctuations in dusty plasmas.

As mentioned above, the electromagnetic fluctuations are closely related to the thermodynamic and structural properties of continuous media. Particularly, the correlation functions of particle densities determine both static and dynamic form factors of the system and thus the spectrum of electromagnetic wave scattering by plasmas [2, 3, 10]. The kinetic coefficients in the Fokker-Planck equation [3, 10, 39] and the collision integrals in the kinetic equations are also determined by particle phase density fluctuations. It is also obvious that fluctuations of the electromagnetic field play the role of the Langevin sources in the Brownian motion of charged particles in plasmas [2, 40]. Therefore, the calculation of the electric field fluctuations in a dusty plasmas [19] is important for describing various specific processes in such plasmas. For example, the intensity of grain diffusion generated by fluctuations determines, to a considerable extent, the processes of formation and melting of plasma crystals [41, 42].

The theory of fluctuations in ordinary collisional plasmas is well developed [4, 7, 13, 43, 44]. The problem of generalization of this theory to the case of dusty plasmas has some issues that remain open. To solve the problem under consideration two approaches are usually used. The first one is based on the description of grain charge dynamics using the charging equation and the explicit representation of the charging currents in terms of fluctuations of plasma particle distribution functions and charging cross sections [26, 45–47]. The disadvantage of this approach is that it requires a phenomenological description of plasma particle collisions with grains (both elastic and inelastic) and additional calculations of "shadow" and bombardment forces (generated by scattering and absorption of plasma particles by a grain in the presence of another grain) to describe consistently the collective grain-grain interaction, if necessary. In the second approach the grain charge is treated as an independent dynamic variable and thus the grain distribution function depends not only on the coordinate and velocity, but also on the grain charge [18, 21, 48–53]. This approach provides an opportunity to work out a consistent kinetic theory of fully ionized dusty plasmas and, in principle, to find collision terms for all particle species. However, this approach faces serious problems with the generalization to the case of collisional plasmas. The main problem here is that calculations of the quantities describing the grain dynamics are expressed in terms of charging cross sections that are known for collisionless plasmas only. At the same time, phenomenological approximations for such cross sections (see, for example, [52] and related references cited therein) have a limited range of applications. Actually, the same problem arises in the first approach, but in that case, one can use semi-phenomenological approximations for charging currents [54, 55] that take into account the influence of collisions and, at the same time, are in a good agreement with the results of experiments and numerical calculations [56] for arbitrary collision frequencies and other plasma parameters. This means that we can avoid calculating charging cross sections if the equations of grain charging dynamics can be formulated in terms of charging currents. Since the appropriate formulation can be easily done, the first approach looks more suitable for the generalization of the theory of electromagnetic fluctuations in collisionless dusty plasmas to the case of collisional plasma background. The above-mentioned problems with unknown charging cross sections for collisional dusty plasmas are the reason why the consistent calculations of fluctuation spectra are related to collisionless [26, 45, 46, 48] or weakly collisional regimes [50–53, 57].

The purpose of the present contribution is to give a consistent linear kinetic description of electromagnetic processes in a collisional weakly ionized dusty plasma with regard to the absorption of electrons and ions by grains and grain charge fluctuations. The main point is that electric field perturbations influence the charging currents and hence produce the grain charge fluctuations. Such a self-consistent influence gives an additional contribution to the dielectric response of the dusty plasma that determines both dispersion and damping of waves and electric field fluctuation spectra in dusty plasmas [58, 59].

The contribution is organized as follows. We start from the equations for fluctuations of distribution functions (Sect. 2). Then we formulate the equations for grain charging dynamics (Sect. 3). In Sect. 4 we analyze the dependencies of the stationary grain charge and charging collision frequencies on plasma parameters. The calculations of the dielectric response functions and detailed analysis of ion-acoustic waves in both cases of ordinary and dusty plasmas are presented in Sects. 5 and 6. Correlation functions of the electron density fluctuations are calculated and analyzed in Sect. 7.

2 Fluctuations of the Distribution Function

The consistent description of fluctuations in a dusty plasma requires considering the charge density fluctuations related to electrons ($\alpha = e$) and ions ($\alpha = i$)

$$\delta\rho_\alpha(\mathbf{r}, t) = e_\alpha \delta n_\alpha(\mathbf{r}, t), \quad \alpha = e, i, \tag{1}$$

as well as the grain charge density fluctuations that may be divided into two terms. The first one is due to variations of the grain number density and the second one is associated with the fluctuations of the grain charge

$$\delta\rho_g(\mathbf{r}, t) = e_g \delta n_g(\mathbf{r}, t) + n_g \delta e_g(\mathbf{r}, t), \tag{2}$$

where e_g is the stationary grain charge and n_g is the mean number density of grains. It should be noted that such representation is valid for fluctuations that satisfy the condition $n_g R^3 \gg 1$. Here R is the spatial scale of perturbation.

The number density fluctuations of charged particles have the form

$$\delta n_\alpha(\mathbf{r}, t) = n_\alpha \int d\mathbf{v} \delta f_\alpha(\mathbf{r}, \mathbf{v}, t), \quad \alpha = e, i, g, \tag{3}$$

where $\delta f_\alpha(\mathbf{r}, \mathbf{v}, t)$ are the fluctuations of the distribution function of the corresponding particle species. In the case of electrons or ions, these may be found in the same way as in ordinary plasmas [44], but regarding the collisions of electrons and ions with grains in addition to collisions with neutrals.

Fluctuations of electron and ion distribution functions satisfy the equation [44]

$$\left\{\frac{\partial}{\partial t} + \mathbf{v}\frac{\partial}{\partial \mathbf{r}}\right\} \delta f_\alpha(X, t) + \nu_\alpha \left\{\delta f_\alpha(X, t) - f_{0\alpha}(\mathbf{v}) \int d\mathbf{v} \delta f_\alpha(X, t)\right\}$$
$$= -\frac{e_\alpha}{m_\alpha} \frac{\partial \delta\phi(\mathbf{r}, t)}{\partial \mathbf{r}} \frac{\partial f_{0\alpha}(\mathbf{v})}{\partial \mathbf{v}}, \tag{4}$$

where $X = (\mathbf{r}, \mathbf{v})$, $\delta\phi(\mathbf{r}, t)$ is the fluctuation of the electrostatic potential, $f_{0\alpha}(\mathbf{v})$ is the unperturbed distribution function (usually it is the Maxwell distribution), and $\nu_\alpha = \nu_{\alpha n} + \nu_{\alpha g}$, where $\nu_{\alpha n}$ and $\nu_{\alpha g}$ are the effective collision frequencies between particles of α species with neutrals and grains.

In the case under consideration, Eq. (4) is justified under following assumptions. The electrons and ions absorbed by the grain recombine on its surface and form the neutral gas atoms (molecules) that evaporate into the surrounding plasma and may be ionized again due to collisions or to external ionization sources. This assumption makes it possible to use the Bhatnagar-Gross-Krook (BGK) collision integral [60] in Eq. (4). It should be noted that Eq. (4) follows also from the equations for the microscopic phase densities of electrons and ions smoothed over physically infinitesimal time τ_{ph} [21] in the appropriate approximation.

Further, the equation for the grain charge fluctuations is formulated. The averaging (smoothing over physically infinitesimal time $\tau_{|} text ph$) of the equation for the microscopic phase density of grains yields a kinetic equation with the collision integral that may be expressed in terms of correlation functions of microscopic quantities [21]. For the sake of simplicity, we use the BGK collision integral in the manner similar to the case of plasma particles. The linearized equation for the fluctuations of the grain distribution function reproduces Eq. (4) with $\alpha = g$, where ν_g is the effective collision frequency between grains and other particles. This equation, however, does not take into account the fluctuations of charging collision frequencies.

The formal solution of Eq. (4) is given by

$$\delta f_\alpha(X,t) = \delta f_\alpha^{(0)}(X,t) - \frac{e_\alpha}{m_\alpha}\int_{-\infty}^{t} dt' \int dX' W_\alpha(X,X';t-t') \frac{\partial \delta\phi(\mathbf{r}',t')}{\partial \mathbf{r}'} \frac{\partial f_{0\alpha}(\mathbf{v}')}{\partial \mathbf{v}'}, \tag{5}$$

where $\delta f_\alpha^{(0)}(X,t)$ is the general solution of the homogenous Eq. (4), i.e., it is the fluctuation of the distribution function in the system without self-consistent interaction through the fluctuation electric field. The second term in (5) is the particular solution of Eq. (4). The function $W_\alpha(X,X';t-t')$ also satisfies the homogenous Eq. (4), but with the initial condition $W_\alpha(X,X';0) = \delta(X-X')$. Hence, it is the probability density of particle transition from the phase point X' to the phase point X during the time interval $t-t'$ for particles of α-species.

As is easy to see, $\delta f_\alpha^{(0)}(X,t)$ play the role of Langevin sources of the electric field fluctuations in the system. Their correlation functions are given by [2, 4, 10]

$$\begin{aligned}\langle \delta f_\alpha^{(0)}(X,t)\delta f_{\alpha'}^{(0)}(X',t')\rangle = \frac{\delta_{\alpha\alpha'}}{n_\alpha}\big\{ & f_{\alpha'}(X',t')W_\alpha(X,X';t-t')\theta(t-t') \\ & + f_\alpha(X,t)W_\alpha(X',X;t'-t)\theta(t'-t)\big\}.\end{aligned} \tag{6}$$

3 Equations of the Grain Charge Dynamics

The grain charge is determined by the electron ($\alpha = e$) and ion ($\alpha = i$) charging currents

$$\frac{\partial e_g(\mathbf{r},t)}{\partial t} = I_{\text{ch}} = \sum_{\alpha=e,i} I_{\text{ch}}^\alpha\left(n_\alpha(\mathbf{r},t), e_g(\mathbf{r},t)\right). \tag{7}$$

For the sake of simplicity, we assume that charging currents depend only on the number density n_α and temperature of plasma particles, as well as on the grain charge. This assumption is justified for the strongly collisional regime. It is not consistently justified for weakly collisional regime, but we will use it. Its correctness may be confirmed by the comparison with the results of kinetic calculations.

For small fluctuations of number density $n_\alpha(\mathbf{r},t) = n_\alpha + \delta n_\alpha(\mathbf{r},t)$ from the average value n_α and small fluctuations of the grain charge $e_g(\mathbf{r},t) = e_g + \delta e_g(\mathbf{r},t)$ from its stationary value that is determined by the condition of zero total charging current, i.e.,

$$I_{\text{ch}}^e(n_e, e_g) + I_{\text{ch}}^i(n_i, e_g) = 0, \tag{8}$$

one obtains from (7) an equation for $\delta e_g(\mathbf{r},t)$

$$\frac{\partial \delta e_g(\mathbf{r},t)}{\partial t} + \nu_{\text{ch}}\delta e_g(\mathbf{r},t) = \sum_{\alpha=e,i} \frac{\partial I^{\alpha}_{\text{ch}}(n_\alpha, e_g)}{\partial n_\alpha}\delta n_\alpha(\mathbf{r},t), \tag{9}$$

where the charging frequency ν_{ch} is given by

$$\nu_{\text{ch}} = \sum_{\alpha=e,i} \nu^{\alpha}_{\text{ch}}, \quad \nu^{\alpha}_{\text{ch}} = -\frac{\partial I^{\alpha}_{\text{ch}}(n_\alpha, e_g)}{\partial e_g}. \tag{10}$$

Now, the explicit forms of charging currents $I^{\alpha}_{\text{ch}}(n_\alpha, e_g)$ in a collisional plasma are needed. Since the mean free path of electrons l_e is usually about two orders higher than l_i in gas discharge plasma, we use the expression

$$I^{e}_{\text{ch}} = e_e n_e \sqrt{8\pi} a^2 v_{Te} \exp(-\alpha) \tag{11}$$

that is obtained in the orbit motion limited (OML) approximation, i.e., the collisions of electrons with neutrals are neglected.

For ionic charging current we use the interpolation formula [54] that reproduces with high accuracy the results of kinetic calculations [56]

$$I^{i}_{\text{ch}} = e_i n_i \sqrt{8\pi} a^2 v_{Ti} \frac{I^{\text{WC}} I^{\text{SC}}}{I^{\text{WC}} + I^{\text{SC}}}, \tag{12}$$

where

$$I^{\text{WC}} = 1 + \alpha\tau + 0.1(\alpha\tau)^2 \lambda_D / l_i, \tag{13}$$

$$I^{\text{SC}} = \sqrt{2\pi}\alpha\tau l_i / a, \quad \tau = T_e / T_i. \tag{14}$$

Here $\alpha = e_e \phi_s / T_e$ (not to be confused with the subscript α that denotes the plasma particle species), ϕ_s is the surface potential, a is the grain radius, $k_D^2 = k_{De}^2 + k_{Di}^2$, $k_{D\alpha}^2 = 4\pi e_\alpha^2 n_\alpha / T_\alpha$, $\lambda_D = 1/k_D$ is the Debye length, $v_{T\alpha} = \sqrt{T_\alpha / m_\alpha}$ is the plasma particle thermal velocity, $l_i = v_{Ti}/\nu_i$ is the ion mean free path, ν_i is the collision frequency of ions with other particles. WC stands for weakly collisional and SC is for strongly collisional.

It is reasonable to assume [56, 61–64] that the electrostatic potential near the grain is described by the Derjaguin-Landau-Verwey- Overbeek (DLVO) potential

$$\phi(r) = \frac{e_g}{r}\frac{\exp[-k_D(r-a)]}{1 + ak_D}, \tag{15}$$

then

$$\alpha = \frac{e_e e_g}{aT_e(1 + ak_D)}. \tag{16}$$

The space-time Fourier transform (FT)

$$f_{\mathbf{k}\omega} = \int_{-\infty}^{\infty} dt \int d\mathbf{r}\, f(\mathbf{r}, t) \exp(i\omega t - i\mathbf{k}\mathbf{r}) \tag{17}$$

of Eq. (2) for the grain charge fluctuations along with (9) gives

$$\delta\rho_{g\mathbf{k}\omega} = e_g \delta n_{g\mathbf{k}\omega} + \frac{i n_g}{\omega + i\nu_{\text{ch}}} \sum_{\alpha=e,i} \frac{I_{\text{ch}}^{\alpha}}{n_\alpha} \delta n_{\alpha\mathbf{k}\omega}. \tag{18}$$

It was taken into account in (18) that according to (11) and (12)

$$\frac{\partial I_{\text{ch}}^{\alpha}(n_\alpha, e_g)}{\partial n_\alpha} = \frac{I_{\text{ch}}^{\alpha}}{n_\alpha}. \tag{19}$$

We substitute Eq. (5) in (3) and after FT obtain

$$\begin{aligned} \delta n_{\alpha\mathbf{k}\omega} &= n_\alpha \int d\mathbf{v} \delta f_{\alpha\mathbf{k}\omega}^{(0)}(\mathbf{v}) + i\frac{e_\alpha n_\alpha}{m_\alpha} \int d\mathbf{v} \int d\mathbf{v}' W_{\alpha\mathbf{k}\omega}(\mathbf{v}, \mathbf{v}')\mathbf{k} \frac{\partial f_{0\alpha}(\mathbf{v}')}{\partial \mathbf{v}'} \delta\phi_{\mathbf{k}\omega} \\ &= \delta n_{\alpha\mathbf{k}\omega}^{(0)} - \frac{k^2}{4\pi e_\alpha} \chi_\alpha(\mathbf{k}, \omega) \delta\phi_{\mathbf{k}\omega}, \quad \alpha = e, i, g, \end{aligned} \tag{20}$$

where $\chi_\alpha(\mathbf{k}, \omega)$ is the dielectric susceptibility of the plasma particle subsystem.

Further, we substitute formula (20) with $\alpha = g$ in the first term of (18) and formula (20) with $\alpha = e, i$ in the second one. Thus we obtain

$$\begin{aligned} \delta\rho_{g\mathbf{k}\omega} = \delta\rho_{g\mathbf{k}\omega}^{(0)} &- \frac{k^2}{4\pi} \chi_g(\mathbf{k}, \omega) \delta\phi_{\mathbf{k}\omega} \\ &+ \frac{i}{\omega + i\nu_{\text{ch}}} \sum_{\alpha=e,i} \nu_{\alpha g} \delta\rho_{\alpha\mathbf{k}\omega}^{(0)} \\ &- \frac{k^2}{4\pi} \frac{i}{\omega + i\nu_{\text{ch}}} \sum_{\alpha=e,i} \nu_{\alpha g} \chi_\alpha(\mathbf{k}, \omega) \delta\phi_{\mathbf{k}\omega}, \end{aligned} \tag{21}$$

where

$$\nu_{\alpha g} = \frac{n_g I_{\text{ch}}^{\alpha}}{e_\alpha n_\alpha} \tag{22}$$

is the frequency of plasma particle collisions with grains.

Equation (21) may be rewritten to be similar to (20):

$$\delta\rho_{g\mathbf{k}\omega} = \delta\tilde{\rho}_{g\mathbf{k}\omega}^{(0)} - \frac{k^2}{4\pi} \tilde{\chi}_g(\mathbf{k}, \omega) \delta\phi_{\mathbf{k}\omega}, \tag{23}$$

where

$$\delta\tilde{\rho}^{(0)}_{g\mathbf{k}\omega} = \delta\rho^{(0)}_{g\mathbf{k}\omega} + \frac{i}{\omega + i\nu_{\mathrm{ch}}} \sum_{\alpha=e,i} \nu_{\alpha g}\delta\rho^{(0)}_{\alpha\mathbf{k}\omega}, \tag{24}$$

$$\tilde{\chi}_g(\mathbf{k}, \omega) = \chi_g(\mathbf{k}, \omega) + \frac{i}{\omega + i\nu_{\mathrm{ch}}} \sum_{\alpha=e,i} \nu_{\alpha g}\chi_\alpha(\mathbf{k}, \omega). \tag{25}$$

Thus,

$$\delta\phi_{\mathbf{k}\omega} = \frac{4\pi\,\delta\rho^{(0)}_{\mathbf{k}\omega}}{k^2\varepsilon(\mathbf{k}, \omega)}, \tag{26}$$

where

$$\delta\rho^{(0)}_{\mathbf{k}\omega} = \sum_{\alpha=e,i,g} \delta\rho^{(0)}_{\alpha\mathbf{k}\omega} + \frac{i}{\omega + i\nu_{\mathrm{ch}}} \sum_{\alpha=e,i} \nu_{\alpha g}\delta\rho^{(0)}_{\alpha\mathbf{k}\omega}, \tag{27}$$

$$\varepsilon(\mathbf{k}, \omega) = 1 + \sum_{\alpha=e,i,g} \chi_\alpha(\mathbf{k}, \omega) + \frac{i}{\omega + i\nu_{\mathrm{ch}}} \sum_{\alpha=e,i} \nu_{\alpha g}\chi_\alpha(\mathbf{k}, \omega). \tag{28}$$

We see that the dielectric permittivity of a dusty plasma in the present description differs from the dielectric permittivity given by the multi-component model

$$\varepsilon_{\mathrm{mc}}(\mathbf{k}, \omega) = 1 + \sum_{\alpha=e,i,g} \chi_\alpha(\mathbf{k}, \omega) \tag{29}$$

by the presence of the last term in (28) that is the renormalized susceptibility of grains generated by the charging processes. Furthermore, in calculations of the plasma particle dielectric response, collisions with grains should be taken into account on an equal footing with the collisions with neutrals, i.e., $\nu_\alpha = \nu_{\alpha n} + \nu_{\alpha g}$.

4 Stationary Grain Charge, Charging and Collision Frequencies

Stationary grain charge is determined by the condition of zero value of the total current on the grain surface (8). Using (11) and (12) one obtains equation for α given by

$$\frac{n_e}{n_i}\mu \exp(-\alpha) = \frac{I^{\mathrm{WC}} I^{\mathrm{SC}}}{I^{\mathrm{WC}} + I^{\mathrm{SC}}}, \tag{30}$$

where $\mu = v_{Te}/v_{Ti} = \sqrt{\tau m_i/m_e}$.

The ratio n_e/n_i in Eq. (30) is found from the condition of quasineutrality that in the case of dusty plasmas is given by

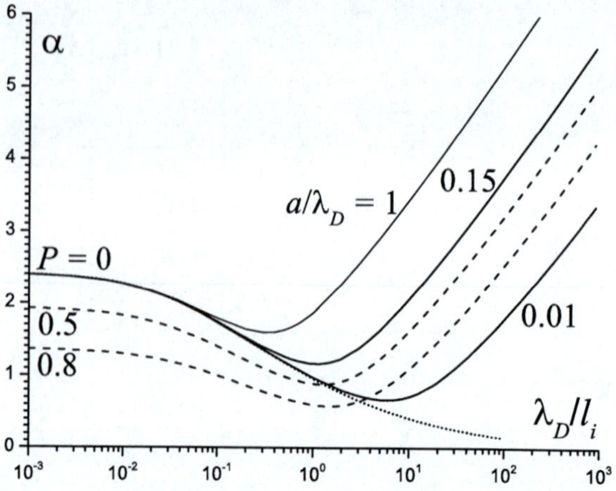

Fig. 1 Dependence of $\alpha = e_e\phi_s/T_e$ on the inverse ion mean free path in a non-isothermal ($\tau = 100$) argon plasma. Solid lines: $a/\lambda_D = 1, 0.15, 0.01$, $P = 0$; dashed lines: $a/\lambda_D = 0.15$, $P = 0.5, 0.8$; the dotted line is the solution of Eq. (34)

$$e_e n_e + e_i n_i + e_g n_g = 0. \tag{31}$$

For singly charged ions ($e_e = -e_i$)

$$\frac{n_e}{n_i} = 1 - P, \quad P = \frac{e_g n_g}{e_e n_i}, \tag{32}$$

where P is the Havnes parameter that describes the portion of the electron charge collected by the dust.

In the weakly collisional regime, the ion current is simplified to

$$I^i_{\rm ch} = e_i n_i \sqrt{8\pi} a^2 v_{Ti} I^{\rm WC}, \tag{33}$$

that yields an equation for α given by

$$\frac{n_e}{n_i}\mu \exp(-\alpha) = I^{\rm WC}. \tag{34}$$

It does not depend on a/λ_D in contrast to Eq. (30).

The dependence of the stationary value of α on the inverse ion mean free path in a non-isothermal ($\tau = 100$) argon plasma is presented in Fig. 1. For a large mean free path l_i and $P = 0$, α tends to the value ≈ 2.4 that reproduces the value given by the OML approximation. An increase in the mean free path leads, at first, to a decrease in α. The dotted line in Fig. 1 is the solution of Eq. (34). It shows that the applicability of formula (33) for the ion current depends on the ion mean free path and on the grain size. A further increase in ion collisionality leads to the increase in α, where, at some point, it starts to depend on the grain size a/λ_D. An increase in the Havnes parameter P causes a decrease in α (dashed lines in Fig. 1). Since $\lambda_{De} = \lambda_D\sqrt{1 + \tau/(1 - P)}$, then for $\tau = 100$ and $P = 0$ $\lambda_{De} \approx 10\lambda_D$. The curve for

$a/\lambda_D = 0.15$ ($\lambda_{De}/a \approx 66.7$) matches the corresponding curve in Ref. [54]. When α is known, we can find the stationary charging currents from (11) and (12).

Let us find the charging frequencies (10) by taking the derivatives of currents (11) and (12)

$$\nu_{\rm ch}^e = I_{\rm ch}^e \frac{\alpha}{e_g}, \tag{35}$$

$$\nu_{\rm ch}^i = -I_{\rm ch}^i \frac{(2I^{\rm WC} - 2 - \alpha\tau)I^{\rm SC} + (I^{\rm WC})^2}{e_g(I^{\rm WC} + I^{\rm SC})I^{\rm WC}}. \tag{36}$$

Charging frequencies in the normalized form are given by

$$\frac{\nu_{\rm ch}^e}{\omega_{pi}} = \frac{\mu}{\sqrt{2\pi\tau}} \frac{k_{Di}}{k_D} \frac{(1-P)\exp(-\alpha)}{1+\lambda_D/a}. \tag{37}$$

$$\frac{\nu_{\rm ch}^i}{\omega_{pi}} = -\frac{1}{\sqrt{2\pi\tau}} \frac{k_{Di}}{k_D} \frac{I^{SC}}{1+\lambda_D/a} \frac{(2I^{\rm WC} - 2 - \alpha\tau)I^{\rm SC} + (I^{\rm WC})^2}{\alpha(I^{\rm WC} + I^{\rm SC})^2}, \tag{38}$$

where $k_{Di}/k_D = \sqrt{\tau}/\sqrt{1-P+\tau}$ for $\tau = 100$, $k_{Di}/k_D \approx 1$.

Figure 2 shows that charging frequencies are smaller than the ionic plasma frequency $\nu_{\rm ch}^\alpha \lesssim \omega_{pi}$ for the parameters under consideration and depend nonmonotonically on the inverse ion mean free path. The behavior of the electron charging frequency is explained using formula (37). A decrease in α leads to an increase in the charging frequency, that is why the frequency maximum coincides approximately with the minimum of α (see Fig. 1). A decrease in a/λ_D causes a decrease in α, but since λ_D/a is contained in the denominator of (37), a decrease in the grain size leads to a decrease in charging frequency. Concerning the influence of the Havnes parameter on the charging frequency, the dependence of k_{Di}/k_D on P may be disregarded in the strongly non-isothermal plasma and the dependence of α on P is manifested in

Fig. 2 Normalized electron (dashed lines) and ion (solid lines) charging frequencies $\nu_{\rm ch}^\alpha/\omega_{pi}$ versus the inverse ion mean free path in a non-isothermal ($\tau = 100$) argon plasma for $a/\lambda_D = 1, 0.15, 0.01$

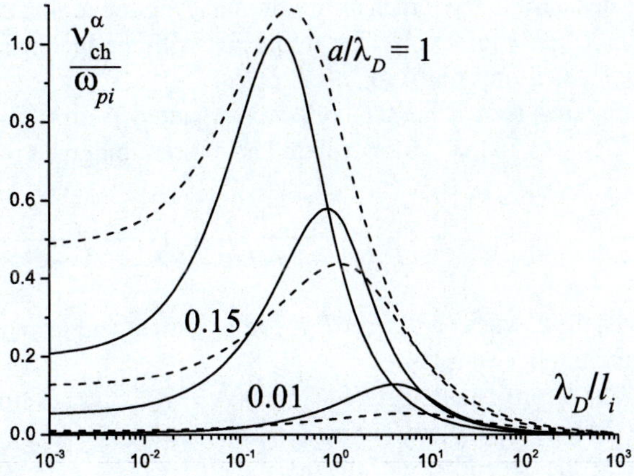

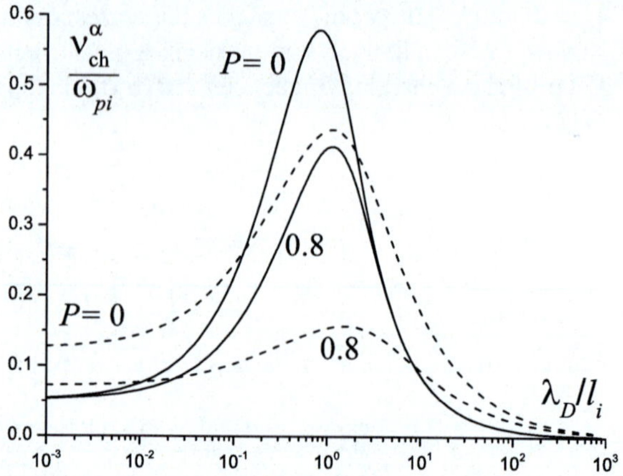

Fig. 3 Normalized electron (dashed lines) and ion (solid lines) charging frequencies $\nu^{\alpha}_{\mathrm{ch}}/\omega_{pi}$ versus the inverse ion mean free path in a non-isothermal ($\tau = 100$) argon plasma for $a/\lambda_D = 0.15$ and $P = 0, 0.8$

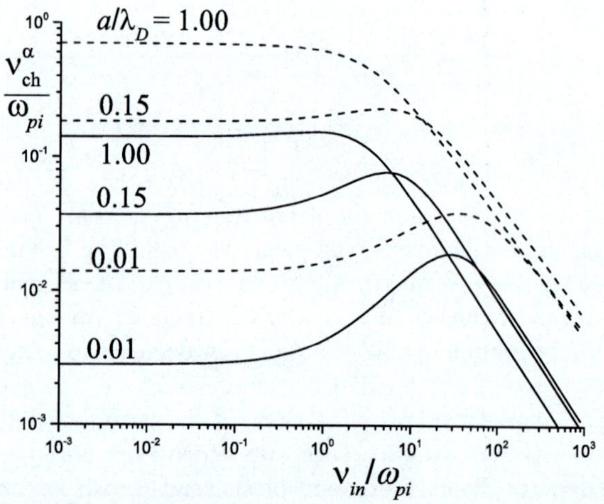

Fig. 4 Electron (dashed lines) and ion (solid lines) charging frequencies $\nu^{\alpha}_{\mathrm{ch}}/\omega_{pi}$ versus the ion-neutral collision frequency in the isothermal ($\tau = 1$) argon plasma for $a/\lambda_D = 1, 0.15, 0.01$

the decrease in the ion charging frequency in the region of its maximum (see Fig. 3). The decrease in the electron frequency with the increase in P is more pronounced, since it is proportional to $1 - P$ (37).

Figure 4 shows that the collision frequency is also lower than ω_{pi}.

Using (32) and (16), we obtain an expression given by

$$\frac{n_g}{k_D^3} = \frac{P}{4\pi\alpha k_D a(1 + k_D a)(1 - P + \tau)}. \tag{39}$$

Thus, the Havnes parameter P together with other plasma parameters determines the grain number density.

The mean distance between grains $\Delta = n_g^{-1/3}$ varies from $\approx$ 3 to 20 Debye lengths for the parameters under consideration (see Fig. 5). According to (39), $\Delta \sim \alpha^{1/3}$ that is why the dependence of the mean distance between grains on the inverse ion mean

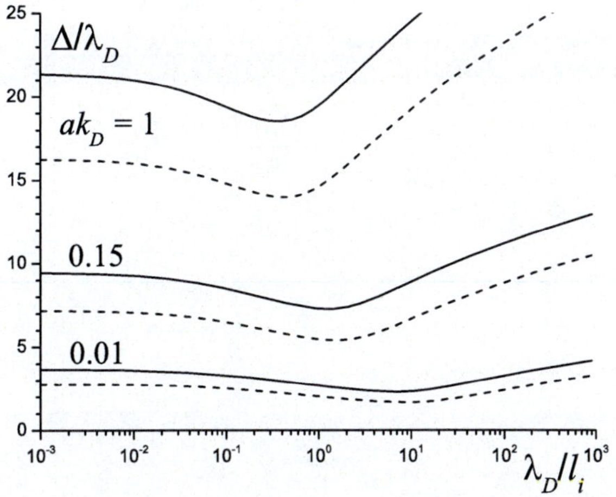

Fig. 5 Mean distance between grains versus the inverse ion mean free path in a non-isothermal ($\tau = 100$) argon plasma for $a/\lambda_D = 1, 0.15, 0.01$, $P = 0.5$ (solid lines) and $P = 0.8$ (dashed lines)

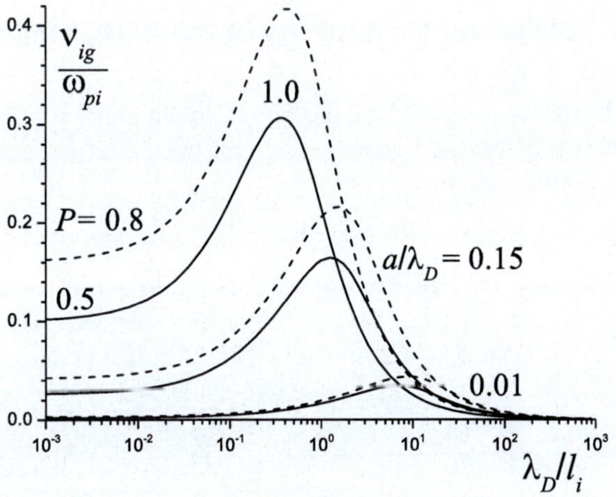

Fig. 6 Normalized ion collision frequency ν_{ig}/ω_{pi} versus the inverse ion mean free path in a non-isothermal ($\tau = 100$) argon plasma for $P = 0.5$ (solid lines) and $P = 0.8$ (dashed lines), $a/\lambda_D = 1, 0.15, 0.01$

free path has the minimum that corresponds to the minimum of the dimensionless surface potential α (see Fig. 1).

Now, let us calculate the frequency of collisions between plasma particles and grains using formula (22). The simple relation $\nu_{ig} = (1 - P)\nu_{eg}$ holds in the stationary state, that is why only ion frequencies are shown in Figs. 6 and 7. Since the charging currents grow with the growth of the grain size (11) and (12), the charging frequency has similar behavior. As follows from (39), $n_g \sim P$ in a non-isothermal plasma, therefore, dashed lines ($P = 0.8$) are higher than the solid lines ($P = 0.5$) in Figs. 6 and 7.

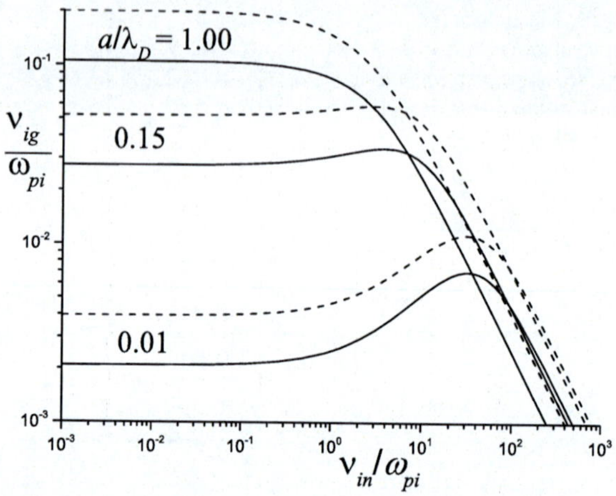

Fig. 7 Ion collision frequency ν_{ig}/ω_{pi} versus the ion-neutral collision frequency in an isothermal ($\tau = 1$) argon plasma for $P = 0.5$ (solid lines) and $P = 0.8$ (dashed lines), $a/\lambda_D = 1, 0.15, 0.01$

5 Dielectric Permittivity in the BGK Model

The dispersion and damping rate of longitudinal waves in a dusty plasma may be studied in terms of the expression for the dielectric permittivity (28) by solving the dispersion equation

$$\varepsilon(\mathbf{k}, \omega) = 0. \tag{40}$$

Obviously, explicit expressions for the dielectric susceptibility in the collisional plasma are required. We use the results obtained [65] from the kinetic equations with the Bhatnagar-Gross-Krook (BGK) collision integral [60], i.e.,

$$\chi_\alpha(\mathbf{k}, \omega) = \frac{k_{D\alpha}^2}{k^2} \frac{(\omega + i\nu_\alpha) W(z_\alpha)}{\omega + i\nu_\alpha W(z_\alpha)}, \tag{41}$$

where $z_\alpha = (\omega + i\nu_\alpha)/kv_{T\alpha}$ and $W(z)$ is the plasma dispersion function [2]

$$W(z) = 1 - ze^{-z^2/2} \int_0^z dy e^{y^2/2} + i\sqrt{\frac{\pi}{2}} ze^{-z^2/2}. \tag{42}$$

In the normalized form we have

$$\chi_i(\mathbf{k}, \omega) = \frac{1}{\tilde{k}^2} \frac{(\tilde{\omega} + i\tilde{\nu}_i) W(z_i)}{\tilde{\omega} + i\tilde{\nu}_i W(z_i)} \tag{43}$$

and

$$\chi_e(\mathbf{k}, \omega) = \frac{1 - P}{\tau \tilde{k}^2} \frac{(\tilde{\omega} + i\tilde{\nu}_e) W(z_e)}{\tilde{\omega} + i\tilde{\nu}_e W(z_e)}, \tag{44}$$

where

$$\tilde{k} = \frac{k}{k_{Di}}, \quad \tilde{\omega} = \frac{\omega}{\omega_{pi}}, \quad \tilde{\nu}_\alpha = \frac{\nu_\alpha}{\omega_{p\alpha}}, \quad z_i = \frac{\tilde{\omega} + i\tilde{\nu}_i}{\tilde{k}}, \quad z_e = \frac{\tilde{\omega} + i\tilde{\nu}_e}{\mu\tilde{k}}. \tag{45}$$

Thus, the presence of dust in the plasma leads not only to the appearance of new terms in the expression for the dielectric permittivity (the last term in formula (28) and $\chi_g(\mathbf{k}, \omega)$ in the second term) but also to the change of $\chi_e(\mathbf{k}, \omega)$ associated with the electron density decrease with respect to the ion density ($\chi_e(\mathbf{k}, \omega)$ is proportional to $1 - P$).

6 Ion-Acoustic Waves

Since the charging frequency $\nu_{\mathrm{ch}}^{\alpha}$ is of the order of magnitude of the ion plasma frequency (see Fig. 2), the grain charge fluctuations can influence the propagation of ion-acoustic waves in dusty plasmas. The study of such an influence is given in what follows.

It is known [65] that in plasma without dust, ion-acoustic waves exist only in strongly non-isothermal case $\tau \gg 1$ and their phase velocity belongs to the range $v_{Ti} \ll \omega/k \ll v_{Te}$. The solution of dispersion equation is given by [58]

$$\omega(\mathbf{k}) = \sqrt{\frac{\omega_{pi}^2}{1 + k_{De}^2/k^2} - \frac{\nu_i^2}{4}} - i\frac{\nu_i}{2}. \tag{46}$$

For $\nu_i = 0$ and $k \ll k_{De}$ Eq. (46) turns into the well-known spectrum for ion-acoustic waves in collisionless plasma

$$\omega_{\mathbf{k}} = \frac{\omega_{pi}}{k_{De}} k = \sqrt{\tau \frac{n_i}{n_e}} v_{Ti} k. \tag{47}$$

The damping of ion-acoustic waves is caused by the resonance interaction of the wave field with ions [65] (the Landau damping) as well as by the collisions of ions, whose contribution to the damping rate is approximately equal to $\Delta\gamma_{\mathbf{k}} = -\nu_i/2$ [which coincides with the imaginary part of the formula (46)]. This estimate is consistent with the results of the numerical solution of the dispersion equation [44, 58].

Now, let us discuss the electron-neutral collision frequency ν_{en}, which is present in formula (44). The simple relation $\nu_{\alpha n} = v_{T\alpha}\sigma_\alpha n_n$, where σ_α is the scattering cross section, is valid for a weakly ionized plasma. Thus, the electron collision frequency is related to ion frequency through the following expression $\nu_e = \nu_i \mu (\sigma_e/\sigma_i)$. The scattering cross section of Ar^+ ions with an energy of 0.1 eV on Ar atoms is about 157 Å^2, and it decreases with the increase in energy (Ref. [66], table 7). The scattering cross section of electrons on argon atoms has the minimum of $\approx 0.1\text{Å}^2$ for the energy of ≈ 0.2 eV, and for ≈ 2.5 eV it is $\approx 3\,\text{Å}^2$ (Ref. [67], Fig. 4). Therefore, we can assume that $\sigma_e/\sigma_i \approx 0.02$ for a non-isothermal plasma, and this value was used in our computations. Since electron collisions almost do not affect the dispersion and damping of ion-acoustic waves, the estimate for the ratio of scattering cross sections can be used without compromising the accuracy of the computations, or the value of ν_{en} can even be taken as zero.

As it was mentioned above, collisions of plasma particles with grains should be included into the effective collision frequencies along with collisions with neutrals, i.e., $\nu_\alpha = \nu_{\alpha n} + \nu_{\alpha g}$. The inverse ion mean free path in a strongly non-isothermal plasma is given by

$$\frac{\lambda_D}{l_i} = \frac{k_{Di}}{k_D}\tilde{\nu}_{in} \approx \frac{\nu_{in}}{\omega_{pi}} = \tilde{\nu}_{in}, \tag{48}$$

and Fig. 6 shows how $\tilde{\nu}_{ig}$ is commensurate with $\tilde{\nu}_{in}$. The main contribution to the effective frequencies is given by the collisions with grains for $\tilde{\nu}_{in} \ll 1$. In the opposite case $\tilde{\nu}_{in} \gg 1$, the main contribution is obviously given by the collisions with neutrals.

Both spectra and damping of ion-acoustic waves in dusty plasmas with regard to the grain charge fluctuations (solid lines in Fig. 8a, b) are studied using the numerical solution of the dispersion equation (40) with respect to the complex frequency $\omega(\mathbf{k}) = \omega_\mathbf{k} + i\gamma_\mathbf{k}$, where the dielectric permittivity is given by expression (28). Since the plasma frequency of grains is much smaller than the ion plasma frequency, the motion of grains does not affect the propagation of ion-acoustic waves. Hence, $\chi_g(\mathbf{k}, \omega)$ is disregarded in (28).

The possibility of the existence of ion-acoustic waves in isothermal collisionless dusty plasmas was discussed in Ref. [27]. It follows from (47) that the phase velocity of ion-acoustic waves is equal to $\omega/k = v_{Ti}\sqrt{\tau n_i/n_e}$ for $k \ll k_{De}$. This means that in ordinary isothermal plasma phase the velocity is equal to ion thermal velocity ($\omega/k = v_{Ti}$), and there occurs strong Landau damping on ions. It decreases in isothermal dusty plasmas under the condition $n_e \ll n_i$. In the case of a non-isothermal ($\tau = 100$) collisional plasma that is considered in the present contribution the phase velocity (the eigenfrequency) also increases. The curve corresponding to $P = 0.8$ is higher than the curve for $P = 0$ in Fig. 8a. This figure shows that the dispersion in dusty plasmas is also well-described by formula (46) (the dashed line) that is obtained for ordinary plasmas but with regard to the change in n_e/n_i. Thus, the grain charge fluctuations and the increase in ν_{in} due to the collisions with grains do not affect the dispersion of ion-acoustic waves for the parameters under consideration.

Since the phase velocity is much higher than the thermal velocity for $\tau = 100$ and the Landau damping is small for $k \ll k_{De}$, the additional increase in the phase velocity due to the decrease in n_e/n_i makes no effect on the wave damping. In contrast, the grain charge fluctuations and the increase in ν_{in} due to the collisions with grains produce a considerable increase in the absolute value of the damping rate (per order for $\tilde{k} \approx 0.02$) and a decrease in the $\omega_\mathbf{k}/|\gamma_\mathbf{k}|$ ratio accordantly (see the insert in Fig. 8a). The dashed line in the insert is given by formula (46).

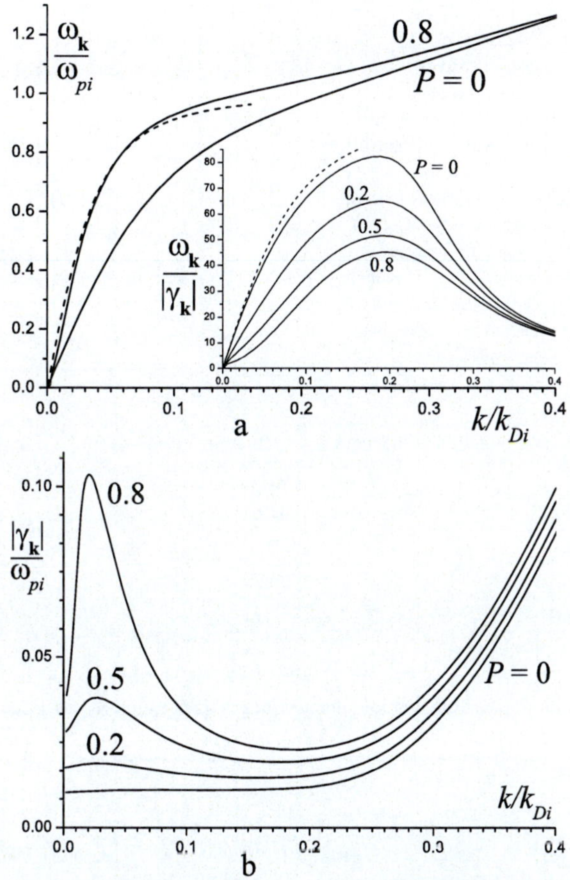

Fig. 8 Eigenfrequencies $\omega_{\mathbf{k}}$ (**a**) and absolute values of damping rates $|\gamma_{\mathbf{k}}|$ (**b**) of ion-acoustic waves in a non-isothermal ($\tau = 100$) argon plasma as the result of the numerical solution of the dispersion equation $\varepsilon(\mathbf{k}, \omega_{\mathbf{k}} + i\gamma_{\mathbf{k}}) = 0$ with the dielectric permittivity (28), (41) versus the wave number for $\nu_{in} = 0.02\omega_{pi}$, $ak_D = 0.15$, $P = 0, 0.2, 0.5, 0.8$ (solid lines). The dashed lines are given by formula (46), and the insert shows the ratio $\omega_{\mathbf{k}}/|\gamma_{\mathbf{k}}|$

Figure 9 shows that the increase in the wave damping due to collisions between ions and grains is approximately described by the formula $\gamma_{\mathbf{k}} = -(\nu_{in} + \nu_{ig})/2$ (dashed lines) for small P. The values of $\tilde{\nu}_{ig}$ and $(\tilde{\nu}_{in} + \tilde{\nu}_{ig})/2$ are presented in Table 1. The same values that correspond to Fig. 8 are presented in Table 2.

Let us consider ion-acoustic waves in a dusty plasma with a different ratio of electron to ion temperatures, namely $\tau = 10$. In this case, the influence of grains on the wave dispersion and the damping rate (see Fig. 10) is similar to the previous case $\tau = 100$ (Fig. 8). The increase in eigenfrequency and damping rate absolute value is observed in the domain of $\tilde{k} \lesssim 0.2$. But there is a considerable difference between these cases, namely the damping rate in the dusty plasma is lower than in the ordinary plasma for $\tilde{k} \gtrsim 0.2$. As it was mentioned above, damping due to the resonant interaction of waves with ions is significant for $\tilde{k} \gtrsim 0.2$. The increase in the phase velocity in a dusty plasma leads to the decrease in the resonant damping. The insert in Fig. 10a shows that the maximum of the $\omega_{\mathbf{k}}/\gamma_{\mathbf{k}}$ ratio shifts toward higher values of k. The value of the maximum itself slightly grows. It differs from the case

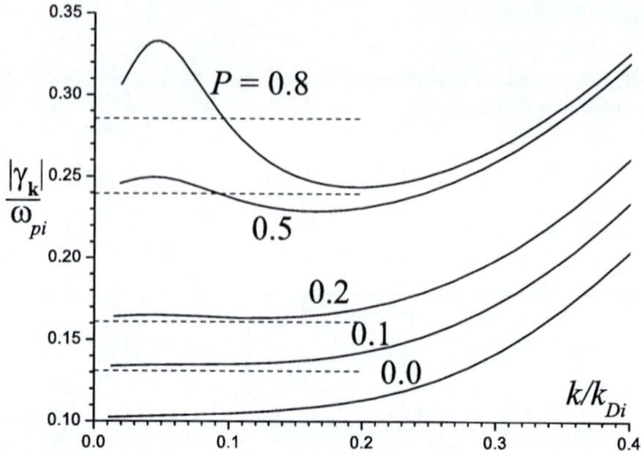

Fig. 9 Absolute values of damping rates $|\gamma_{\mathbf{k}}|$ of ion-acoustic waves in a non-isothermal ($\tau = 100$) argon plasma as the result of the numerical solution of the dispersion equation $\varepsilon(\mathbf{k}, \omega_{\mathbf{k}} + i\gamma_{\mathbf{k}}) = 0$ with the dielectric permittivity (28), (41) versus the wave number for $\nu_{in} = 0.2\omega_{pi}$, $ak_D = 1$, $P = 0, 0.1, 0.2, 0.5, 0.8$. The dashed lines are given by $(\tilde{\nu}_{in} + \tilde{\nu}_{ig})/2$

Table 1 Ion collision frequency with grains of radius $a = \lambda_D$ for $\tilde{\nu}_{in} = \nu_{in}/\omega_{pi} = 0.2$

P	$\tilde{\nu}_{ig}$	$(\tilde{\nu}_{in} + \tilde{\nu}_{ig})/2$
0.1	0.062	0.131
0.2	0.122	0.161
0.5	0.279	0.240
0.8	0.371	0.286

Table 2 Ion collision frequency with grains of radius $a = 0.15\lambda_D$ for $\tilde{\nu}_{in} = \nu_{in}/\omega_{pi} = 0.02$

P	$\tilde{\nu}_{ig}$	$(\tilde{\nu}_{in} + \tilde{\nu}_{ig})/2$
0.2	0.0146	0.0173
0.5	0.0351	0.0276
0.8	0.0521	0.0361

$\tau = 100$ (the insert in Fig. 8a) where the presence of grains produced a considerable decrease in $\omega_{\mathbf{k}}/\gamma_{\mathbf{k}}$ ratio, though it remains higher than that for $\tau = 10$.

Finally, we consider ion-acoustic waves in an isothermal ($\tau = 1$) dusty plasma. We see from Fig. 11a that the eigenfrequency of ion-acoustic waves in an isothermal plasma increases with P similarly to the non-isothermal case. Furthermore, the region of small k where the waves do not exist grows wider. The changes in the behavior of damping rates that are observed upon the decrease in τ from 100 to 10 become more pronounced for $\tau = 1$ (see. Fig. 11b), in particular, this concerns the decrease in the Landau damping. The insert in Fig. 11a shows that the maximum of the ratio $\omega_{\mathbf{k}}/|\gamma_{\mathbf{k}}|$ shifts to the higher values of k, and the value of the maximum itself considerably increases though remains lower than that for $\tau = 10$ and 100.

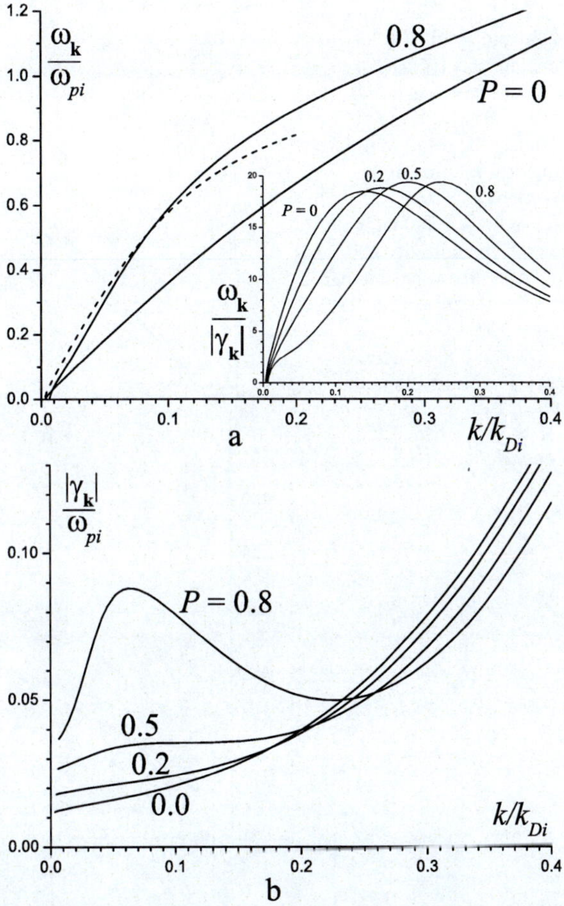

Fig. 10 Eigenfrequencies $\omega_{\mathbf{k}}$ (**a**) and absolute values of damping rates $|\gamma_{\mathbf{k}}|$ (**b**) of ion-acoustic waves in a non-isothermal ($\tau = 10$) argon plasma as the result of the numerical solution of the dispersion equation $\varepsilon(\mathbf{k}, \omega_{\mathbf{k}} + i\gamma_{\mathbf{k}}) = 0$ with the dielectric permittivity (28), (41) versus the wave number for $\nu_{in} = 0.02\omega_{pi}$, $ak_D = 0.15$, $P = 0, 0.2, 0.5, 0.8$ (solid lines). The dashed line is given by formula (46), and the insert shows the ratio $\omega_{\mathbf{k}}/|\gamma_{\mathbf{k}}|$

7 Fluctuation Spectra

In this section we calculate the electron density correlation function $\langle \delta\rho_e^2 \rangle_{\mathbf{k}\omega}$. We start from the electron density fluctuations and it follows from Eqs. (20) and (26) that

$$\delta\rho_{e\mathbf{k}\omega} = \delta\rho^{(0)}_{e\mathbf{k}\omega} - \frac{\chi_e(\mathbf{k}, \omega)}{\varepsilon(\mathbf{k}, \omega)} \delta\rho^{(0)}_{\mathbf{k}\omega}, \tag{49}$$

using (27) we get

$$\delta\rho_{e\mathbf{k}\omega} = \delta\rho^{(0)}_{e\mathbf{k}\omega} - \frac{\chi_e(\mathbf{k}, \omega)}{\varepsilon(\mathbf{k}, \omega)} \left[\sum_{\alpha=e,i} \left(1 + \frac{i\nu_{\alpha g}}{\omega + i\nu_{\text{ch}}} \right) \delta\rho^{(0)}_{\alpha\mathbf{k}\omega} + \delta\rho^{(0)}_{g\mathbf{k}\omega} \right]. \tag{50}$$

Finally,

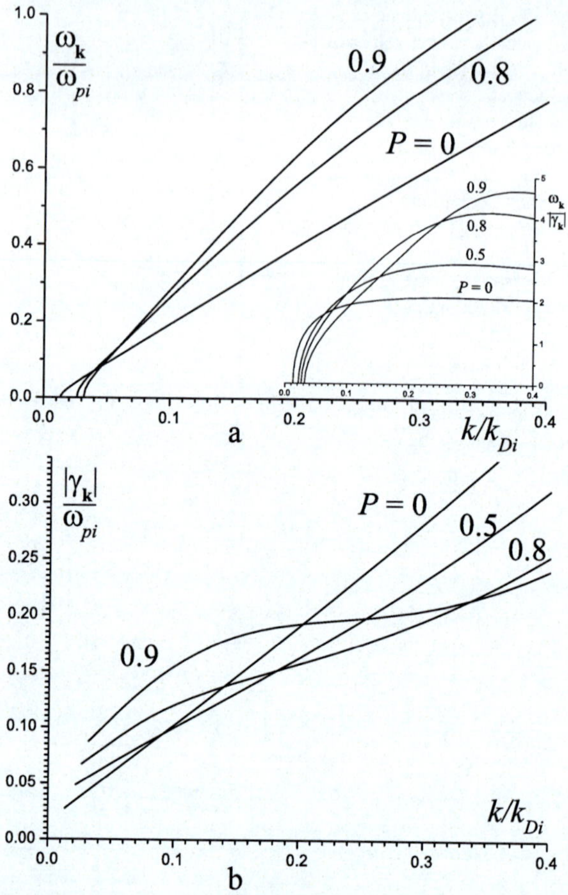

Fig. 11 Eigenfrequencies $\omega_{\mathbf{k}}$ (**a**) and absolute values of damping rates $|\gamma_{\mathbf{k}}|$ (**b**) of ion-acoustic waves in isothermal ($\tau = 1$) argon plasma as the result of the numerical solution of the dispersion equation $\varepsilon(\mathbf{k}, \omega_{\mathbf{k}} + i\gamma_{\mathbf{k}}) = 0$ with the dielectric permittivity (28), (41) versus the wave number for $\nu_{in} = 0.02\omega_{pi}$, $ak_D = 0.15$, $P = 0, 0.5, 0.8, 0.9$ (solid lines). The insert shows the ratio $\omega_{\mathbf{k}}/|\gamma_{\mathbf{k}}|$

$$\begin{aligned}\langle\delta\rho_e^2\rangle_{\mathbf{k}\omega} &= \left|1 - \frac{\chi_e(\mathbf{k},\omega)}{\varepsilon(\mathbf{k},\omega)}\left(1 + \frac{i\nu_{eg}}{\omega + i\nu_{\text{ch}}}\right)\right|^2 \langle\delta\rho_e^{(0)2}\rangle_{\mathbf{k}\omega} \\ &+ \left|\frac{\chi_e(\mathbf{k},\omega)}{\varepsilon(\mathbf{k},\omega)}\left(1 + \frac{i\nu_{ig}}{\omega + i\nu_{\text{ch}}}\right)\right|^2 \langle\delta\rho_i^{(0)2}\rangle_{\mathbf{k}\omega} \\ &+ \left|\frac{\chi_e(\mathbf{k},\omega)}{\varepsilon(\mathbf{k},\omega)}\right|^2 \langle\delta\rho_g^{(0)2}\rangle_{\mathbf{k}\omega} \qquad (51)\end{aligned}$$

the electron density correlation function is expressed in terms of $\chi_\alpha(\mathbf{k}, \omega)$ and $\langle\delta\rho_\alpha^{(0)2}\rangle_{\mathbf{k}\omega}$

$$\langle\delta\rho_\alpha^{(0)2}\rangle_{\mathbf{k}\omega} = e_\alpha^2 n_\alpha \int d\mathbf{v} \int d\mathbf{v}' W_{\alpha\mathbf{k}\omega}(\mathbf{v}, \mathbf{v}') f_{0\alpha}(\mathbf{v}') + \text{c.c.} \qquad (52)$$

in the equilibrium state [2, 4, 10]

$$\langle \delta\rho_\alpha^{(0)2} \rangle_{\mathbf{k}\omega} = \frac{T_\alpha k^2}{2\pi\omega} \mathrm{Im}\chi_\alpha(\mathbf{k}, \omega). \tag{53}$$

Disregarding the grain charge variations, i.e., putting the frequencies ν_{ch} and $\nu_{\alpha g}$ equal to zero in (51), we obtain

$$\langle \delta\rho_e^2 \rangle_{\mathbf{k}\omega} = \left| 1 - \frac{\chi_e(\mathbf{k}, \omega)}{\varepsilon_{\mathrm{mc}}(\mathbf{k}, \omega)} \right|^2 \langle \delta\rho_e^{(0)2} \rangle_{\mathbf{k}\omega} + \left| \frac{\chi_e(\mathbf{k}, \omega)}{\varepsilon_{\mathrm{mc}}(\mathbf{k}, \omega)} \right|^2 \langle \delta\rho_i^{(0)2} \rangle_{\mathbf{k}\omega} + \left| \frac{\chi_e(\mathbf{k}, \omega)}{\varepsilon_{\mathrm{mc}}(\mathbf{k}, \omega)} \right|^2 \langle \delta\rho_g^{(0)2} \rangle_{\mathbf{k}\omega} \tag{54}$$

the electron density correlation function in the multi-component model, where $\varepsilon_{\mathrm{mc}}(\mathbf{k}, \omega)$ is defined by (29).

The ion density correlation function is found in a similar manner.

We performed the calculations for argon dusty plasmas with various values of the grain size ak_D, Havnes parameter P, and ion-neutral collision frequency ν_{in} for both isothermal $\tau = 1$ and non-isothermal $\tau > 1$ plasmas. Since the plasma frequency of grains is much lower than the ion plasma frequency, the motion of grains makes no effect on the fluctuations in this frequency domain and the last terms in (51) and (54) are disregarded as well as $\chi_g(\mathbf{k}, \omega)$ in (28) and (29).

We can highlight the main factors included in our description that influence the fluctuation spectra of the electron density in dusty plasmas, i.e.,

- The grain charge variations that are described by the last term in (28) and corresponding terms in (51) that contain ν_{ch} and $\nu_{\alpha g}$.
- The plasma particle collisions with neutrals and grains that define the plasma particle effective collision frequencies $\nu_\alpha = \nu_{\alpha n} + \nu_{\alpha g}$ that influence the susceptibility of electrons and ions (41).
- The decrease in the electron to ion density ratio in dusty plasmas that influences the stationary grain charge α [see Eq. (30)] and the electron susceptibility (44).

In order to clarify the individual input of the second factor we start from considering the electron density correlation function given by (54) and $\nu_\alpha = \nu_{\alpha n}$ in (41) i.e., we start with studying the influence of ion-neutral collisions on the electron density fluctuations in ordinary plasma. Figure 12 shows that the intensity of fluctuations in a non-isothermal ($\tau = 100$) plasma is sensitive to ion collisions with neutrals. The fluctuations are considerably suppressed even for small values of ν_{in}/ω_{pi}.

Concerning the last factor, it was shown in Sect. 5 that the decrease in the n_e/n_i results in the increase in dust ion-acoustic wave eigenfrequency (see Fig. 8) and therefore the fluctuation maxima are also shifted toward higher frequencies. This effect is illustrated by Fig. 13 where the fluctuation spectra are obtained using the expressions for ordinary collisional plasma but with $n_e/n_i = 1 - P$. Besides the fluctuation maximum shift, its decrease is also observed. It can be explained by the decrease in $\langle \delta\rho_e^{(0)2} \rangle_{\mathbf{k}\omega}$ via the decrease in the electron susceptibility (44), since it is proportional to $1 - P$.

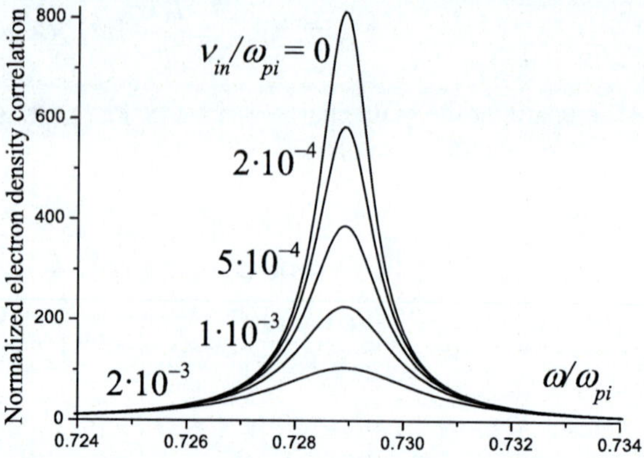

Fig. 12 Normalized electron density correlation spectra $\langle \delta n_\alpha^2 \rangle_{\mathbf{k}\omega} \omega_{pi}/n_i$ in ordinary non-isothermal ($\tau = 100$) argon plasma for $\nu_{in}/\omega_{pi} = 0, 2 \cdot 10^{-4}, 5 \cdot 10^{-4}, 1 \cdot 10^{-3}, 2 \cdot 10^{-3}$, $k/k_{Di} = 0.1$

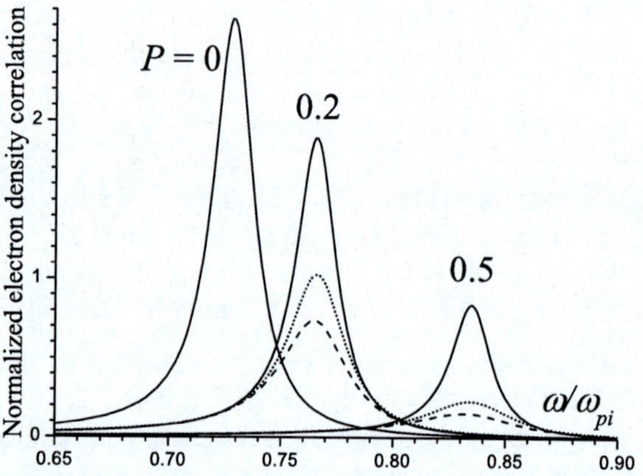

Fig. 13 Normalized electron density correlation spectra $\langle \delta n_\alpha^2 \rangle_{\mathbf{k}\omega} \omega_{pi}/n_i$ in ordinary non-isothermal ($\tau = 100$) argon plasma for $\nu_{in}/\omega_{pi} = 0.02$, $k/k_{Di} = 0.1$, $P = 0, 0.2, 0.5$. The dashed line corresponds to (54) and plasma particle susceptibility (41) including collisions with grains $\nu_\alpha = \nu_{\alpha n} + \nu_{\alpha g}$, $a/\lambda_D = 0.15$. Dotted line corresponds to (51)

In order to see the isolated effect of the dust charge variability (first factor), we, initially, "turn on" both second and third factors. Namely, the collisions of plasma particles with grains are included into the effective collision frequency $\nu_\alpha = \nu_{\alpha n} + \nu_{\alpha g}$ in (41), but for the electron density correlation function formula (54) is still used (the dashed line in Fig. 13). For $a/\lambda_D = 0.15$ and $P = 0.2, 0.5$ the ion-grain collision frequencies equal to $\nu_{ig}/\omega_{pi} \approx 0.015$ and 0.035 respectively, thus they are of the order of the ion-neutral collision frequency $\nu_{in}/\omega_{pi} = 0.02$. As should be expected, the increase in the ion collisionality leads to the decrease in fluctuation intensity. Finally, we include into consideration the grain charge variation (the dotted line in Fig. 13), which means that we calculate the electron density correlation function using formula (51) with the dielectric permittivity (28). One may conclude that variations of the grain charge lead to the enhancement of the electron density fluctuations as compared to the dashed line, but they are considerably suppressed as compared to the results of the multi-component description (the solid line).

The fluctuation spectra in a strongly non-isothermal ($\tau = 100$) plasma are presented in Fig. 14 show that both positions and intensities of the maxima depend on

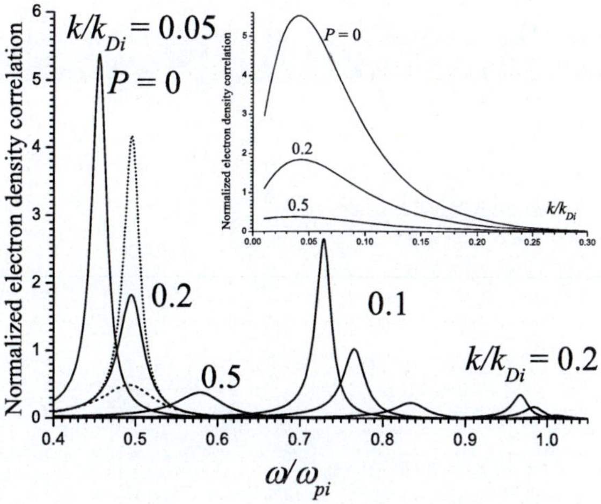

Fig. 14 Normalized electron density correlation spectra $\langle \delta n_{\alpha}^2 \rangle_{\mathbf{k}\omega} \omega_{pi}/n_i$ in a non-isothermal ($\tau = 100$) argon plasma for $\nu_{in} = 0.02\omega_{pi}$, $a/\lambda_D = 0.15$, $k/k_{Di} = 0.05, 0.1, 0.2$, $P = 0, 0.2, 0.5$, $a/\lambda_D = 0.01$ (doted line) and $a/\lambda_D = 1$ (dashed line). The insert shows the maximum value of the normalized electron correlation function versus k/k_{Di}

the wave number k/k_{Di} and coincide with the eigenfrequency of ion-acoustic waves in a collisional dusty plasma (see Fig. 8). The presence of grains leads to the shift of fluctuation maxima toward higher frequencies and to the decrease in fluctuation intensity. It was already mentioned that the increase in the eigenfrequency of the ion-acoustic wave is caused by the decrease in the electron to ion density ratio n_e/n_i. This assumption is confirmed by the curves in Fig. 14 corresponding to $k/k_{Di} = 0.05$, $P = 0.2$ and different values of the grain size $a/\lambda_D = 0.01, 0.15, 1$. The values of collision ν_{ig} and charging ν_{ch}^{α} frequencies for $a/\lambda_D = 0.01$ are much lower than for $a/\lambda_D = 1$ (see Figs. 4 and 7). As a consequence the fluctuations are less suppressed in the presence of small grains than of a big one, but the value of the shift depends almost entirely on the Havnes parameter.

The insert in Fig. 14 shows that the highest intensity of fluctuations occurs for $k/k_{Di} \approx 0.05$ and the increase in the Havnes parameter leads to a decrease in the fluctuation intensity in all wave number domains under consideration.

The transformation of the fluctuation spectra in an ordinary non-isothermal plasma ($\tau = 10$) with the increase in the ion collisionality is shown in Fig. 15. With growing ν_{in}/ω_{pi}, the fluctuations decrease, but then the fluctuation maxima grow at $\omega = 0$. The presence of dust grains also increases the ion effective collision frequency $\nu_i = \nu_{in} + \nu_{ig}$ due to collisions between ions and grains. For example, $\nu_{ig}/\omega_{pi} = 0.028$ for $P = 0.5$, $ak_D = 0.15$ and $\nu_{in}/\omega_{pi} = 0.02$, thus $\nu_i/\omega_{pi} = 0.048$ and fluctuations are also suppressed. But the maximum value of the electron density correlation in a dusty plasma is approximately equal to that in an ordinary plasma with $\nu_{in} = 0.1$. It means that the influence of dust on the fluctuation spectra should not be described only by the increase in the ion collisionality. Also, the maximum is shifted toward higher frequency. As we have already mentioned, the increase in the fluctuation frequency is provided by the decrease in n_e/n_i.

The fluctuation spectra in an isothermal plasma (see Fig. 16) differ from those in a non-isothermal one: the maxima are broader and located at lower frequencies. The

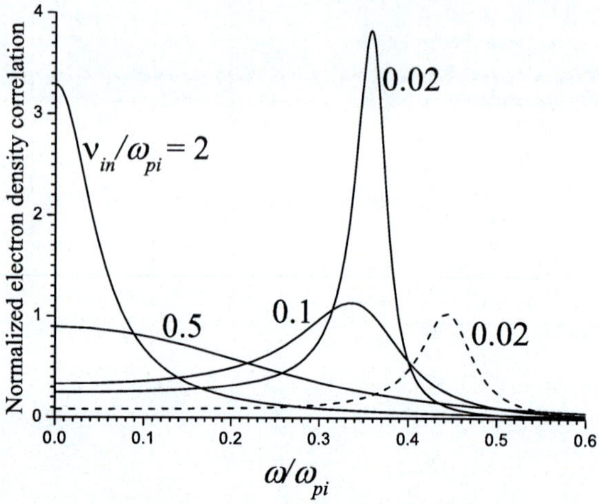

Fig. 15 Normalized electron density correlation spectra $\langle \delta n_\alpha^2 \rangle_{\mathbf{k}\omega} \omega_{pi}/n_i$ in a non-isothermal ($\tau = 10$) argon plasma for $k/k_{Di} = 0.1$, $\nu_{in}/\omega_{pi} = 0.02, 0.1, 0.5, 2$, $P = 0$ (solid lines) and $P = 0.5$, $a/\lambda_D = 0.15$ (dashed line)

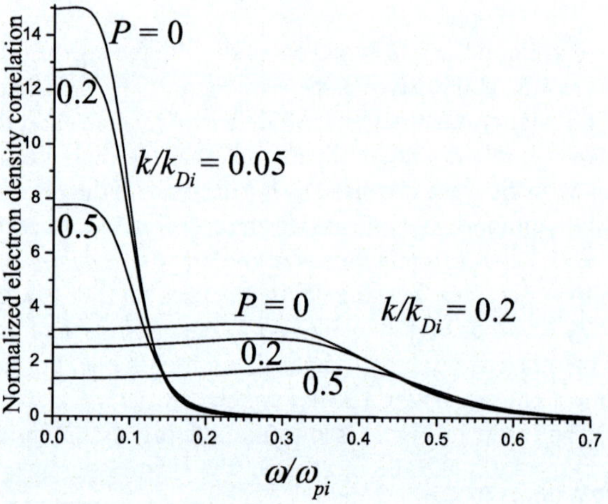

Fig. 16 Normalized electron density correlation spectra $\langle \delta n_\alpha^2 \rangle_{\mathbf{k}\omega} \omega_{pi}/n_i$ in an isothermal ($\tau = 1$) argon plasma for $\nu_{in} = 0.02\omega_{pi}$, $a/\lambda_D = 0.15$, $k/k_{Di} = 0.05, 0.1, 0.2$ and $P = 0, 0.2, 0.5$

presence of grains suppresses the fluctuations but not so efficiently as in the case of non-isothermal plasmas. Figure 17 illustrates the influence of the grain size on the fluctuations in an isothermal plasma. Since ν_{ig} and ν_{ch} for $a/\lambda_D = 0.01$ are much lower than for $a/\lambda_D = 1$ (see Figs. 4 and 7), the presence of grains of different sizes change the fluctuation spectra differently, even if the Havnes parameter is the same.

As it was mentioned above, we have omitted the last term in (51) and $\chi_g(\mathbf{k}, \omega)$ in (28) in our calculations. It means that we considered immovable grains. That is the reason why the maxima corresponding to the collective fluctuations of the grains (dust-acoustic resonances) are not observed in the region of the grain plasma frequency $\omega_{pg} \ll \omega_{pi}$ in Figs. 15, 16, 17.

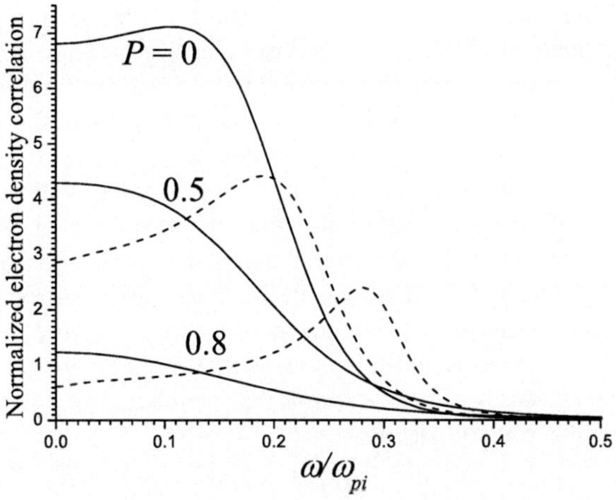

Fig. 17 Normalized electron density correlation spectra $\langle \delta n_{\alpha}^{2} \rangle_{\mathbf{k}\omega} \omega_{pi}/n_i$ in an isothermal ($\tau = 1$) argon plasma for $\nu_{in} = 0.02\omega_{pi}$, $k/k_{Di} = 0.1$, $P = 0, 0.5, 0.8$, $a/\lambda_D = 0.01$ (dashed lines) and $a/\lambda_D = 1$ (solid lines)

8 Conclusions

The expression for the dielectric permittivity of a collisional weakly ionized dusty plasma is obtained within the kinetic description. The account of grain charge fluctuations leads to the renormalization of the grain susceptibility. Both electron and ion susceptibilities are influenced by the increase in electron and ion collision frequencies due to their collisions with grains.

The charging and collision frequencies are analyzed in a wide range of ion collisionality. It was found that both frequencies are lower than or about the ion plasma frequency $\lesssim \omega_{pi}$ for the gas discharge plasma parameters.

The dispersion of ion-acoustic waves is mostly affected by the decrease in the electron to ion density ratio n_e/n_i in dusty plasmas, specifically the eigenfrequency and the phase velocity grow with the decrease in n_e/n_i. The growth of the phase velocity leads to the decrease in the Landau damping that is observed for $k/k_{Di} \gtrsim 0.2$ in isothermal and weakly non-isothermal plasma. The collisions between plasma particles and grains as well as charge fluctuations provide additional mechanisms for the wave energy dissipation and lead to a considerable growth in the absolute value of the damping rate.

Electron density correlation spectra are strongly affected by the presence of grains with variable charges. The main factors of this influence are the decrease in the electron to ion density ratio, the increase in ion collisionality due to collisions with grains, and grain charge variations.

In the case of non-isothermal plasmas the positions of the ion-acoustic resonances and their intensities depend on the grain density (Havnes parameter). The decrease in the electron to ion density ratio n_e/n_i in a dusty plasma leads to the shift of the fluctuation maxima toward higher frequencies and to the decrease in fluctuations due to the decrease in electron susceptibility. The increase in ion effective collision frequency additionally suppresses the electron density correlations. The variations of the grain charges themselves enhance the electron density fluctuations, but, in summary, the fluctuations in dusty plasmas are considerably suppressed. This effect depends on the grain size and is more pronounced for bigger grains since the ion-grain collision frequency is proportional to the square of the grain radius.

The presence of grains increases the effective ion collision frequency, but the resulting influence on the fluctuation spectra is different from that in the ordinary plasma with the same value of ion-neutral collision frequency. Particularly, the increase in the ion-neutral collisionality in an ordinary plasma can lead to the suppression of the ion-acoustic resonance and the growth of the maximum near the zero frequency that is not observed in dusty plasmas. In the case of isothermal plasmas the presence of grains results in even more crucial changes. For high densities (the Havnes parameter exceeds 0.5) of small ($a/\lambda_D = 0.01$) grains the ion-acoustic maxima become visible.

References

1. N.N. Bogoliubov, *Problems of a Dynamical Theory in Statistical Physics* (North-Holland, Amsterdam, 1962)
2. S. Ichimaru, *Basic Principles of Plasma Physics: A Statistical Approach* (Benjamin, 1973)
3. A.I. Akhiezer, I.A. Akhiezer, R.V. Polovin, A.G. Sitenko, K.N. Stepanov, *Plasma Electrodynamics: Linear Theory* (Pergamon Press, 1975)
4. Y.L. Klimontovich, *Kinetic Theory of Nonideal Gas and Nonideal Plasma* (Pergamon Press, 1982)
5. Y.L. Klimontovich, *Statistical Physics* (Harwood Academic Publ, New York, 1986)
6. A.G. Sitenko, *Electromagnetic Fluctuations in Plasmas* (Academic Press, 1967)
7. R.H. Williams, W.R. Chappell, Phys. Fluids **14**(3), 591 (1971)
8. A.G. Zagorodnii, Y.L. Klimontovich, I. Yakimenko, Theor. Math. Phys. **75**(1), 377 (1988)
9. Y.L. Klimontovich, A.Y. Shevchenko, I.P. Yakimenko, A.G. Zagorodny, Contrib. Plasma Phys. **29**(6), 551 (1989)
10. Y.L. Klimontovich, H. Wilhelmsson, I.P. Yakimenko, A.G. Zagorodny, Phys. Rep. **175**(5–6), 263 (1989)
11. V.V. Belyi, Y.L. Klimontovich, V.P. Nalivaiko, Fizika Plazmy **8**, 1063 (1982)
12. A. Sitenko, A. Zagorodny, P. Schram, Phys. B: Cond. Matt. **228**(1), 125 (1996)
13. I. Holod, A. Zagorodny, J. Weiland, Phys. Rev. E **71**, 046401 (2005)
14. V. Belyi, Y. Kukharenko, J. Wallenborn, Contrib. Plasma Phys. **42**(1), 3 (2002)
15. V.V. Belyi, Sci. Rep. **8**(1), 1 (2018)
16. V.V. Belyi, I.V. Paiva-Veretennicoff, Theor. Math. Phys. **62**(2), 196 (1985)
17. V.V. Belyi, EPL (Europh. Lett.) **111**(4), 40011 (2015)
18. V.N. Tsytovich, O. Havnes, Comments Plasma Phys. Contr. F. **15**(5), 267 (1993)
19. V.E. Fortov, A.V. Ivlev, S.A. Khrapak, A.G. Khrapak, G.E. Morfill, Phys. Rep. **421**(1–2), 1 (2005)
20. C. Cui, J. Goree, IEEE T. Plasma Sci. **22**(2), 151 (1994)
21. P.P.J.M. Schram, A.G. Sitenko, S.A. Trigger, A.G. Zagorodny, Phys. Rev. E **63**(1), 016403 (2000)
22. M.R. Jana, A. Sen, P.K. Kaw, Phys. Rev. E **48**(5), 3930 (1993)
23. G.E. Morfill, E. Grün, T.V. Johnson, Planet. Space Sci. **28**(12), 1087 (1980)
24. O. Havnes, T. Aslaksen, F. Melandsø, T. Nitter, Phys. Scripta **45**(5), 491 (1992)
25. R.K. Varma, P.K. Shukla, V. Krishan, Phys. Rev. E **47**(5), 3612 (1993)

26. A.G. Sitenko, A.G. Zagorodny, Y.I. Chutov, P.P.J.M. Schram, V.N. Tsytovich, Plasma Phys. Contr. F. **38**(12A), A105 (1996)
27. P.K. Shukla, V.P. Silin, Phys. Scripta **45**(5), 508 (1992)
28. A. Barkan, N. D'angelo, R.L. Merlino, Planet. Space Sci. **44**(3), 239 (1996)
29. R.L. Merlino, A. Barkan, C. Thompson, N. D'angelo, Phys. Plasmas **5**(5), 1607 (1998)
30. N. d'Angelo, Planet. Space Sci. **42**(6), 507 (1994)
31. S.V. Vladimirov, K.N. Ostrikov, M.Y. Yu, Phys. Rev. E **60**(3), 3257 (1999)
32. A.V. Ivlev, D. Samsonov, J. Goree, G. Morfill, V.E. Fortov, Phys. Plasmas **6**(3), 741 (1999)
33. J. Vranjes, B.P. Pandey, S. Poedts, Phys. Plasmas **9**(4), 1464 (2002)
34. H.J. Lee, Phys. Plasmas **11**(2), 849 (2004)
35. J.X. Ma, M.Y. Yu, Phys. Plasmas **1**(11), 3520 (1994)
36. R.K. Varma, Phys. Plasmas **7**(9), 3505 (2000)
37. A.I. Momot, A.G. Zagorodny, Phys. Scripta **71**(5), 543 (2005)
38. M.R. Rouhani, M. Jamshidi, H.H. Pajouh, Astrophys. Space Sci. **357**(1), 91 (2015)
39. A. Sitenko, V. Malnev, *Plasma Physics Theory* (CRC Press, 1994)
40. Y.L. Klimontovich, *The Statistical Theory of Nonequilibrium Processes in Plasma* (MIT, Cambridge, MA, 1967)
41. H. Thomas, G.E. Morfill, V. Demmel, J. Goree, B. Feuerbacher, D. Möhlmann, Phys. Rev. Lett. **73**(5), 652 (1994)
42. J.H. Chu, I. Lin, Phys. Rev. Lett. **72**(25), 4009 (1994)
43. A.G. Sitenko, A.A. Gurin, J. Exp. Theor. Phys. **22**, 1089 (1966)
44. A.I. Momot, A.G. Zagorodny, Phys. Plasmas **18**(10), 102110 (2011)
45. A.G. Sitenko, A.G. Zagorodny, V.N. Tsytovich, A.I.P. Conf, Proc. **345**(1), 311 (1995)
46. V.P. Kubaichuk, A.G. Zagorodny, Phys. Scripta **60**(6), 549 (1999)
47. U. De Angelis, G. Capobianco, C. Marmolino, C. Castaldo, Plasma Phys. Contr. F. **48**(12B), B91 (2006)
48. V.N. Tsytovich, U. de Angelis, Phys. Plasmas **6**(4), 1093 (1999)
49. A.G. Zagorodny, P.P.J.M. Schram, S.A. Trigger, Phys. Rev. Lett. **84**, 3594 (2000)
50. S. Ratynskaia, M. De Angeli, U. De Angelis, C. Marmolino, G. Capobianco, M. Lontano, E. Lazzaro, G. Morfill, G. Gervasini, Phys. Rev. Lett. **99**(7), 075002 (2007)
51. S. Ratynskaia, M. De Angeli, E. Lazzaro, C. Marmolino, U. De Angelis, C. Castaldo, A. Cremona, L. Laguardia, G. Gervasini, G. Grosso, Phys. Plasmas **17**(4), 043703 (2010)
52. P. Tolias, S. Ratynskaia, U. de Angelis, Phys. Plasmas **18**(7), 073705 (2011)
53. P. Tolias, S. Ratynskaia, U. de Angelis, Phys. Rev. E **85**, 026408 (2012)
54. S.A. Khrapak, G.E. Morfill, Phys. Plasmas **15**(11), 114503 (2008)
55. S.A. Khrapak, P. Tolias, S. Ratynskaia, M. Chaudhuri, A. Zobnin, A. Usachev, C. Rau, M.H. Thoma, O.F. Petrov, V.E. Fortov, G.E. Morfill, EPL (Europh. Lett.) **97**(3), 35001 (2012)
56. I.L. Semenov, A.G. Zagorodny, I.V. Krivtsun, Phys. Plasmas **19**(4), 043703 (2012)
57. V.N. Tsytovich, U. de Angelis, A.V. Ivlev, G.E. Morfill, Phys. Plasmas **12**(8), 082103 (2005)
58. A.I. Momot, A.G. Zagorodny, O.V. Momot, Phys. Plasmas **25**(7), 073706 (2018)
59. A.I. Momot, A.G. Zagorodny, O.V. Momot, Phys. Rev. E **99**, 013206 (2019)
60. P.L. Bhatnagar, E.P. Gross, M. Krook, Phys. Rev. **94**(3), 511 (1954)
61. A.V. Filippov, A.G. Zagorodny, A.F. Pal., A.N. Starostin, A.I. Momot, JETP Lett. **86**(12), 761 (2008)
62. S.A. Khrapak, B.A. Klumov, G.E. Morfill, Phys. Rev. Lett. **100**(22), 225003 (2008)
63. A.G. Zagorodny, A.I. Momot, A.V. Filippov, A.F. Pal, A.N. Starostin, Ukr. J. Phys. **54**(11), 1089 (2009)
64. A.I. Momot, A.G. Zagorodny, EPL (Europhys. Lett.) **114**, 65004 (2016)
65. A.F. Aleksandrov, L.S. Bogdankevich, A.A. Rukhadze, *Principles of Plasma Electrodynamics* (Springer, 1978)
66. A.V. Phelps, J. Phys. Chem. Ref. Data **20**(3), 557 (1991)
67. H.B. Milloy, R.W. Crompton, J.A. Rees, A.G. Robertson, Aust. J. Phys. **30**(1), 61 (1977)

Fluctuations and Kinetics in Non-linear Non-equilibrium Systems

Space-Time Dynamics of High-Q Optical Resonators

F. Tabbert, S. V. Gurevich, K. Panajotov, and M. Tlidi

Abstract We consider high-Q resonators subjected to an optical injection. We focus on the formation of stationary dissipative periodic and localized structures by using the well-known Lugiato-Lefever equation. We construct their bifurcation diagrams in the case of homogeneous injection in a three-dimensional setting where the transport process is provided by 2D diffraction and 1D dispersion. After discussing the mechanism of formation of an aperiodic but localized state, we investigate the effect of the inhomogeneity on the formation of localized structures. In the regime of small but positive inhomogeneity, a significant increase of the single localized peak stability domain is observed. We have also investigated the case of a negative inhomogeneity and constructed the bifurcation diagrams.

Foreword

We dedicate this chapter to the memory of our friend Slava Belyi (1945–2020). The tragic death of Slava has deeply saddened and affected numerous friends and colleagues.

F. Tabbert · S. V. Gurevich
Institute for Theoretical Physics, University of Munster, Wilhelm-Klemm-Str. 9, 48149 Munster, Germany
e-mail: gurevics@uni-muenster.de

K. Panajotov
Vrije Universiteit Brussel, Department of Applied Physics & Photonics, Pleinlaan 2, 1050 Brussels, Belgium
e-mail: kpanajot@b-phot.org

M. Tlidi (✉)
Faculté des Sciences, Université Libre de Bruxelles (ULB), Campus Plaine CP. 231, 1050 Bruxelles, Belgium
e-mail: mtlidi@ulb.ac.be

L. Brenig et al. (eds.), *Nonequilibrium Thermodynamics and Fluctuation Kinetics*, Fundamental Theories of Physics 208,
https://doi.org/10.1007/978-3-031-04458-8_12

1 Introduction

There has been intense research on high-Q optical cavities for applications in frequency comb generation [1–3]. A frequency comb is a broad spectrum composed of equidistant narrow lines. It should be noted that T. Hänsch and J. Hall [4] have obtained the Nobel Prize in Physics in 2005 for their pioneering work on optical frequency comb technology. Optical cavities, which allow for light confinement, constitute an important part in laser physics and photonic devices [5]. Based on Fabry-Perot cavities, edge-emitting, and vertical-cavity surface-emitting lasers are widely used in the industry. More recent high-Q resonators based on whispering-gallery modes allowed for light confinement in a small volume and high finesse. Quality factors reaching 4×10^8 have been realized using toroid microcavities on a chip at 1.550 nm [6]. In the near-infrared, $Q \sim 10^8$ has been demonstrated in toroidal microresonators for cavity quantum electrodynamics [7] and in fused-silica microspheres [8]. Optical frequency combs generated by microresonators have revolutionized many fields of science and technology, such as high-precision spectroscopy, metrology, and photonic analog-to-digital conversion [5]. Most of these combs, having a 100 dB dynamic range and a frequency range larger than one octave, are theoretically described with a very good precision by the Lugiato-Lefever equation (LLE [9]). The LLE is a basic model for the investigation of the dynamics of microresonator-based optical frequency combs which were experimentally evidenced for the first time in [3]. These Kerr combs are expected to revolutionize the generation of ultra-stable lightwave and microwave signals for aerospace engineering, optical communication networks, and microwave photonic systems. This issue has witnessed a constant increase in the number of publications with an explosive growth of the localized structures (LSs) theme, as evidenced by recent review papers [10, 11].

Most of the works on passive cavities employ resonant structures to increase the local-field intensity and hence the nonlinear optical effects that characterize the effective material response to intensive optical field. In Kerr media, pattern formation originates from the third-order nonlinearity via a process called modulational instability. Frequency combs generated in Kerr microcavity resonators [10, 11] are the spectral content of the stable localized patterns occurring in a pinning range of the parameter space where the system exhibits multistability [12]. Localized structures and localized patterns consist of stable single or more peaks on top of a low-intensity background [12]. Moreover, LSs appear in other classes of systems such as population dynamics, chemistry, nonlinear optics, and laser physics (see overviews on this issue [13–22]).

In another line of research, it has been shown analytically and experimentally that a Kerr resonator driven by an inhomogeneous Gaussian pumping beam supports stable localized structures in a bistable regime [23]. These types of LSs connect the two homogeneous solutions and become stable, thanks to the well-known mechanism based on front interaction. Front dynamics have been investigated in synchronously pumped passive all-fiber resonators [24, 25]. These structures occur in a regime far from any modulational instability assuming that the system operates in a direct

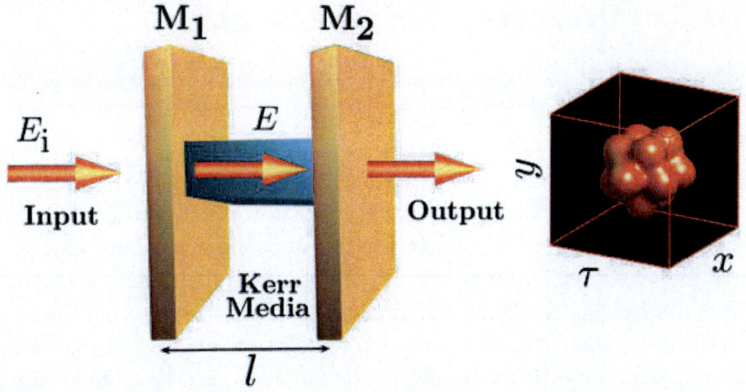

Fig. 1 Schematic setup of an optical diffractive and dispersive Fabry-Perot cavity filled with a Kerr medium and coherently driven by an external field E_i. The input-output mirrors $M_{1,2}$ are separated by a distance l and are identical. The field output intensity shows isosurface associated with the formation of three-dimensional localized structures involving six light bullets. Figure reproduced from [28]

dispersion regime. The trajectory of the position of the LS is derived from the LLE and its hyperbolic tangent analytical expression perfectly fits the experimental data [23]. The experimental analysis supported by numerical simulations has indicated that LSs do not necessarily stabilize at minima or maxima of the injection [26]. A phase-modulated injection has been also applied to a Kerr resonator [27]. This effect allows for the stabilization of a single peak LS out of a multistability regime [27].

This chapter is organized as follows. After an introduction, we present in Sect. 2 an analysis of the homogeneous, periodic stationary solutions in the classical LLE with homogeneous injection, where we are going to mostly summarize previous results [28] on diffractive and dispersive Kerr resonators submitted to an optical injection (see Fig. 1). We discuss in Sect. 3 the conditions under which localized structures and soliton frequency combs are formed in high-Q resonators. In Sect. 4, we focus on localized structures and we investigate their homoclinic snaking bifurcation diagram in the presence of inhomogeneous pumping. We conclude in Sect. 5.

Our aim in this contribution is not to review but to highlight some important properties of the LLE and its diverse applicability in the fields of photonics and nonlinear sciences. Therefore, this introduction is meant for active research physicists and engineers working in nonlinear science and nonlinear photonics wishing to have a quick overview of recent developments in terms of the applicability of the LLE concerning soliton frequency combs.

2 Space-Time Dynamics of High-Q Resonators

2.1 The Mean-Field Model

High-Q Kerr resonators are widely described by the paradigmatic LLE [9]. This model is used to investigate the dynamical properties of laser fields confined in nonlinear optical resonators. The LLE model was derived using the mean-field approach, and external power can be coupled into the cavity only if the system is close to resonance. This implies that both the linear cavity detuning and the nonlinear cavity phase shift must be much smaller than unity. In addition, we assume that the cavity is much shorter than the diffraction, dispersion and nonlinearity spatial scales. Furthermore, a single longitudinal mode operation is assumed [9]. In its general form the LLE reads [29]

$$\frac{\partial E}{\partial t} = i\left(\alpha\nabla_\perp^2 + \beta\frac{\partial^2}{\partial\tau^2}\right)E - (1+i\theta)E + i|E|^2E + E_i. \tag{1}$$

Here $E(x, y, t, \tau)$ is the normalized slowly varying envelope of the electric field that circulates within the cavity and E_i is the amplitude of the injected field which is real and positive in order to fix the origin of the phase. The time variable t corresponds to the slow evolution of E over successive round-trips. τ accounts for the fast dynamics. The parameter θ is the cavity detuning with respect to the injected field. β is the chromatic dispersion coefficient. The diffraction process acting in the transverse plane (x, y) is described by the Laplace operator $\nabla_\perp^2 = \partial^2/\partial x^2 + \partial^2/\partial y^2$. Diffraction α and dispersion β coefficients can have the same or opposite signs. The diffraction coefficient can be negative if the cavity is operating in a self-imaging configuration or when using a left-handed material. The dispersion coefficient β can be either positive or negative depending on whether the dispersion is anomalous or normal, respectively.

The mean-field approximation used to derive the LLE Eq. (1) has been compared with the propagation model supplemented by cavity boundary conditions for both quadratic [30, 31] and Kerr [32] cavities. This model has also been successfully applied to describe all-fiber cavities [33] and resonators filled with left-handed materials [34]. It has been considered the basic model for the study of microresonator-based optical frequency combs [35] (see also the review by Lugiato et al. [36] and [37, 38]).

The LLE Eq. (1) possesses homogeneous stationary solutions E_{hs}, which can be easily calculated by setting the spatial and temporal derivatives in Eq. 1 to zero. This yields the input-output characteristics (Fig. 2)

$$|E_i|^2 = |E_{hs}|^2\left(1 + (\theta - |E_{hs}|^2)^2\right). \tag{2}$$

The implicit expression given in Eq. (2) can be split-up in two expressions for the real and the imaginary parts of the solution, respectively:

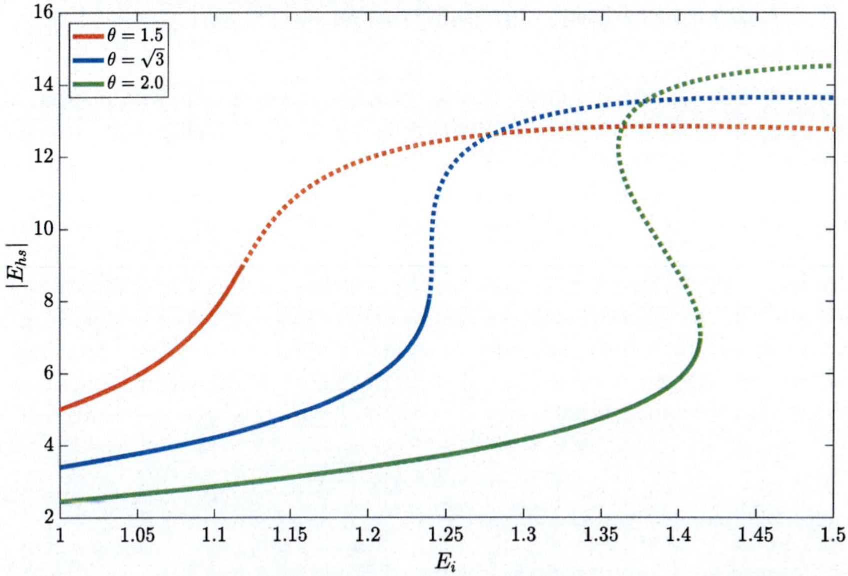

Fig. 2 Input-output characteristics: intracavity field amplitude $|E_{hs}|$ as a function of the injected field amplitude E_i without considering spatial and temporal effects, i.e., neglecting the diffractive and dispersion terms. Dotted lines represent unstable solutions, whereas solid lines depict stable solutions. The transition from a monostable ($\theta = 1.5$, red line) to a bistable regime ($\theta = 2.0$, green line) in a cusp bifurcation is clearly visible. For $\theta = 2.0$ there exists a region of parameters where three different stationary solutions exist, with the middle one being unstable. The blue line depicts the response curve at the critical value of $\theta = \sqrt{3}$, i.e., the border between the monostable and the bistable regime. When spatial modulation is considered, the monostable (red curve) and critical (blue curve) lose stability in a modulational instability. The bistable curve (green) loses stability

$$\mathrm{Re}(E_{hs}) = \frac{E_i}{1 + (|E_{hs}|^2 - \theta)^2}, \tag{3}$$

$$\mathrm{Im}(E_{hs}) = \frac{E_i(|E_{hs}|^2 - \theta)}{1 + (|E_{hs}|^2 - \theta)^2}. \tag{4}$$

There exists a critical point marking the onset of a hysteresis loop when both the first and the second derivatives of E_i with respect to $|E_{hs}|^2$ vanish. This corresponds to the inflection point of the intensity response curve corresponding to the critical detuning for the onset $\theta = \theta_c = \sqrt{3}$. For $\theta < \sqrt{3}$ ($\theta > \sqrt{3}$) the transmitted intensity $|E_{hs}|^2$ as a function of the input intensity E_i^2 is monostable (bistable). The transition from a single solution to three possible solutions is sometimes referred to as a cusp bifurcation due to the cusp-like shape of the response curve. When neglecting dispersion and diffraction terms in Eq. (1), the linear stability analysis shows that two of these states are stable, which is why this coexistence is often referred to as optical bistability.

2.2 Modulational Instability and Pattern Formation

The linear stability analysis of the homogeneous solutions Eq. (2) with respect to perturbations of the form $\exp(i\mathbf{q}.\mathbf{r} + \sigma t)$ with $\mathbf{r} = (x, y, \tau)$ and $\mathbf{q} = (q_x, q_y, q_\tau)$ leads to a characteristic equation

$$\sigma^2 + 2\sigma + 1 + (I_{hs} - \theta)(I_{hs} - \theta) + \left[q^2 - 2(2I_{hs} - \theta)\right] q^2 = 0, \tag{5}$$

where $I_{hs} = |E_{hs}|^2$. Equation (5) is obtained by assuming the following scaling $(x, y) \rightsquigarrow \sqrt{\alpha}(x, y)$ and $\tau \rightsquigarrow \sqrt{\beta}\tau$. In this section we assume that α and β are both positive. The homogeneous stationary solution undergoes a symmetry-breaking instability when the eigenvalue σ vanishes for a finite wavenumber [9]. Above this instability point, there exist a finite band of unstable Fourier modes $q_- < q < q_+$, with

$$q_\pm^2 = 2I_s - \theta \pm \sqrt{I_s - 1}, \tag{6}$$

which are linearly unstable and trigger the spontaneous evolution of the intracavity field towards a stationary, periodic solution that occupies the whole space available in the (x, y, τ) Euclidean space. The threshold associated with the symmetry breaking instability is obtained when $q_- = q_+$. The critical intracavity field intensity is $I_{hsc} = |E_{hsc}|^2 = 1$ and the corresponding critical injected field intensity at the onset of the instability is $I_{ic} = E_{ic}^2$. At this bifurcation point, the critical wavelength of the pattern which emerges from the symmetry-breaking or modulational instability is $\Lambda_c = 2\pi/q_c$ with $q_c = \sqrt{2 - \theta}$. At the threshold associated with the modulational instability, the homogeneous stationary solution becomes unstable with respect to modes satisfying the relation, $q_x^2 + q_y^2 + q_\tau^2 = 2 - \theta$. There exists a finite band of unstable Fourier modes lying on a sphere with radius $q_c = \sqrt{2 - \theta}$. In the absence of a source of anisotropy such as a birefringence or walk-off, the Fourier modes are arbitrarily directed in the space (q_x, q_y, q_τ) and there is no preferred direction. Although a large number of unstable modes is excited along arbitrary directions, a periodic dissipative structure is selected. In the linear regime, these three-dimensional structures can be approximated by a linear superposition of p pairs of opposite wave vectors q_j lying on the critical sphere of radius q_c as $E(\mathbf{r}, t) = E_s + \mathbf{e}\sum_{j=1}^{p} A_j \exp(i\mathbf{q_j}.\mathbf{r}) + c.c.$, where $c.c.$ denotes the complex conjugate and $\mathbf{e}$ the eigenvector of the corresponding Jacobian matrix associated with the zero eigenvalue, i.e., $\sigma = 0$. The lamellae and rhombic structures are characterized by $p = 1$ and $p = 2$, respectively, and the 3D hexagons or hexagonally packed cylinders correspond to $q = 3$ with $\sum_{j=1}^{3} \mathbf{q_j} = 0$. The face-centered-cubic (fcc) lattice and the quasiperiodic crystals are obtained for $p = 4$ and $p = 5$, respectively. The body-centered-cubic (bcc) lattice corresponds to $n = 6$. The bcc optical crystals are characterized by six pairs of wavevectors whose coordinates are $q_c(\pm 1, \pm 1, 0)/\sqrt{2}$, $q_c(\pm 1, 0, \pm 1)/\sqrt{2}$, and $q_c(0, \pm 1, \pm 1)/\sqrt{2}$. These wavevectors form a regular octahedron of eight faces in the form of equilateral triangles.

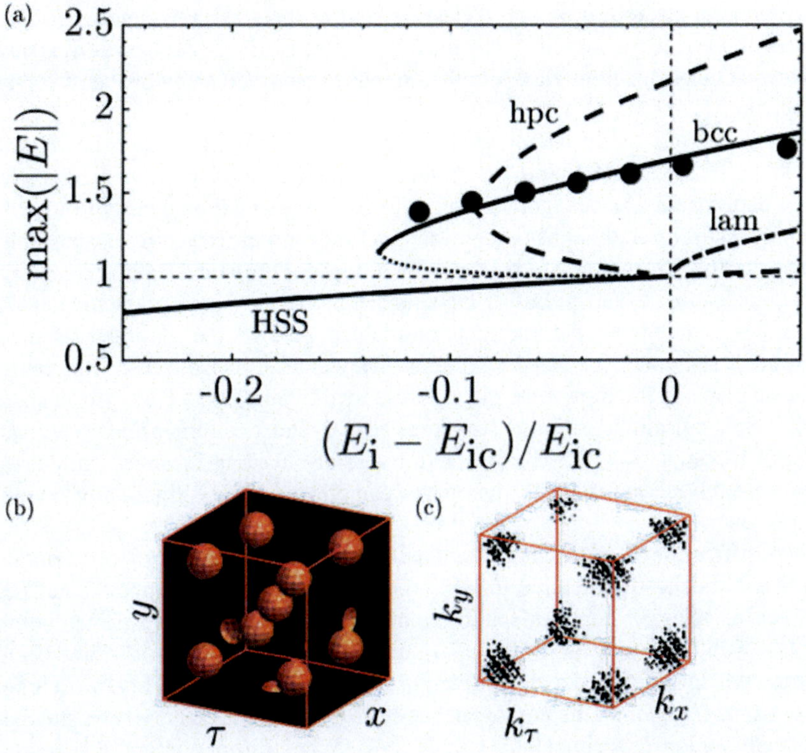

Fig. 3 (a) 3D Bifurcation diagram associated with three-dimensional periodic dissipative structures obtained in the weakly nonlinear regime ($\theta = 0.7 < 41/30$). Broken lines correspond to unstable solutions, and the continuous line corresponds to stable branches of solutions. The black dots along the body-centered cubic (bcc) branch denote the maximum amplitude values associated with the bcc solutions obtained by numerical simulations of the 3D LLE Eq. 1. The lamellae and face-centered cubic crystals, denoted respectively by lam and fcc, are unstable. The HSS denotes the homogeneous steady states which are stable for $E_i < E_{ic}$ and unstable for $E_i > E_{ic}$. (b) Isosurface with the bcc solutions obtained from numerical simulations, and (c) the Fourier spectrum of the bcc. Parameters are $E_i = 1.05$ and $\theta = 0.7$. Figure reproduced from [45]

A weakly nonlinear analysis allows determining the three-dimensional solutions of amplitude equations. The relative stability analysis has been performed in [29, 39] showing that the bcc crystals are the most stable solutions over all others. Note that in two-dimensional settings, hexagonal structures are the most stable structures [40]. The bifurcation diagram displayed in Fig. 3(a) summarizes the results of the weakly nonlinear analysis and the relative stability analysis. We plot the amplitude of some of the above-mentioned three-dimensional solutions as a function of the injected field with respect to the relative distance from the modulational critical threshold. Note that the analytical solutions shown in the 3D bifurcation diagram Fig. 3(a) had not been checked numerically [29] until recently. More recently, numerical simulations of the 3D LLE using a fourth-order exponential time differencing Runge-Kutta method for the temporal discretization [41, 42] and a Fourier spectral method with periodic boundary conditions for the spatial discretization of the LLE [43, 44] has been carried out. The results of this numerical simulation indeed provided an evidence of stable bcc structures [28, 45].

When the chromatic dispersion is neglected, i.e., $\beta = 0$, it has been shown that the coupling between diffraction and nonlinearity may lead to appearance of stable periodic patterns such as hexagons that emerge from a modulational instability [46]. The relative stability analysis has shown that rolls and rhombic structures are unstable with respect to perturbations in the form of hexagonal structures [40]. Secondary bifurcations leading to self-pulsating hexagonal patterns have been predicted in [47, 48]. Besides periodic patterns, the same mechanism predicts the possible existence of aperiodic, stationary localized structures. They consist of isolated or randomly distributed peaks in the transverse plane surrounded by regions in the uniform state [12, 49]. Self-pulsating localized structures in one and two-dimensional settings have been investigated [50]. The interaction of self-pulsating localized structures that emit weakly decaying dispersive waves can allow for the formation of bound states [51–54].

When diffraction is neglected, i.e., $\alpha = 0$, polarization properties of periodic [55–57] and localized structures [58–61, 63, 64] have also been investigated. In large intensity regimes, complex spatio-temporal dynamics such as spatio-temporal chaos [65–68], rogue waves [66, 69, 70], and chimera states [71, 72] have been demonstrated. In regimes devoid of Turing or modulational instability, switching wave or fronts connecting the homogeneous steady states of LLE have been studied theoretically and/or experimentally [24, 73, 74]. When diffraction and dispersion have a comparable influence, the spatio-temporal dynamics of the nonlinear Kerr cavity can be described by the three-dimensional LLE [29, 75, 76].

3 Localized Solutions of the Lugiato-Lefever Equation: Homoclinic Snaking Bifurcation

We have shown how different dissipative structures can be excited in high-Q resonators using the LLE in a weakly nonlinear regime where the modulational instability appears supercritically, i.e., $\theta < 41/30$. In this section, we explore the regime where this bifurcation appears subcritically for $\theta > 41/30$. This inversion of the bifurcation associated with periodic solutions induces a regime where a single homogeneous steady state coexists with periodic structures. In this hysteresis loop, we can generate aperiodic and localized states with appropriate initial conditions [77]. Since LSs can only occur in a region where a stable homogeneous solution coexists with stable patterned solutions, we will restrict ourselves to the latter subcritical case. It is worth noting that the distinction between a super- and subcritical bifurcation does not coincide with the above-mentioned differentiation between the mono- and bistable regime, i.e., localized solutions can exist in both regimes. After discussing

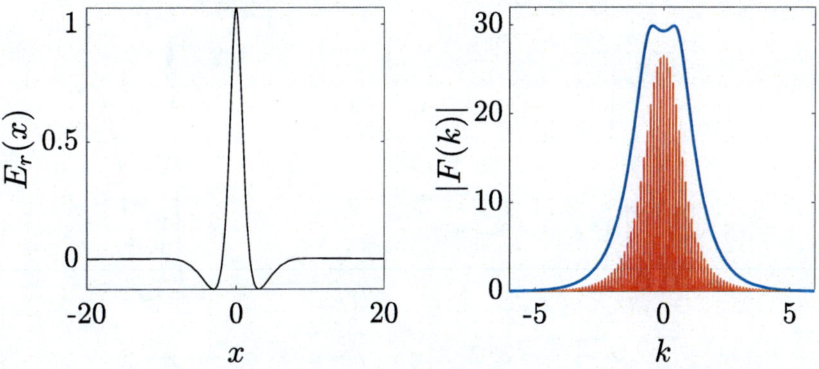

Fig. 4 Left: Real part of the solution profile of a single LS for $\theta = 1.7$, $E_i = 1.2$. Right: Absolute value (blue) and real part (red) of the Fourier transform of the LS. The real part consists of a sequence of equidistant peaks (also called frequency comb). The imaginary part is not depicted for the sake of better visualization; however, it exhibits the same comb-like structure

both homogeneous and periodic solutions of the LLE, we will now focus on the formation of LSs, and we choose the parameters from the paper by Scroggie et al. [12]. The first theoretical prediction of localized structures has been realized in the monostable regime for $\theta = 1.7$ and $E_i = 1.2$. An example of a single peak LS is shown in Fig. 4 (Left). The investigation of LSs has recently sparked new interest since a direct link between LSs of the temporal LLE and Kerr frequency combs (KCs) has been established [36]. It was shown that the frequency spectrum of a temporal LS created, e.g., in a microresonator, consists of a sequence of equidistant peaks spanning over a broad frequency band. An example of a LS together with the corresponding KC can be found in Fig. 4. In this parameter region, localized solutions do exist. Figure 5 shows two examples of LS in this region, one for smaller θ with very pronounced oscillatory tails and one for larger θ where the oscillatory tails are less pronounced, yet still present. The presence (or absence) of oscillatory tails is of great importance for the bifurcation structure of LS since the form of the tails predicts the possibility of bound states between two or more LSs.

We are now going to analyze the bifurcation structure of LSs in this regime by employing numerical continuation algorithms provided by the MATLAB package *pde2path*. As a measure, we use the averaged L_1 norm

$$L_1 = \int d\xi |\mathrm{Re}(E - \overline{E})|.$$

$\overline{E}$ denotes the mean value of the electric field $E(\xi)$ averaged over the domain size, i.e., homogeneous solutions possess the norm $L_1 = 0$.

In Fig. 6, the emergence of a single LS (green line) bifurcating from the first periodic branch with an odd number of peaks (red line) is depicted. The fact that this solution branches off from the periodic branch close to the Turing point is due

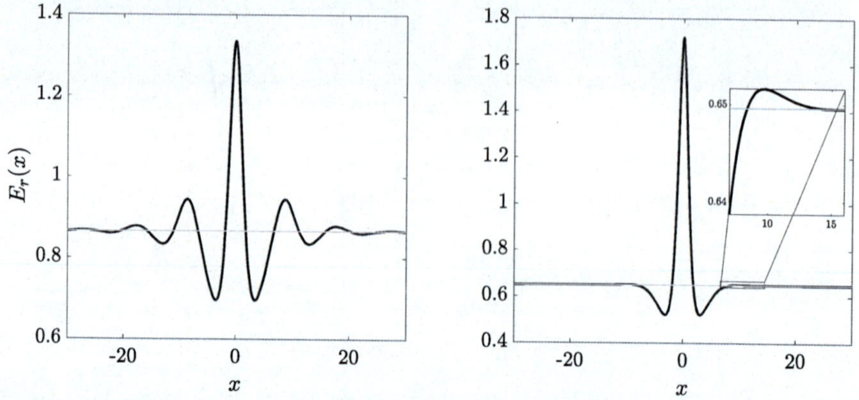

Fig. 5 Real part of the solution profile of a single LS for $\theta = 1.5$, $E_i = 1.114$ (left), and $\theta = 1.7$, $E_i = 1.2$ (right). The amplitude of oscillatory tails decreases drastically with increasing θ. However, one side-maximum is still present in the case of $\theta = 1.7$ (see inset)

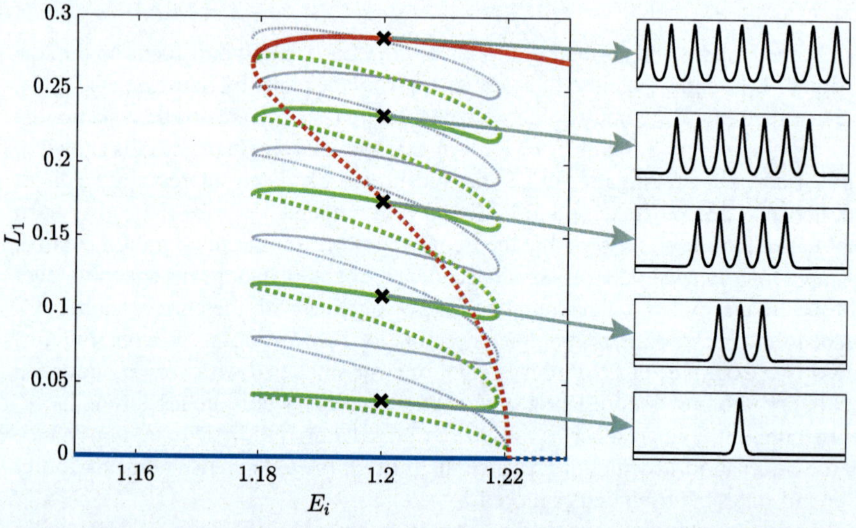

Fig. 6 Left: Emergence of a single LS and odd numbers of bound LSs in homoclinic snaking for $\theta = 1.7$ with E_i as the continuation parameter and with the L_1 norm as a measure. Shortly after the periodic nine-peak solution (red line) bifurcates from the homogeneous solution (blue line) at the Turing point, a single LS solution (green line) branches off of the periodic solution. In a sequence of folds, the solution gains stability and then gains two extra peaks. This pattern continues until the domain is filled with nine peaks and the branch of LSs reconnects with the periodic branch. The thin gray lines in the background depict the even branch of LSs which are thoroughly discussed in Fig. 7. Right: Solution profiles of the LSs at the marked positions illustrating the transition from a single LS to a domain-filling pattern by addition of peaks

to a finite size effect [78]. On an infinite domain one would expect both the periodic branch and the LS branch to bifurcate at the same Turing point. The single LS then gains stability in a fold. The solution profile of the single LS is depicted on the bottom right of Fig. 6 as well as the solution profiles of all upper branches. The single LS then loses stability in another fold but gains two additional peaks, then stabilizes again as a three-peak solution. This sequence of folds continues until the peaks fill the entire domain and the branch of LSs reconnects with the first stable periodic branch (in the case of Fig. 6, the same red branch it bifurcated from originally).

This sequence of consecutive folds is typically referred to as homoclinic snaking [79–81] (see overviews on this issue [82], in the theme issue [83], and in the very recent Editorial to homoclinic snaking of 2021 in memory of Patrick Woods [84]). It has been thoroughly studied in the LLE [49, 73, 85]. Besides the described branch consisting of odd numbers of LSs, there exists also an even branch of LSs as depicted in Fig. 7 (gray line in Fig. 6). In this case, a solution consisting of two bound LSs (green line) bifurcates from the first periodic branch that becomes Turing unstable (in this case the orange eight-peak branch). In a sequence of folds, the structure gains additional peaks as in the case of the odd branch until the entire domain is filled and the structure reconnects with the periodic ten-peak branch (red line). This ten-peak solution is the first periodic solution with an even number of peaks that gains stability. The eight-peak solution (orange) becomes Turing unstable before the ten-peak solution; however, it does not reach stability. The fact that the even branch connects to a different periodic branch than the odd LS branch can be attributed to a finite size effect [86]. The even and the odd branch of the homoclinic snaking are connected by so-called ladders [86], i.e., unstable nonsymmetric solution branches connecting the two branches. However, since these branches are completely unstable, they will not be the focus of the following analysis.

With increasing the detuning θ, the Turing bifurcation vanishes by colliding with the right fold of the cusp-like homogeneous branch at $\theta = 2.0$. Localized structures still exist and can be directly traced by direct continuation starting from a parameter regime $\theta < 2$ into the regime $\theta > 2$ [86], i.e., the stable single LS solution below and above $\theta = 2$ are connected. The same does not hold for bound states of two or more LSs because the shape of the LSs changes qualitatively around $\theta \approx 2$. The localized solution at the maximum possible injection E_i at the fold of the single LS depicted in Fig. 6 loses its side maximum at $\theta = 2.08$. Therefore, LSs are not able to form stable bound states but repel each other above this value. The homoclinic snaking as described above therefore breaks down [86].

In [86], this loss of tail was identified as a Belyakov-Devaney transition by analyzing the system of four first-order ODEs describing the spatial dynamics of temporally stationary solutions. The linear stability analysis of the homogeneous solution in this framework shows a Belyakov-Devaney transition at $\theta = 2.0$ at the fold. That is, the system goes from having complex spatial eigenvalues $\lambda_{1,2,3,4} = \pm q_0 \pm iq_0$ to real eigenvalues $\lambda_{1,2} = \pm q_1$ and $\lambda_{3,4} = \pm q_2$. Losing the imaginary part of the spatial eigenvalues is equivalent to a loss of oscillatory tails in the linearized limit. The fact that this transition takes place at $\theta = 2.0$, whereas numerically, reminiscences of the oscillatory tails can be found up to $\theta = 2.08$ is due to nonlinear effects that are not

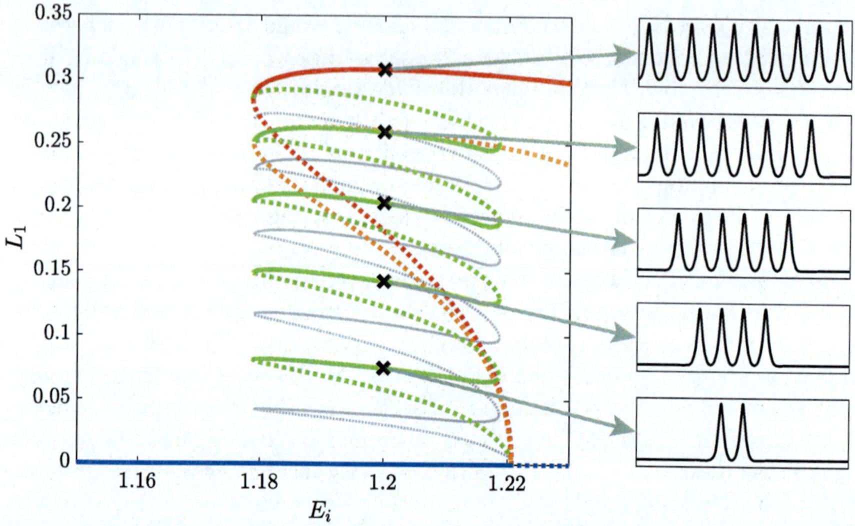

Fig. 7 Left: Emergence of even numbers of bound LSs in homoclinic snaking for $\theta = 1.7$ with E_i as the continuation parameter and with the L_1 norm as a measure. A solution of two bound LS (green line) bifurcates from the eight-peak periodic solution (orange line) which is the first periodic solution with an even number of peaks to branch off of the homogeneous solution (blue line). In a sequence of folds, the solution gains stability and then gains two extra peaks. This pattern continues until the domain is field with ten peaks and the branch of LSs reconnects with the periodic branch of ten peaks, which is the first periodic branch with an even number of peaks that gains stability. The thin gray lines in the background depict the odd branch of LSs which are thoroughly discussed in Fig. 6. Right: Solution profiles of the LSs at the marked positions illustrating the transition from a single LS to a domain-filling pattern by addition of peaks

considered in the linearized framework of [86]. In contrast to the case of $\theta < 2$, the single LS does not bifurcate from the periodic branch close to the Turing bifurcation, the single LS branch for $\theta > 2$ is the first solution branch, bifurcating from the homogeneous solution.

In the same manner, the branch with $k = 2\Delta k$ coarsens to two peaks positioned at a distance of $L/2$. The same goes for higher values of k, where a number of peaks are positioned equidistant on the domain, i.e., a differentiation between a periodic state and a state consisting of fewer equidistant LSs is not systematically possible. Bulks of LSs that are closer together can exist for small values of E_i below the Belyakov-Devaney transition; however, the distance between the LSs diverges in the vicinity of the Belyakov-Devaney transition connecting these states to the branches of equidistant LSs [86]. A branch with the peak number n is also connected to the branch with the peak number $2n$ by a ladder-like branch on which small peaks grow in between every pair of existing peaks. This structure is referred to as foliated snaking. A detailed discussion of this phenomenon is out of the scope of this chapter but can be found in [86].

Going to even larger values of the detuning $\theta > 3$, the LSs are destabilized in an Andronov-Hopf bifurcation leading to breathing LSs and through further instabilities to spatio-temporal chaos [68]. Since the scope of this chapter are the properties of stable LSs, it is necessary to determine where this instability sets in, in order to obtain an overview of the parameter regime where LSs are stable. To that aim, we again deploy numerical continuation algorithms provided by *pde2path*. Instead of using only one continuation parameter while fixing all other parameters, we now use θ as the continuation parameter and E_i as a free parameter that is determined to fulfill an additional condition characterizing a fold or Andronov-Hopf bifurcation. In this manner, it is possible to track the folds and the Andronov-Hopf bifurcation points that delimit the stability of LSs in parameter space [87, 88].

4 Inhomogeneous High-Q Resonators

4.1 The Lugiato-Lefever Model with Inhomogeneous Injection

The Lugiato-Lefever equation in its original form presented in the last section exhibits localized solutions in a broad parameter range. In this chapter, we are going to expand the analysis to the case of inhomogeneous injection. In the classical LLE describing spatial pattern formation, the injected beam is assumed to be a plane wave, i.e., E_i is constant along the coordinates (x, y, τ). The investigation of the effects of an inhomogeneous injection is important for two reasons: on the one hand, small spatial inhomogeneities are unavoidable in any experimental setup, yet they break the translational symmetry, which is typically assumed in theoretical models. This symmetry-breaking effect can have a strong influence on the formation, position, and dynamics of LSs and therefore needs to be investigated. On the other hand, it might prove beneficial to introduce small inhomogeneities artificially. Since the injected beam is comparatively easy to manipulate, this would provide a simple mechanism to alter the properties of the system under consideration.

Although the LLE was intensively studied in the past decades, the effects of inhomogeneous injection have not been as extensively studied and have gotten into the focus of experimental research only recently [23]. In that work, the LLE with a Gaussian pulse is applied to a Kerr cavity containing a liquid crystal. Later, the same idea has been applied to all-fiber cavities [89, 90]. Another example of the study of inhomogeneous injection can be found in Cole et al. [91], where the results suggest that a phase-modulated injection can protect single LS generation by preventing the multistability between different numbers of LSs typically present in the LLE. In this section, we are limiting ourselves to the case of homogeneous injection with an additional inhomogeneous portion in the shape of a Gaussian. For this purpose, we rewrite the LLE in one-dimensional setting as

$$\frac{\partial E(\xi,t)}{\partial t} = E_i + A\exp\left(-\frac{\xi^2}{B}\right) + \left[-(1+i\theta) + \mathrm{i}|E(\xi,t)|^2 + \mathrm{i}\frac{\partial^2}{\partial \xi^2}\right]E(\xi,t). \tag{7}$$

Here, the coordinate ξ can be seen as either one of the transverse spatial coordinates x or y or the fast time τ in the reference frame moving with the group velocity of the light within the cavity while t is the slow time proportional to the round-trip time. In contrast to Hendry et al. [90], who studied purely Gaussian injection, we are going to mainly discuss the effects of small additional inhomogeneities, i.e., E_i is comparably large as in the homogeneous case, whereas the amplitude of the Gaussian typically is $A < 1$. The width of the Gaussian is fixed to a moderate width of $B = 4.0$ comparable to the typical length-scale of the system. Altering this width within a reasonable limit does not affect the results presented here as drastically as altering A and therefore will not be studied in this section. We are also limiting ourselves to a Gaussian shape of the inhomogeneity. In the following we are focusing on the impact introduced by the inhomogeneity in the injected field on the formation of stationary LSs in different parameter regimes, also briefly discussing the impact on homogeneous and periodic solutions. We will construct the bifurcation diagram associated with the formation of LSs. The inhomogeneity allows us to not only control the position of LSs but strongly affects their stability domains. In particular, a new stability domain of a single-peak localized structure appears outside of the region of multistability between multiple peaks of localized states [87, 88]. Finally, in the last subsection, we apply the potential well model to the LLE, which was derived and discussed in the context of the Swift-Hohenberg equation [92] and for the LLE [87]. We close with a full exploration of parameter space, comparing the results in the case of inhomogeneous injection to those in the case of homogeneous injection presented in the previous section.

4.2 Homoclinic Snaking in the Presence of Inhomogeneities

We start with the analysis of the bifurcation structure of Eq. (7) in the regime of homoclinic snaking discussed in the last section. Here, we are going to analyze the emergence of LSs with increasing E_i and discuss the effects of the inhomogeneity compared to the case of homogeneous injection. We are going to show that the inhomogeneity can act either attracting or repelling on the LSs, leading to LSs that are pinned to the inhomogeneity on the center or on the side, respectively. Varying the amplitude A in parameter continuations we are then going to discuss the transitions between the different LSs and identify the bifurcations responsible for these transitions. In this section we investigate the effect of small added inhomogeneities on the solution structure of the LLE in the regime of homoclinic snaking, i.e., in the regime of $\theta < 2$. One of the most obvious consequences of the introduction of an inhomogeneous term is that the perfectly homogeneous solution as described in Eqs. (3–4) vanishes. However, especially in the case of small A we are interested in, one can still identify a solution that we are going to denote as quasi-homogeneous due

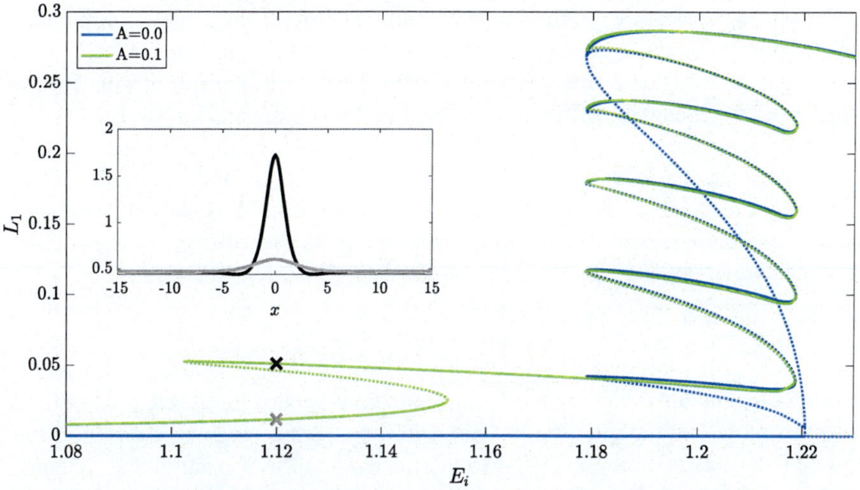

Fig. 8 Homoclinic snaking in the presence of an attracting inhomogeneity for $A = 0.1$, $B = 4.0$, $\theta = 1.7$ (green line) and in the absence of the inhomogeneity (blue line) obtained by numerical continuation on a domain of $L = 100$. The inset depicts two solutions at the positions marked by the crosses. In the presence of the inhomogeneity, no perfectly homogeneous solution exists but a quasi-homogeneous solution (gray line, inset). This quasi-homogeneous solution is connected with the LS solution (black line, inset) via two folds. The region of existence (and stability) of the single LS is drastically increased even by the small inhomogeneity, whereas the higher branches of the snaking diagram remain almost unaltered

to its similarities to the homogeneous solution in the previous section. The quasi-homogeneous solution shows a slight bump or peak (depending on the sign of A) at the position of the inhomogeneity while consisting of the homogeneous solution elsewhere. An example of the quasi-homogeneous solution for $A = 0.1$, $\theta = 1.7$, $E_i = 1.12$ can be found in the inset of Fig. 8 (gray line). The black line depicts a LS at the same position.

Apart from the obvious difference in amplitude, the quasi-homogeneous solution can be qualitatively distinguished from LSs by considering the different underlying mechanisms responsible for the peak formation. The shape and height of the peak (or bump) in the quasi-homogeneous solution are mainly determined by the locally increased (or decreased) injection. Spatial interactions due to the Laplacian or nonlinear effects hardly influence the peak formation. In other words, the value of the quasi-homogeneous solution at the peaks maximum approximately matches the value of the homogeneous solution with increased injection.

Focusing now on the destabilization of the quasi-homogeneous solution one can note that the destabilization with respect to periodic perturbations as described in the previous section is unlikely to happen in the presence of an inhomogeneity, since the solution itself is inhomogeneous, defining a specific length-scale. Instead, the quasi-homogeneous solution loses stability in a fold as depicted in Fig. 8. Following another fold, a stable LS positioned on the center of the inhomogeneity emerges.

Concerning the stability of the LS, two important differences to the case of homogeneous injection have to be pointed out. First of all, the LS in the case of $A = 0.1$ is stable, not neutrally stable, towards perturbations in the form of an infinitesimal translation, i.e., the eigenvalue of the translational mode is non-vanishing and negative. The breaking of the translational symmetry of the system by the inhomogeneity in the case of positive A leads to a pinning of the LS on the center of the inhomogeneity.

Second, the onset of the region of stability of the single LS is shifted drastically towards smaller values of E_i, as can be seen in Fig. 8. This shift is understandable since the overall injection E_{inj} at the peak consists of two contributions:

$$E_{\text{inj.}}(x = 0) = E_i + A, \tag{8}$$

i.e., a smaller amount of homogeneous injection E_i is needed for the creation of LSs. However, more surprisingly, the next fold in the snaking diagram, delimiting the region of stability of a single LS is hardly affected at all by the additional inhomogeneous injection. In Fig. 8, the case of an attracting inhomogeneity ($A = 0.1$, green line) is depicted as well as the previously described case of homogeneous injection ($A = 0.0$, blue line), showing that the fold where a single LS loses stability is not drastically shifted. This difference provides valuable insights into the prerequisites necessary for the formation of LSs. Only the injection (or driving in general) at the maximum of the LS needs to exceed a certain value to create LSs; however, the amount of injection at the sides of the LS seems to determine the end of the region of stability.

Apart from this theoretical insight, this effect can be beneficial for the experimental realization of cavity solitons or Kerr combs since it shows that by deploying slightly inhomogeneous injection, one can drastically widens the region of stability of the desired structures. Furthermore, as Fig. 8 suggests, the upper branches of the snaking diagram are hardly affected at all by the inhomogeneity. At the end of the snaking branch, the LSs even connect to a solution that fills the entire domain with peaks; however, this solution is not perfectly periodic and such a solution does not exist for all parameter values of θ and A. The fact that the upper branches of the homoclinic snaking do not change in the presence of the inhomogeneity leads to the second potentially useful result for experimental realizations: by using slightly inhomogeneous injection, the region where only single LSs are stable gets widened in contrast to the branches of several bound LSs yielding a parameter region where solely the single LS is stable. Avoiding the multistability associated with homoclinic snaking might be a promising way to address single LSs or Kerr combs more easily.

In the case of negative A, the bifurcation diagram in Fig. 9 shows that the norm L_1 as a function of the injected field amplitude E_i undergoes a homoclinic snaking type of bifurcation. The quasi-homogeneous solution (lowest branch, solution profile depicted on the lower left) evolves into a solution with a peak on each side of the inhomogeneity. This branch (blue line) undergoes the classical homoclinic snaking, with additional peaks growing on each side throughout the snaking. Solution profiles at the positions marked by circles are depicted on the left. Shortly before the first fold of the even branch, a single-peak solution (solution profile on the lower right)

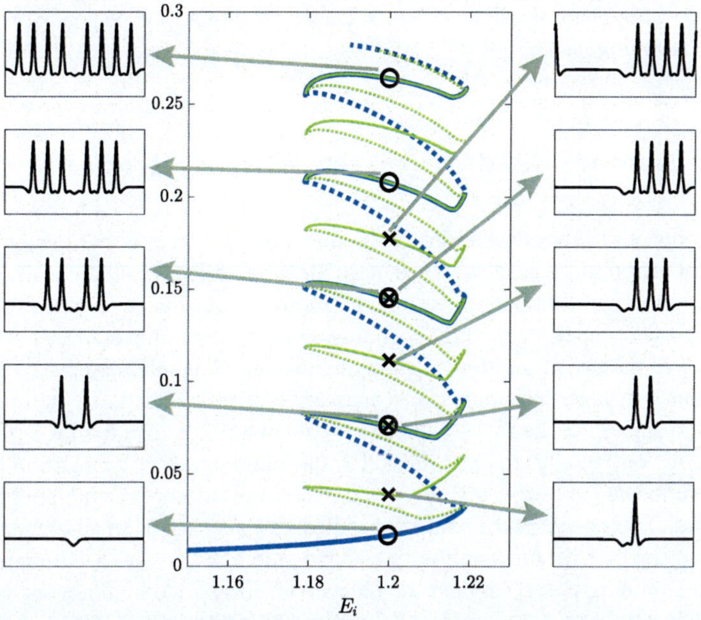

Fig. 9 Bifurcation diagram associated with the formation of localized structures under the effect of inhomogeneity for with a negative amplitude of $A = -0.1$. Other parameters are $\theta = 1.7$ and $B = 0.4$

bifurcates in a subcritical pitchfork bifurcation. This solution undergoes a symmetry-broken snaking where with each turn only one peak is added to the solution profile. Peaks are only added on the far side of the inhomogeneity. Solution profiles at the positions marked by the crosses are depicted on the right. Both the green and the blue branches overlap substantially, since, e.g., a two-peak solution possesses essentially the same L_1-norm, regardless of the position of the peaks. It should be noted that none of the branches connect to a periodic or a quasi-periodic solution. If one continues to follow the branches they wind down again towards lower peak solutions.

Following the green branch, solution profiles at the positions marked with a cross are depicted on the right-hand side of Fig. 9, showing that in this symmetry-broken version of a homoclinic snaking diagram, peaks are not added in pairs to the solution but the solution gains additional peaks one by one. Since the depicted L_1-norm of an n-peak solution does not differ greatly for different peak positions, the green and the blue branches overlap when both branches show an even number of peaks. One can note that similar bifurcations to the one leading to the emergence of the green branch can be found close to every fold of the blue branch as is the case with the so-called ladders [93]. The additional branches emerging from these bifurcations are not depicted in Fig. 9 for the sake of clarity but these branches would also exhibit solutions that show an uneven number of peaks on each side of the inhomogeneity. Therefore, not only the completely symmetric and the completely asymmetric cases as depicted in Fig. 9 are possible, but so are every configuration of peaks on either side of the inhomogeneity.

4.3 The Potential Well Model for the Inhomogeneous LLE

Since the impact of inhomogeneities in the injection on LSs is rather intriguing as discussed in the previous section where both direct numerical simulations of the LL Eq. (7) and numerical continuation techniques indicated that inhomogeneities can act either attracting or repelling the LS depending on the amount of overall injection. To further analyze the transition from an attracting to a repelling inhomogeneity, we use a method that considers the single LS as an over-damped particle in a potential well and has been applied to the Swift-Hohenberg equation [92] and to the LLE [87]. To estimate semi-analytically the force exerted by the inhomogeneity at a position $R = 0$ on a particle located at position R, we perform a convolution of the inhomogeneity with the spatial derivative of the solution profile in the absence of an inhomogeneity. For this purpose, we decompose the right-hand side of Eq. 7 into a homogeneous part N_{hom} containing everything but the delayed terms and the inhomogeneity, and an inhomogeneous part N_{inh} containing the inhomogeneity, i.e.:

$$\partial_t E(\mathbf{x}, t) = N_{\text{hom}}[E] + N_{\text{inh}}[\xi] . \tag{9}$$

We consider the following ansatz

$$E(\xi, t) = E_0(\xi - R(t)) = E_{0\text{h}}(\xi - R(t)) + w(\xi - R(t)), \tag{10}$$

where E_0 and $E_{0\text{h}}$ are the stationary solution of the inhomogeneous and homogeneous system, respectively. The function $w(\xi, t)$ describes the shape deformation of the stationary solution caused by the inhomogeneity. The position $R(t)$ denotes the center of the localized structure. The goal is to derive a differential equation that describes the time evolution of the position $R(t)$. Inserting the ansatz Eq. 10 into Eq. 9 yields:

$$-\frac{\partial R}{\partial t}\partial_\xi E_0[\xi - R(t)] = N_{\text{hom}}[E_0(\xi - R(t))] + N_{\text{inh}}[\xi] . \tag{11}$$

In addition, expanding $N_{\text{hom}}[E_0(\xi - R(t))]$ around $E_{0\text{h}}(\xi - R(t))$ results in:

$$N_{\text{hom}}[E_0] = N_{\text{hom}}[E_{0\text{h}}] + \mathfrak{L}'[E_{0\text{h}}]w + \frac{1}{2}\mathfrak{L}''[E_{0\text{h}}]ww + \frac{1}{6}\mathfrak{L}'''[E_{0\text{h}}]www. \tag{12}$$

Inspection of Eq. 12 reveals that $\langle \varphi_G(\xi - R(t)) | N_{\text{hom}}[E_0(\xi - R(t))] \rangle = 0$, where φ_G is the Goldstone mode of the homogeneous system associated with zero eigenvalue [92]. In fact, $N_{\text{hom}}[E_{0\text{h}}] = 0$, because $R_{0\text{h}}$ is a stationary solution of the homogeneous system. The linear term in w vanishes, because $\mathfrak{L}'$ is a self-adjoint operator. The quadratic and cubic terms vanish, because even and odd functions are multiplied and integrated over the full domain. Projecting $\langle\varphi_G(\xi - R(t))|$ onto Eq. 9 leads to:

$$\frac{\partial R}{\partial t} = \frac{-\langle\varphi_G(\xi)|N_{\text{inh}}[\xi + R(t)]\rangle}{\langle\varphi_G(\xi)|\partial_\xi E_0(\xi)\rangle}. \tag{13}$$

The dynamics of the LS can be reduced to an ODE system with only the variable $R(t)$ which is the distance from the center of mass of the LS to the center of the inhomogeneity.

$$\frac{\partial R}{\partial t} = C \int \left\{ \text{Re}[\partial_\xi E_{\text{hls}}(\xi)]\; A e^{-(\xi+R)^2/B} \right\} d\xi = F(R), \tag{14}$$

where $C^{-1} = -\int \partial_\xi \mathbf{E}_{\text{hls}}(\xi) \cdot \partial_\xi \mathbf{E}_{\text{inhls}}(\xi) d\xi$ and $\mathbf{E}_{\text{hls}}(\xi)$ refers to the stationary LS in the homogeneous case ($A = 0$) written as a vector-function with the real and imaginary part as separate components, whereas $E_{\text{hls}}(\xi)$ refers to the same solution written as a complex scalar. $\mathbf{E}_{\text{inhls}}$ is the stationary centered LS solution in the presence of the inhomogeneity (stable or unstable) also written in vector form. The right hand side of Eq. (14) can be interpreted as the force $F(R)$ exerted by the inhomogeneity on an over-damped particle at a position R representing the LS. More detailed description of the method used to derive Eq. 14 is provided in [94]. The corresponding potential $V(R)$ therefore is defined as:

$$-\partial_R V(R) = F(R), \tag{15}$$

and can be calculated by numerically evaluating the integral in Eq. (14). It has to be noted, however, that the derivation loses its validity for large values of A; since then, the shape deformation due to the inhomogeneity cannot be neglected.

Two examples of the calculated potential $V(R)$ are depicted in Fig. 10, i.e. for $\theta = 3.0$, $A = 0.1$ and $E_i = 1.6$ (left) and $E_i = 2.0$ (right) respectively. The transition from an attracting to a repulsive inhomogeneity, which has been discussed in the previous section can be reproduced by the potential well approach. Furthermore, the center of mass positions of the stable solutions (orange lines), which have been calculated numerically, coincide well with the minima of the potential. The potential well model therefore does not only qualitatively describe the transition from an attracting to a repelling inhomogeneity but can also be deployed to quantitatively estimate the position of LSs in the vicinity of an inhomogeneity.

The potential well model applied to the LLE however has one limitation: the examples chosen in Fig. 10 both showcase where the position of the LLE is determined by its shape, i.e., the LSs pin to the inhomogeneity either with their maximum in the attracting case, or with their first minimum in the repelling case. As described previously, in a small parameter regime in between, they are pinned by certain ideal values E_{ideal} of the injection, which can be between minima and maxima. The potential well model does not reflect this behavior and therefore is not suitable to shed

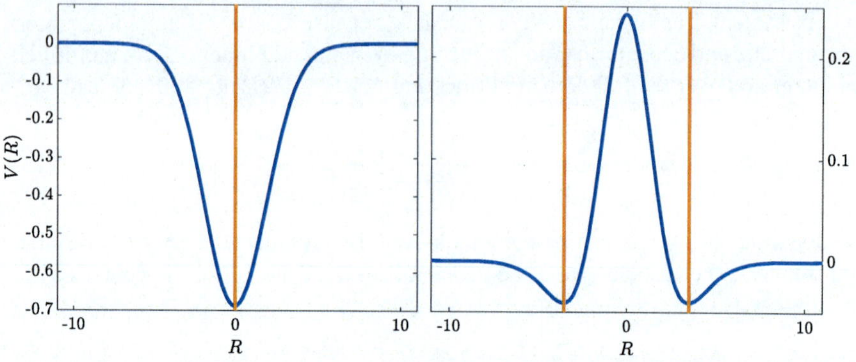

Fig. 10 Potential $V(R)$ induced by an inhomogeneity of $A = 0.1$, $B = 4.0$ for $\theta = 3.0$, $E_i = 1.6$ (left) and $E_i = 2.0$ (right). In agreement with the previous results, the inhomogeneity acts attracting in the case of the smaller injection on the left-hand side and repelling in the case of the larger injection on the right-hand side. The orange lines mark the position of the center of mass obtained from numerical continuation showing a good agreement with the minima of the potential

light on it. It has to be noted though that this parameter regime in the presence of small inhomogeneities is rather narrow, which is why the potential well model in general still can be applied successfully to the LLE.

5 Conclusions

We discussed the formation of soliton frequency combs emanating from the output of high-Q resonators filled with a Kerr media and driven by a coherent injected beam. This simple device is well described by the Lugiato-Lefever equation. In the first part, we have investigated the formation of periodic structures emerging from a three-dimensional modulational instability that occurs in the cavity under the combined influence of 2D diffraction and 1D chromatic dispersion. We showed how the system evolves from the unstable homogeneous state towards the stable three-dimensional body-centered cubic crystals. This 3D periodic structure is the most stable solution over others such as lamellae, face-centered cubic, or quasiperiodic crystals and its bifurcation structure is constructed in the absence of inhomogeneities. These structures consist of regular distribution of localized structures traveling at the group velocity of light within the cavity often called light bullets. In the regime where the modulational instability appears subcritically, another class of aperiodic but localized structures is formed in a regime where homogeneous solutions coexist with a periodic structure in a finite range of parameters called the pinning zone. We have discussed the formation of localized structures by drawing their homoclinic type of bifurcation and have discussed their link to frequency combs. Frequency combs generated in high-Q optical resonators are nothing but the spectral content of the stable localized structure occurring in the cavity.

The second part is more focused on the effect of inhomogeneity. We have considered an inhomogeneous injection in the form of a Gaussian beam. We have shown that for positive values of A, inhomogeneity acts as an attracting force on LS, and that the LS pinned on the center of the inhomogeneity also persists for small negative values of the inhomogeneity. In the case of negative A, the inhomogeneity acts repelling on the center of the localized state. Finally, we have discussed the application of the potential well approach after constructing the homoclinic snaking bifurcation.

Acknowledgements K.P. acknowledges support by the Fonds Wetenschappelijk Onderzoek-Vlaanderen FWO (G0E5819N) and the Methusalem Foundation. M.T acknowledges financial support from the Fonds de la Recherche Scientifique FNRS under Grant CDR no. 35333527 "Semiconductor optical comb generator". A part of this work is supported by the "Laboratoire Associé International" University of Lille - ULB on "Self-organisation of light and extreme events" (LAI-ALLURE).

References

1. T. Udem, R. Holzwarth, T.W. Hänsch, Optical frequency metrology. Nature **416**, 233 (2002)
2. D.J. Jones, S.A. Diddams, J.K. Ranka et al., Carrier-envelope phase control of femtosecond mode-locked lasers and direct optical frequency synthesis. Science **288**, 635 (2000)
3. P. Del'Haye, A. Schliesser, O. Arcizet, T. Wilken, R. Holzwarth, T.J. Kippenberg, Optical frequency comb generation from a monolithic microresonator. Nature **450**, 1214 (2007)
4. T.W. Hänsch, Rev. Modern Phys. **78**, 1297 (2006)
5. T. Fortier, E. Baumann, Commun. Phys. **2**, 1 (2019)
6. T.J. Kippenberg, S.M. Spillane, K.J. Vahala, Appl. Phys. Lett. **85**, 6113 (2004)
7. S.M. Spillane, T.J. Kippenberg, K.J. Vahala, K.W. Goh, E. Wilcut, H.J. Kimble, Phys. Rev. A **71**, 013817 (2005)
8. D.W. Vernooy, V.S. Ilchenko, H. Mabuchi, E.W. Streed, H.J. Kimble, Opt. Lett. **23**, 247 (1998)
9. L.A. Lugiato, R. Lefever, Phys. Rev. Lett. **58**, 2209 (1987)
10. T. Herr, V. Brasch, J.D. Jost, C.Y. Wang, N.M. Kondratiev, M.L. Gorodetsky, T.J. Kippenberg, Nat. Photon. **8**, 145 (2014)
11. T.J. Kippenberg, A. L. Gaeta, M. Lipson, and M. L. Gorodetsky, Science **361**, eaan8083 (2018)
12. A.J. Scroggie, W.J. Firth, G.S. McDonald, M. Tlidi, R. Lefever, L.A. Lugiato, Chaos. Solitons Fractals **4**, 1323 (1994)
13. F. Arecchi, S. Boccaletti, P. Ramazza, Phys. Rep. **318**, 1 (1999)
14. P. Mandel, M. Tlidi, J. Opt. B Quant. Semicl. Opt. **6**, R60 (2004)
15. B.A. Malomed, D. Mihalache, F. Wise, L. Torner, J. Opt. B Quant. Semicl. Opt. **7**, R53 (2005)
16. N. Akhmediev, A. Ankiewicz, *Dissipative Solitons: From Optics to Biology and Medicine*, Vol. 751 (Springer, 2008)
17. T. Ackemann, W. Firth, G.-L. Oppo, Adv. At. Mol. Phys. **57**, 323 (2009)
18. N.N. Rosanov, *Spatial Hysteresis and Optical Patterns* (Springer Science & Business Media, 2013)
19. H. Leblond, D. Mihalache, Phys. Rep. **523**, 61 (2013)
20. M. Tlidi, K. Staliunas, K. Panajotov, A.G. Vladimirov, M.G. Clerc, Philos. Trans. R. Soc. A Math. Phys. Eng. Sci. **372**, 20140101 (2014)
21. M. Tlidi, M.G. Clerc, *Nonlinear Dynamics: Materials, Theory and Experiments*, vol. 173 (Springer, 2016)
22. D. Mihalache, Roman. Rep. Phys. **69**, 403 (2017)
23. V. Odent, M. Tlidi, M.G. Clerc, P. Glorieux, E. Louvergneaux, Phys. Rev. A **90**, 011806(R) (2014)

24. S. Coen, M. Tlidi, P. Emplit, M. Haelterman, Phys. Rev. Lett. **83**, 2328 (1999)
25. Y. Xu, S. Coen, Opt. Lett. **39**, 3492 (2014)
26. I. Hendry, W. Chen, Y. Wang, B. Garbin, J. Javaloyes, G.-L. Oppo, S. Coen, S.G. Murdoch, M. Erkintalo, Phys. Rev. A **97**, 053834 (2018)
27. D.C. Cole, J.R. Stone, M. Erkintalo, K.Y. Yang, X. Yi, K.J. Vahala, S.B. Papp, Optica **5**, 1304 (2018)
28. S.S. Gopalakrishnan, K. Panajotov, M. Taki, M. Tlidi, Phys. Rev. Lett. **126**, 153902 (2021)
29. M. Tlidi, M. Haelterman, P. Mandel, EPL (Europhys. Lett.) **42**, 505 (1998)
30. M. Tlidi, M. Le Berre, E. Ressayre, A. Tallet, L. Di Menza, Phys. Rev. A **61**, 043806 (2000)
31. M. Tlidi, P. Mandel, M. Le Berre, E. Ressayre, A. Tallet, L. Di Menza, Opt. Lett. **25**, 487 (2000)
32. X. Dong, C. Spiess, V.G. Bucklew, W.H. Renninger, Phys. Rev. Res. **3**, 033252 (2021)
33. M. Haelterman, S. Trillo, S. Wabnitz, Opt. Commun. **91**, 401 (1992)
34. P. Kockaert, P. Tassin, G. Van der Sande, I. Veretennicoff, M. Tlidi, Phys. Rev. A **74**, 033822 (2006)
35. Y.K. Chembo, D. Gomila, M. Tlidi, C.R. Menyuk, Eur. Phys. J. D **71**, 299 (2017)
36. L. Lugiato, F. Prati, M. Gorodetsky, T. Kippenberg. Philos. Trans. R. Soc. A Math. Phys. Eng. Sci. **376**, 20180113 (2018)
37. M. Tlidi, M. Clerc, K. Panajotov, Philos. Trans. R. Soc. A Math. Phys. Eng. Sci. **376**, 20180114 (2018)
38. M. Tlidi, M. Clerc, K. Panajotov, Philos. Trans. R. Soc. A Math. Phys. Eng. Sci. **376**, 20180276 (2018)
39. M. Tlidi, J. Opt. B Quant. Semicl. Opt. **2**, 438 (2000)
40. M. Tlidi, R. Lefever, P. Mandel, J. Opt. B Quant. Semicl. Opt. **8**, 931 (1996)
41. L.N. Trefethen, *Spectral Methods in MATLAB* (SIAM, Philadelphia, 2000)
42. S.M. Cox, P.C. Matthews, J. Comput. Phys. **176**, 430 (2002)
43. A.K. Kassam, High order timestepping for stiff semilinear partial differential equations (University of Oxford, 2004)
44. Y. Saad, *Numerical Methods for Large Eigenvalue Problems* (SIAM, Philadelphia, 2011)
45. M. Tlidi, S.S. Gopalakrishnan, M. Taki, K. Panajotov, Chaos. Solitons Fractals **152**, 111364 (2021)
46. W.J. Firth, A.J. Scroggie, G.S. McDonald, L.A. Lugiato, Phys. Rev. A **46**, R3609 (1992)
47. D. Gomila, P. Colet, Phys. Rev. A **68**, 011801 (2003)
48. D. Gomila, P. Colet, Phys. Rev. E **76**, 016217 (2007)
49. D. Gomila, A. Scroggie, W. Firth, Phys. D Nonlinear Phenomena **227**, 70 (2007)
50. W.J. Firth, G.K. Harkness, A. Lord, J.M. McSloy, D. Gomila, P. Colet, J. Opt. Soc. Am. B **19**, 747 (2002)
51. A.G. Vladimirov, J.M. McSloy, D.V. Skryabin, W.J. Firth, Phys. Rev. E **65**, 046606 (2002)
52. D. Turaev, A.G. Vladimirov, S. Zelik, Phys. Rev. Lett. **108**, 263906 (2012)
53. A.G. Vladimirov, S.V. Gurevich, M. Tlidi, Phys. Rev. A **97**, 013816 (2018)
54. A.G. Vladimirov, M. Tlidi, M. Taki, Phys. Rev. A **1083**, 063505 (2021)
55. S. Wabnitz, Phys. Rev. A **38**, 2018 (1988)
56. T. Hansson, M. Bernard, S. Wabnitz, J. Opt. Soc. Am. B **35**, 835 (2018)
57. J. Fatome, B. Kibler, F. Leo, A. Bendahmane, G.-L. Oppo, B. Garbin, S.G. Murdoch, M. Erkintalo, S. Coen, Opt. Lett. **45**, 5069 (2020)
58. E. Averlant, M. Tlidi, K. Panajotov, L. Weicker, Opt. Lett. **42**, 2750 (2017)
59. F. Copie, M.T.M. Woodley, L. Del Bino, J.M. Silver, S. Zhang, P. Del'Haye, Phys. Rev. Lett. **122**, 013905 (2019)
60. R. Suzuki, S. Fujii, A. Hori, T. Tanabe, IEEE Photon. J. **11**, 1 (2019)
61. M. Saha, S. Roy, S.K. Varshney, Phys. Rev. A **101**, 033826 (2020)
62. B. Garbin, J. Fatome, G.-L. Oppo, M. Erkintalo, S.G. Murdoch, S. Coen, Phys. Rev. Lett. **126**, 023904 (2021)
63. B. Kostet, S.S. Gopalakrishnan, E. Averlant, Y. Soupart, K. Panajotov, M. Tlidi, OSA Continuum **4**, 1564 (2021)
64. B. Kostet, Y. Soupart, K. Panajotov, M. Tlidi, Phys. Rev. A **104**, 053530 (2021)

65. Y.K. Chembo, D.V. Strekalov, N. Yu, Phys. Rev. Lett. **104**, 103902 (2010)
66. A. Coillet, Y.K. Chembo, Chaos Interdiscip. J. Nonlinear Sci. **24**, 013113 (2014)
67. M. Anderson, F. Leo, S. Coen, M. Erkintalo, S.G. Murdoch, Optica **3**, 1071 (2016)
68. Z. Liu, M. Ouali, S. Coulibaly, M. Clerc, M. Taki, M. Tlidi, Opt. Lett. **42**, 1063 (2017)
69. N. Akhmediev et al., J. Opt. **18**, 063001 (2016)
70. M. Tlidi, K. Panajotov, Chaos Interdiscip. J. Nonlinear Sci. **27**, 013119 (2017)
71. M.G. Clerc, M.A., Ferré, S., Coulibaly, R.G. Rojas, M. Tlidi, Opt. Lett. **42**, 2906 (2017)
72. M.G. Clerc, S. Coulibaly, M.A. Ferré, M. Tlidi, Chaos Interdiscip. J. Nonlinear Sci. **30**, 043107 (2020)
73. P. Parra-Rivas, D. Gomila, M. Matias, S. Coen, L. Gelens, Phys. Rev. **89**, 043813 (2014)
74. P. Parra-Rivas, M.A. Matías, P. Colet, L. Gelens, D. Walgraef, D. Gomila, Phys. Rev. E **95**, 020201 (2017)
75. P. Tassin, G. Van der Sande, N. Veretenov, P. Kockaert, I. Veretennicoff, M. Tlidi, Opt. Express **14**, 9338 (2006)
76. N. Veretenov, M. Tlidi, Phys. Rev. A **80**, 023822 (2009)
77. M. Tlidi, P. Mandel, R. Lefever, Phys. Rev. Lett. **73**, 640 (1994)
78. A. Bergeon, J. Burke, E. Knobloch, I. Mercader, Phys. Rev. E **78**, 046201 (2008)
79. G.W. Hunt, G.J. Lord, A. Champneys, Comput. Methods Appl. Mech. Eng. **170**, 239 (1999)
80. P. Woods, A. Champneys, Phys. D Nonlinear Phenom. **129**, 147 (1999)
81. P. Coullet, C. Riera, C. Tresser, Phys. Rev. Lett. **84**, 3069 (2000)
82. J. Burke, E. Knobloch, Chaos Interdiscip. J. Nonlinear Sci. **17**, 037102 (2007)
83. M. Tlidi, M. Taki, T. Kolokolnikov, Chaos Interdiscip. J. Nonlinear Sci. **17**, 037101 (2007)
84. A. Champneys, IMA J. Appl. Math. **86**, 845 (2021)
85. M. Tlidi, L. Gelens, Opt. Lett. **35**, 306 (2010)
86. P. Parra-Rivas, D. Gomila, L. Gelens, E. Knobloch, Phys. Rev. E **97**, 042204 (2018)
87. F. Tabbert, T. Frohoff-Hlsmann, K. Panajotov, M. Tlidi, S.V. Gurevich, Phys. Rev. A **100**, 013818 (2019)
88. F. Tabbert, S.V. Gurevich, K. Panajotov, M. Tlidi, Chaos Solitons Fractals **152**, 111317 (2021)
89. Y. Xu, S. Coen, Opt. Lett. **39**, 3492 (2014)
90. I. Hendry, W. Chen, Y. Wang, B. Garbin, J. Javaloyes, G.-L. Oppo, S. Coen, S. Stuart, M. Erkintalo, Phys. Rev. A **97**, 053834 (2018)
91. D.C. Cole, J.R. Stone, M. Erkintalo, K.Y. Yang, X. Yi, K.J. Vahala, S.B. Papp, Optica **5**, 1304 (2018)
92. F. Tabbert, K. Panajotov, M. Tlidi, S.V. Gurevich, Phys. Rev. E **95**, 032213 (2017)
93. J. Burke, E. Knobloch, Phys. Rev. E **73**, 056211 (2006)
94. F. Tabbert, Ph.D. Thesis, University of Munster (2019)

Optimization in Engineering Processes: An Application of a Generalized Fluctuation–Dissipation Theorem

Annie Steinchen

Dedicated to the memory of Slava Belyi
A missed colleague
Recently deceased from Covid-19

Abstract The generalization of the fluctuation–dissipation theorem is presented as a tool to optimize transport processes beyond the linear laws of irreversible thermodynamics. The stability criterion accounting for non-local effects enlightens the prominent role of the generalized quality factor Q and of the phase shift between forces and fluxes and allows us to understand the microscopic response of complex systems to fluctuations. It is the purpose of the present small review paper to show example systems whose size is of the order of magnitude of the mean free path of the heat carriers as it is in nanoscopic devices.

1 Introduction

During the last century, the linear laws of irreversible thermodynamics as first introduced by L. Onsager [1] have been widely used by engineers [2] and the minimum of entropy production [3–5] valid for these linear phenomenological laws was used as a variational principle ruling the optimization of energy exchanges even out of its application validity.

As already shown by many physicists for more than 50 years, when the phenomenological laws between forces and fluxes deviate from the linearity [6, 7], new variational principles are governing the regression of the fluctuations [8] so that these last variational principles are now introduced to optimize processes in many complex systems.

When extending irreversible thermodynamics to non-linear phenomena undergoing stochastic variations of the forces and fluxes, Nicolis [9] has derived an

A. Steinchen (✉)
MADIREL, Université Aix-Marseille, Marseille, France
e-mail: annie.steinchen@gmail.com

L. Brenig et al. (eds.), *Nonequilibrium Thermodynamics and Fluctuation Kinetics*, Fundamental Theories of Physics 208,
https://doi.org/10.1007/978-3-031-04458-8_13

adequate definition of entropy and entropy production accounting for the effect of the fluctuations that reduce to Prigogine's formalism of classical irreversible thermodynamics in the mean-field limit.

In his two last papers, Belyi [10, 11] shows that the spectral function of the fluctuations is determined not only by dissipation but also by the derivatives of the dispersion. New non-local effect originating from an additional phase shift between the force and the response of the system is expected to be responsible for the properties of nanomaterials such as carbon nanotubes [12] or graphene films, commonly used now in advanced technologies.

Non-Fourier heat transfer and thermal waves [13] are indeed often reported [14] in new devices designed for microelectronic cooling.

The extended thermodynamics has been applied recently by Jou et al. to heat transfer in nanotubes [15, 16] and has been reviewed recently by G. Lebon [17].

Recently, Dong et al. [18] on the basis of the thermomass model of Cao and Guo [19] gave a generalized expression of the entropy production for non-linear heat transfer.

Because of semantic problems, it is often difficult to compare engineers' and physicists' approaches, and to give a sense of the word "optimization" in terms of stability, regression of fluctuations, and variational principles, it was for me a nightmare that I try to evacuate in the next section.

2 Generalized Entropy Production in Non-Fourier Heat Transfer

Historically, the linear law for heat conduction was shown to fail in solid dielectrics at low temperatures. The search for a generalized Fourier equation has led Cattaneo [20] and Vernotte to introduce their famous "telegrapher equation" that describes the finite propagation speed of the second sound phonon gas and provides information on the phonon scattering in solid-state heat transfer processes.

The Cattaneo–Vernotte equation introduces in addition to the classical Fourier law for the heat flow q, an inertial term $\tau \partial_t q$ where τ is the heat flux relaxation time

$$\tau \partial_t q + q = -\lambda \nabla T \tag{1}$$

where λ is the heat conductivity (generally temperature-dependent).

Together with the energy conservation law, the following temperature equation is obtained, at constant τ and λ called the telegrapher equation by analogy with the propagation of a signal along an electric wire

$$\tau \frac{\partial^2 T}{\partial t^2} + \frac{\partial T}{\partial t} = \chi \nabla^2 T \tag{2}$$

where $\chi = \lambda/cv$ is the heat diffusivity.

As emphasized by Lebon [17], in nanomaterials, the mean free path of the heat carriers is of the order of magnitude of the size of the system so that neither the classical law of Fourier nor even Cattaneo's equation is applicable so that the foundations of classical non-equilibrium thermodynamics have been revisited. The role of boundary conditions is also of fundamental importance in nanodevices, because in such systems, bulk and boundary effects may be of the same order of magnitude. Due to their mesoscopic sizes, the nanosystems make the bridge between the macroscopic description of irreversible thermodynamics and the microscopic approach of statistical mechanics. The competition between the diffusive and ballistic modes of heat transfer is described by a parameter linked to the size L of the system: the Knudsen number $Kn = 1/L$. For large values of L, $Kn << 1$, the regime is diffusive while when L is of the same order of magnitude or smaller than the mean free path of heat carriers, $Kn \geq 1$ the ballistic regime dominates.

In the extended irreversible thermodynamics approach, the dissipative fluxes are introduced as new variables of the generalized non-equilibrium entropy [6]. This leads to the definition of thermodynamic forces and fluxes in order to maintain the requirement of positiveness of the entropy production.

The non-equilibrium entropy is smaller than the equilibrium entropy depending on the heat flux as an additional independent variable and reads

$$s(e, q) = s_{eq}(e, 0) - \frac{\tau}{(2\lambda T^2)q^2} \tag{3}$$

In other words, the equilibrium entropy contains in addition to the entropy of non-equilibrium, a term due to the entropy created to reach equilibrium $\tau/(2\lambda T^2)q^2$.

By a Legendre transform, the non-equilibrium entropy may be written as the function of the temperature gradient

$$s * (e, \nabla T) = s(e, q) - \frac{\tau \nabla T}{T^2} \tag{4}$$

It is interesting to note that Cattaneo's relation is now derivable from a potential identified as s^* $(e, \nabla T)$; indeed, $\partial s/\partial \boldsymbol{q} = 0$ at the Fourier heat flux $\boldsymbol{q} = -\lambda \nabla T$, i.e., at the local equilibrium value of $\boldsymbol{q}$ instead of the global equilibrium value $\boldsymbol{q} = 0$ for $s(e, \boldsymbol{q})$ chosen as the reference entropy. As shown by Lebon [17], the consequences of this new non-equilibrium presentation are that it avoids the paradox of infinite velocity of propagation of high-frequency signals and that it provides explicit formulations of non-equilibrium entropy and temperature.

The non-local effects in small systems, peculiarly in non-Fourier heat transfers, give a concrete example of the fluctuation–dissipation–dispersion relation for slow processes as recently derived by Belyi [11].

Like the non-Joule dispersion contribution characterized by a non-local effect due to an additional phase shift between the force and the response of the system, in nanomaterials non-Fourier heat transfer, there exists a similar balance between the two governing parameters mainly the heat capacity and the heat conductivity as between electric capacitance and inductance in electric oscillation in plasma investigated by Belyi [10] in its generalization of the Callen–Welton formula to systems with slowly varying parameters.

The analogy appears straightforward and coherent with the recent ballistic-diffusion model introduced by Chen [21, 22] and used for optimization purposes by numerous Chinese engineering teams, for example, Xu M. and Hu H. [23] or Cao [24].

3 Conclusions

The nanoscopic systems are fascinating objects of research both in engineering and in fundamental physics. There is a need to match both approaches in order to design the most efficient nanodevices for cooling microelectronic devices. Optimization in engineering aims to avoid the dangerous effects of amplification of temperature fluctuations that could occur when the spectral function of the fluctuations is determined not only by the dissipation but also by the derivatives of the dispersion.

The dichotomy between microscopic and macroscopic descriptions is particularly important in these mesoscopic systems that require the use of a multiscale approach. Moreover, as the surface properties become dominant, all the governing parameters have to be renormalized, the phonon–boundary interaction being sometimes prominent. Even the definition of temperature in these systems is still to be clarified. A very stimulating area of research!

References

1. L. Onsager, Phys. Rev. **37**, 450 (1931)
2. A. Bejan, Entropy generation minimization: the new thermodynamics of finite size devices and finite time processes. J. Appl. Phys. **79**, 1191–1218 (1996)
3. I. Prigogine, Etude thermodynamique des phénomènes irréversibles, Desoer, Liège (1947)
4. P. Glansdorff, I. Prigogine, in *Thermodynamics of Structure, Stability and Fluctuations* (Wiley, New York, 1971)
5. D. Kondepudi, I. Prigogine, *Modern Thermodynamics from Heat Engines to Dissipative Structures* (Wiley, Chichester, 1998)
6. D. Jou, J. Casas-Vazquez, G. Lebon, *Extended Irreversible Thermodynamics* (Springer, Berlin-Heidelberg, 1993)
7. M. Velarde, Toward a non-equilibrium non-linear thermodynamics, in *Dynamics of Multiphase Flows Across Interfaces*, ed. by A. Steinchen, Lecture Notes in Physics (Springer, 1996), pp. 253–267
8. J. Verhas, Gyarmaty's variational principle of dissipative processes. Entropy **16**, 2362–2383 (2014)
9. G. Nicolis, Y. De Deker, Stochastic approach to irreversible thermodynamics. Chaos **27**, 104615 (2017)
10. V.V. Belyi, Fluctuations out of equilibrium. Phil. Trans. R. Soc. A **376**, 20170383 (2018)

11. V.V. Belyi, Fluctuation-dissipation-dispersion relation for slow processes and quality factor for oscillation systems (2019)
12. A.P. Paz, J.M. Garcia-Lastra, T. Markussen, K.S. Thygesen, A. Rubio, Carbon nanotubes as heat dissipaters in microelectronics. Eur Phys J B **86**, 234–247 (2013)
13. S. Huberman, R.A. Duncan, K. Chen, B. Song, V. Chiloyan, Z. Ding, A.A. Maznev, G. Chen, K.A. Nelson, Observation of second sound in graphite at temperatures above 100 K. Science **364**, 374–379 (2019)
14. M. Wang, N. Wang, Z.-Y. Guo, Non-Fourier heat conductions in nanomaterials. J. Appl. Phys. **110**, 064310 (2011)
15. D. Jou, J. Casas-Vázquez, G.G. Lebon, M, A phenomenological scaling approach for heat transport in nano-systems., Appl. Math. Lett. 18 (2005) 963–967
16. D. Jou, A. Sellito, F.X. Alvarez, Heat waves and phonon–wall collisions in nanowires. Proc. R. Soc. A **467**, 2520–2533 (2011)
17. G. Lebon, Heat conduction at micro and nanoscales: a review through the prism of Extended Irreversible Thermodynamics. J. Non-Equilib. Thermodyn. **39**(1), 35–59 (2014)
18. Y. Dong, B.-Y. Cao, Z.-Y. Guo, General expression for entropy production in transport processes based on the thermomass model. Phys. Rev. E **85**, 061107(1–7) (2012)
19. B.-Y. Cao, Z.-Y. Guo, Equation of motion of a phonon gas and non-Fourier heat conduction. J. Appl. Phys. **102**, 053503 (2007)
20. C. Cattaneo, A form of hea t conduction which eliminates the paradox of instantaneous propagation. Comptes Rendus Acad. Sci. Paris **247**, 431–433 (1958)
21. G. Chen, Ballistic-diffusion heat-conduction equations. Phys. Rev. Lett. **86**, 2297–2300 (2001)
22. G. Chen, Ballistic-diffusionequations for transient heat conduction from nano to macroscales. ASME J. HeatTransf. **124**, 320–329 (2002)
23. M. Xu, H. Hu, A ballistic-diffusive heat conduction model extracted from Boltzmann transport equation. Proc. R. Soc. A **467**, 1851–1864 (2011)
24. Y.-C. Hua, C. B-Y., Ballistic-diffusive heat conduction in multiply-constrained nanostructures. Int. J. Therm. Sc. **101**, 126–132 (2016)

Co-evolutionary Complex Networks

Ioannis Antoniou, Evangelos Ioannidis, and Nikos Varsakelis

Abstract Co-evolutionary networks are mathematical graphs which are not passive media of communication channels, simply supporting the spread of information. The channels are actually redefined according to the spread of information among the nodes. Co-evolutionary networks are learning. We report results on both passive spread and active co-evolution in real networks, namely the influence of communication policies among agents, the impact of "experts" and "change agents", hidden influence, the spread of false beliefs, conflict, and the emergence of polarization and centralization.

1 Introduction

When Slava sent to one of us (IA) his work on the generalized fluctuation–dissipation theorem [1] and subsequent work [2], we realized that detailed balance was under discussion [3]. In the graph theoretic representation, detailed balance means locally that the in- and out-flows are balanced for each node. This of course is true for transport networks, but not for communication networks. Thermodynamics, reformaluted in the language of Networks [4–6], reveals the geometry of macroscopic relations. Diffusion in networks is a discrete process formulated as the spread of local (node) properties in the network via the links. The key advantage of network

I. Antoniou (✉) · E. Ioannidis
Faculty of Sciences, School of Mathematics, Aristotle University of Thessaloniki, 54124 Thessaloniki, Greece
e-mail: iantonio@math.auth.gr

E. Ioannidis
e-mail: ioannidek@math.auth.gr

N. Varsakelis
Faculty of Economic and Political Sciences, School of Economics, Aristotle University of Thessaloniki, 54124 Thessaloniki, Greece
e-mail: barsak@econ.auth.gr

L. Brenig et al. (eds.), *Nonequilibrium Thermodynamics and Fluctuation Kinetics*, Fundamental Theories of Physics 208,
https://doi.org/10.1007/978-3-031-04458-8_14

theory [7, 8] is the explicit incorporation of interdependence among the elementary constituents of real systems. Interdependence is the mechanism conditioning the emergence of structures far from equilibrium as emphasized by Ilya Prigogine [9]. We review recent results concerning non-equilibrium spread in both passive and active complex networks. The key idea underlying our work is Prigogine's formulation of traffic. Prigogine distinguished cars from molecules by the fact that molecules just collide and spread in a passive way, while car traffic is conditioned by active selections among possible links [10–12]. This intelligent active transport is different from the conventional energy-consuming active transport inspired by biology [13]. The spread dynamics in communication networks is formulated in Sect. 2. The influence of communication policies is discussed in Sect. 3 and the impact of "experts" in Sect. 4. Attitude Change and Hidden Influences are addressed in Sect. 5 and the spread of false beliefs in Sect. 6. Finally, we discuss the emergence of Polarization (Sect. 7) and centralization (Sect. 8) in Co-Evolutionary Networks. Co-evolution means that space is not just a passive medium of communication channels supporting the spread, but a distributed medium of learning according to the spread of information.

2 Spread Dynamics

We denote by $\psi_\kappa(t)$ the activation of agent κ at time t. Activation means any relevant qualification of each agent κ, such as knowledge level, attitude, opinion, and state of health. The spread dynamics equation of the activations $\psi_\kappa(t)$ in a network of N agents is a generalization of the discrete diffusion equation in networks:

$$\psi_\kappa(t+1) = \psi_\kappa(t) + \sum_{\nu=1}^{\mathrm{N}} \Phi_{\nu\kappa}(t) \tag{1}$$

$\sum_{\nu=1}^{\mathrm{N}} \Phi_{\nu\kappa}(t)$ is the influence function of agent κ during time interval $(t, t+1]$.

$\Phi_{\nu\kappa}(t)$ is the influence function of agent κ from agent ν during time interval $(t, t+1]$.

The influence of agent κ effectively sums the influences $\Phi_{\nu\kappa}$ from all agents of the network. The specific form of the influence function depends on the nature of the activation as well as on the modeling assumptions. In the case of activation spread, the influence function is

$$\Phi_{\lambda\kappa}(t) = [\![\mathfrak{D}_\kappa^{\mathrm{s}}(\mathrm{t}) = \lambda]\!] \cdot \mathrm{w}_{\lambda\kappa} \cdot [\![\varepsilon_{\lambda\kappa}^{\min} \leq \psi_\lambda(\mathrm{t}) - \psi_\kappa(\mathrm{t}) \leq \varepsilon_{\lambda\kappa}^{\max}]\!] \cdot (\psi_\lambda(\mathrm{t}) - \psi_\kappa(\mathrm{t})) \tag{2}$$

the term $[\![Q]\!]$ is the Iverson bracket which converts Boolean values to numbers 0, 1:

$$[\![Q]\!] = \begin{cases} 1, & \textit{if } Q \textit{ is True} \\ 0, & \textit{if } Q \textit{ is False} \end{cases}$$

$\psi_\kappa(t)$ is the activation of agent κ at time t, taking values in the interval $[-1, 1]$ assessing the degree of positiveness or negativeness.

$w_{\lambda\kappa}$ is the weight of influence of agent λ on agent κ, for activation change, taking values in the interval $[-1, 1]$.

$\mathfrak{D}^{\mathbb{s}}_\kappa(t)$ is the selection of the in-neighbor λ by agent κ for communication, specified by the corresponding selection probability $p^{\mathbb{s}}_{\lambda\kappa}$. The bracket $[\![\mathfrak{D}^{\mathbb{s}}_\kappa(t) = \lambda]\!]$ guarantees that only agent λ was selected by agent κ at time t. Here, it is implicitly assumed that **only one** agent is selected at a time. The selection probability $p^{\mathbb{s}}_{\lambda\kappa}$ reflects the adopted selection rule ($\mathbb{s}$) [14–16].

$\varepsilon^{min}_{\lambda\kappa}$ and $\varepsilon^{max}_{\lambda\kappa}$ are the "confidence bounds" for influence [15]. The bracket $[\![\varepsilon^{min}_{\lambda\kappa} \le \psi_\lambda(t) - \psi_\kappa(t) \le \varepsilon^{max}_{\lambda\kappa}]\!]$ guarantees that activation influence may take place only if $\varepsilon^{min}_{\lambda\kappa} \le \psi_\lambda(t) - \psi_\kappa(t) \le \varepsilon^{max}_{\lambda\kappa}$. This actually is a filtering rule ($\mathbb{f}$), excluding (filtering out) agents inappropriate for communication. The assumption that the influence of agent λ on agent κ is proportional to the difference in activations $\psi_\lambda(t) - \psi_\kappa(t)$ is common in social influence models [14, 15]. This assumption is actually a discrete form of diffusion.

3 Communication Policies in Organizational Knowledge Networks

In knowledge networks, the activation $\psi_\kappa(t)$ of each agent κ is its knowledge level, taking values in the interval $[0, 1]$. The knowledge transfer efficiency from agent λ to agent κ defines the weight $w_{\lambda\kappa}$ of the communication channel $\lambda \to \kappa$, taking values in the interval $[0, 1]$. Knowledge transfer between two agents may take place only from the agent with higher knowledge to the agent with lower knowledge, i.e. agent κ may upgrade his/her knowledge level by the selected in-neighbor agent λ, only if $\psi_\lambda(t) - \psi_\kappa(t) > 0$. The opposite direction does not contribute to the knowledge of the selecting agent κ. This assumption is effectively a filtering rule ($\mathbb{f}$) over the set of in-neighbors, regarding the possibility for knowledge upgrade of the selecting agent κ:

$$[\![\psi_\lambda(t) - \psi_\kappa(t) > 0]\!]$$

Each agent in real networks adopts some communication policy [14], which specifies (a) the selection rule for communication, and (b) the prioritization of selection versus filtering. There are 4 main selection rules, namely (i) random selection, (ii) selection taking into account the incoming weights of the in-neighbors, (iii) selection taking into account the knowledge levels of the in-neighbors, and (iv) selection taking into account both the knowledge levels and the incoming weights of the in-neighbors. We present two extreme communication policies, namely

(A) The conventional communication policy $\mathbb{f}(\mathbb{s} = r)$ where filtering is implemented after selection ($\mathbb{fs}$ prioritization). The selections of agents are assumed random ($\mathbb{s} = r$) with selection probability:

$$p_{\lambda\kappa}^{\mathbb{s}=r} = \frac{[\![w_{\lambda\kappa} \neq 0]\!]}{\sum_{\nu=1}^{N} [\![w_{\nu\kappa} \neq 0]\!]}$$

(B) The communication policy $(\mathbb{s} = w\mathfrak{K})\mathbb{f}$ where filtering is implemented before selection ($\mathbb{sf}$ prioritization). The selections ($\mathbb{s} = w\mathfrak{K}$) of agents take into account the knowledge transfer efficiency (w) of each communication channel as well as the knowledge level ($\mathfrak{K}$) of each communicating agent. In this case, the selection probability is

$$p_{\lambda\kappa}^{\mathbb{s}=w\mathfrak{K}} = \frac{w_{\lambda\kappa} \cdot \psi_{\lambda}(t)}{\sum_{\nu=1}^{N} (w_{\nu\kappa} \cdot \psi_{\nu}(t))}$$

The influence function (2) of the spread dynamics equation (1) takes two forms, corresponding to the $\mathbb{fs}$ prioritization and $\mathbb{sf}$ prioritization:

$$\Phi_{\nu\kappa}^{\mathbb{fs}}(t) = [\![\psi_{\nu}(t) - \psi_{\kappa}(t) > 0]\!] \cdot [\![\mathfrak{D}_{\kappa}^{\mathbb{s}}\left(\psi(t), w_{\mu\kappa}\right) = \nu]\!] \cdot w_{\nu\kappa} \cdot [\psi_{\nu}(t) - \psi_{\kappa}(t)] \tag{3}$$

$$\Phi_{\nu\kappa}^{\mathbb{sf}}(t) = [\![\mathfrak{D}_{\kappa}^{\mathbb{s}}\left(\psi(t), [\![\psi_{\mu}(t) - \psi_{\kappa}(t) > 0]\!] \cdot w_{\mu\kappa}\right) = \nu]\!] \cdot w_{\nu\kappa} \cdot [\psi_{\nu}(t) - \psi_{\kappa}(t)] \tag{4}$$

Faster knowledge attainment within organizations leads to improved innovation, and therefore competitive advantage. Interventions on the organizational network may be risky or costly or time-demanding. We have investigated [14] several communication policies in knowledge networks, which reduce the knowledge attainment time without interventions. We present (Fig. 1) the impact of the selection rule and prioritization on the emerging knowledge dynamics for two real organizational networks, namely a Consulting Company (CC) and a Manufacturing Company (MC).

We see clearly from Fig. 1 that the most efficient communication policy $(\mathbb{s} = w\mathfrak{K})\mathbb{f}$ results in significant improvement of knowledge dynamics, compared to the conventional $\mathbb{f}(\mathbb{s} = r)$ policy. For the CC network, the improvement is about a 58% decrease in the Knowledge Attainment Time, while for the MC network the corresponding decrease is 74%. The impact resulting from changing the communication policy from $\mathbb{f}(\mathbb{s} = r)$ to $(\mathbb{s} = w\mathfrak{K})\mathbb{f}$ is stronger for larger network sizes [14].

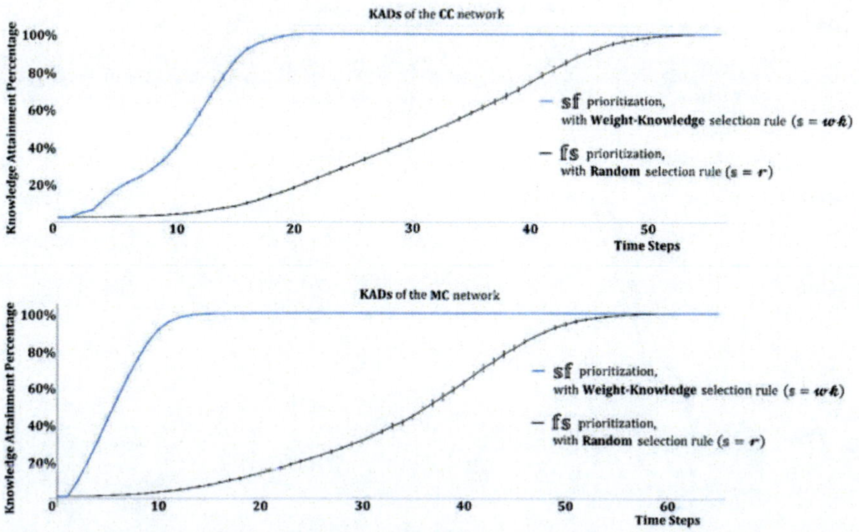

Fig. 1 Knowledge Attainment Diagrams (KAD) of the CC network (up) and the MC network (down), comparing the $(\mathbb{s} = w\kappa)\mathbb{f}$ communication policy versus the conventional $\mathbb{f}(\mathbb{s} = r)$ policy. The vertical axis is the percentage of the agents of the network, who have attained the maximal knowledge due to spread, and the horizontal axis is time

4 Experts in Organizational Knowledge Networks

We have seen (Sect. 3) that "selecting" after "filtering" ($\mathbb{sf}$ prioritization) is more effective compared to conventional $\mathbb{fs}$ prioritization. It is well-known that the presence of experts (highly knowledgeable agents) in the network generally improves the performance of knowledge networks. We investigate the impact of prioritization combined with experts' positioning in the network. We discuss the simplest random $(\mathbb{s} = r)$ selection rule for communication with both prioritizations $\mathbb{f}(\mathbb{s} = r)$ and $(\mathbb{s} = r)\mathbb{f}$. The influence function (2) of the spread dynamics equation (1) takes the corresponding forms:

$$\Phi_{\nu\kappa}^{\mathbb{f}(\mathbb{s}=r)}(t) = [\![\psi_\nu(t) - \psi_\kappa(t) > 0]\!] \cdot [\![\mathfrak{D}_\kappa^{\mathbb{s}=r}\big(\psi(t), w_{\mu\kappa}\big) = \nu]\!] \cdot w_{\nu\kappa} \cdot [\psi_\nu(t) - \psi_\kappa(t)] \quad (5)$$

$$\Phi_{\nu\kappa}^{(\mathbb{s}=r)\mathbb{f}}(t) = [\![\mathfrak{D}_\kappa^{\mathbb{s}=r}\big(\psi(t), [\![\psi_\mu(t) - \psi_\kappa(t) > 0]\!] \cdot w_{\mu\kappa}\big) = \nu]\!] \cdot w_{\nu\kappa} \cdot [\psi_\nu(t) - \psi_\kappa(t)] \quad (6)$$

The resulting knowledge attainment diagrams (KADs) for 5 representative positionings of the experts are presented in Fig. 2.

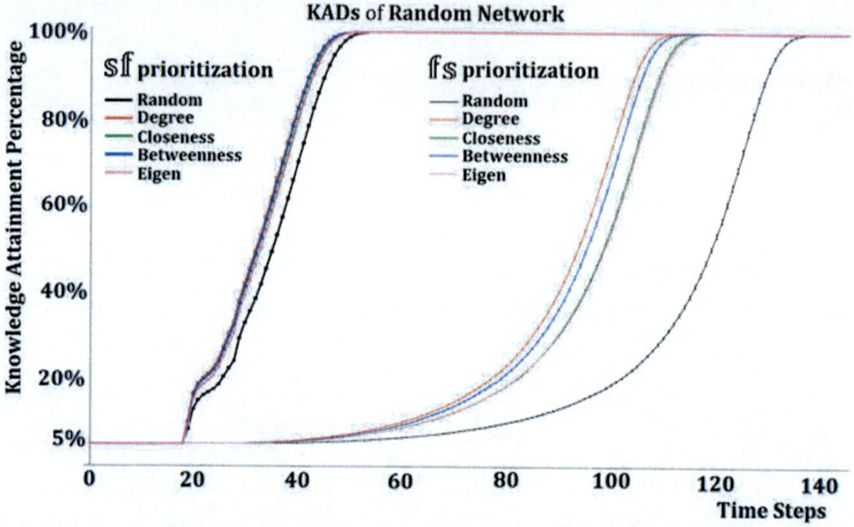

Fig. 2 Knowledge Attainment Diagrams (KAD) of Random Network (5% experts, weights = 0.2, size = 20 N, where N = 200 agents). The 5 policies for positioning experts (random, degree, closeness, betweenness, and eigencentrality) are realized with the conventional prioritization "selection" before "filtering" (fs prioritization—thin lines—right side), as well as with the reverse prioritization "selection" after "filtering" (sf prioritization—thick lines—left side)

Knowledge attainment (Fig. 2) is significantly faster, when agents implement "Selection" after "Filtering" (sf), compared to the conventional wisdom of "Selection" before "Filtering" (fs). This effect depends neither on the policy for positioning experts nor on the structure of the network [17]. Most importantly, the best knowledge attainment performance resulting from the conventional fs prioritization can never match the performance of any knowledge attainment resulting from the superior sf prioritization. The reason for this fact is that sf prioritization ensures that every "selection" realized after "filtering" results in some knowledge upgrade with certainty. In other words, there are no "unfruitful" selections of agents for knowledge upgrade, and therefore there is no waste of valuable time. On the contrary, with the fs prioritization, selections of no knowledge upgrade are possible. The structure of the network (random, small-world, scale-free) is less significant, if agents are intelligent [17].

5 Attitude Change and Hidden Influence in Organizational Networks

The attitude of each agent is the activation $\psi_\kappa(t)$, taking values in the interval $[-1, 1]$ assessing the degree of positiveness or negativeness. The weight $w_{\lambda\kappa}$ of influence of agent λ on agent κ, for attitude change, takes values in the interval [0, 1]. Agents resisting attitude change may "filter" relevant social influences of their in-neighbors [15]. The selecting agent κ may be influenced by the selected in-neighbor λ, only if the distance of their attitudes is not greater than the "tolerance" value $\varepsilon_{\lambda\kappa}$:$|\psi_\lambda(t) - \psi_\kappa(t)| \le \varepsilon_{\lambda\kappa}$. This criterion for social influence is known as "bounded confidence" [15]. We assume that $\varepsilon_{\lambda\kappa} = 2 \cdot w_{\lambda\kappa}$, taking values $0 \le \varepsilon_{\lambda\kappa} \le 2$ for weight values $0 \le w_{\lambda\kappa} \le 1$:

$$[\![|\psi_\lambda(t) - \psi_\kappa(t)| \le \varepsilon_{\lambda\kappa}]\!]$$

We investigate 3 selection rules (Table 1) for communication in attitude networks: (i) selection by randomness, (ii) selection by high link weight, and (iii) selection by high link weight and high degree centrality.

The term $\mathbb{d}_\lambda^{out}$ is the out-strength centrality, taking values in the interval [0, 1], defined as

$$\mathbb{d}_\lambda^{out} = \frac{deg_\lambda^{out}}{\underset{\xi=1,2,\dots,N}{max}\left\{deg_\xi^{out}\right\}} = \frac{\sum_{\mu=1}^{\mathrm{N}} w_{\lambda\mu}}{\underset{\xi=1,2,\dots,N}{max}\left\{\sum_{\mu=1}^{\mathrm{N}} w_{\xi\mu}\right\}}$$

The selections of agents are realized with the conventional prioritization (filtering after selection) $\mathbb{fs}$. The influence function (2) of the spread dynamics equation (1) takes the form:

$$\Phi_{\lambda\kappa}^{\mathbb{fs}}(t) = [\![\mathfrak{D}_\kappa^{\mathbb{s}}(t) = \nu]\!] \cdot w_{\nu\kappa} \cdot [\![|\psi_\nu(t) - \psi_\kappa(t)| \le \varepsilon_{\nu\kappa}]\!] \cdot (\psi_\nu(t) - \psi_\kappa(t)) \quad (7)$$

We investigate the change of attitude of the agents of the network. Initially, all agents have more or less the same attitude. We investigate selected Change Management Strategies, aiming to influence the agents toward a "desired" attitude ψ^{target}.

Table 1 Selection rules

Selection rule $\mathbb{s}$	Selection probability $p_{\lambda\kappa}^{\mathbb{s}}$
Random $\mathbb{s} = r$	$p_{\lambda\kappa}^{r} = \frac{[\![w_{\lambda\kappa} \neq 0]\!]}{\sum_{\nu=1}^{\mathrm{N}} [\![w_{\nu\kappa} \neq 0]\!]}$
Incoming weight $\mathbb{s} = w$	$p_{\lambda\kappa}^{w} = \frac{w_{\lambda\kappa}}{\sum_{\nu=1}^{N} w_{\nu\kappa}}$
Incoming weight and out-strength $\mathbb{s} = w\mathbb{d}$	$p_{\lambda\kappa}^{w\mathbb{d}} = \frac{w_{\lambda\kappa} \cdot \mathbb{d}_\lambda^{out}}{\sum_{\nu=1}^{N} (w_{\nu\kappa} \cdot \mathbb{d}_\nu^{out})}$

The Change Management Strategies are implemented by engaging certain agents initiating "change" by purposely disseminating messages, opinions, values, and behaviors. These professionals are known as "change agents" [18]. The objective of change agents is to influence the agents of the network, so that their attitude becomes eventually close to the desired attitude ψ^{target}. As change agents are in charge of bringing about change within the network, their dynamics is distinguished from the dynamics of the other (regular) agents. Change agents ω adjust their attitudes intentionally toward the desired attitudeψ^{target}, taking into account the attitude of their out-neighbors. The attitude of each change agent ω should remain "close enough" to the average attitude $\overline{\psi}_{\omega}^{out}(t)$ of his/her out-neighbors, due to their "bounded confidence". The average attitude of the out-neighbors of the change agent ω at time t is

$$\overline{\psi}_{\omega}^{out}(t) = \frac{\sum_{\nu=1}^{N} [\psi_{\nu}(t) \cdot [\![w_{\omega\nu} > 0]\!]]}{\sum_{\nu=1}^{N} [\![w_{\omega\nu} > 0]\!]}$$

The attitude of each change agent ω at time $t+1$ is a shift in the average attitude $\overline{\psi}_{\omega}^{out}(t)$ toward the desired attitude ψ^{target}, weighted by a "slicing" parameter β with values $0 \le \beta \le 1$. Therefore, the attitude dynamics equation for each change agent ω is

$$\psi_{\omega}(t+1) = \overline{\psi}_{\omega}^{out}(t) + \beta \cdot \left(\psi^{target} - \overline{\psi}_{\omega}^{out}(t)\right) \tag{8}$$

Change agents are actually triggering a "change cascade" in the network. "Thin slicing" guarantees that influence is not "blocked" due to the "tolerance" limitations. The term $\beta \cdot \left(\psi^{target} - \overline{\psi}_{\omega}^{out}(t)\right)$ in Eq. (8) is the "slice" of influence, used by the change agent ω, to adapt to "tolerance" limitations of out-neighbors. Therefore, the metaphor "thin slicing" means that the "slicing" parameter β is small enough (up to 15% [15]). Change agents "adapt" their attitude to their local environment as "chameleons" for maximal influence (hidden influence). The action of change agents, in this perspective, is *internal locally adaptive control*. Results are obtained by simulating the solutions (change management scenaria) on 4 real organizational networks [15]:

An organizational network of a "Consulting Company" (CC) of size 46.
An organizational network of a "Manufacturing Company" (MC) of size 77.
A partnership network of a "Law Firm" (LF) of size 71.
An organizational network of an "IT Department" (IT) of size 56.

In order to investigate Change Management Strategies, we estimate the relative decrease (%) of the average Change Adoption Time (CAT) in the case where change agents are engaged based on some centrality, and communication among agents is non-random ($\mathbb{s} = w$ or $\mathbb{s} = wd$, Table 1). We compare (Fig. 3) the impact of the Change Management Strategies with the case of no planning for initiating change

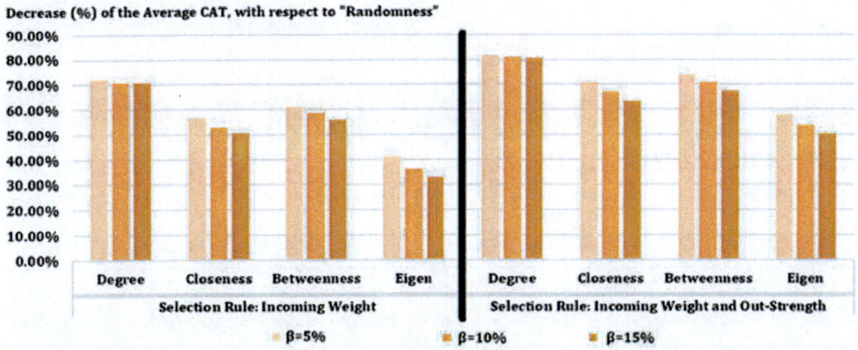

Fig. 3 Relative decrease (%) of the average Change Adoption Times (CAT) with respect to random engagement of change agents and "random" selection rule ($\mathbb{s} = r$), for the IT network with 10% change agents. The 4 engagement policies of change agents based on high centrality (degree, closeness, betweenness, and eigencentrality) are realized with "incoming weight" selection rule ($\mathbb{s} = w$) (left) and with "incoming weight and out-strength" selection rule ($\mathbb{s} = w\mathrm{d}$) (right). Results are presented for slicing parameter $\beta = 5, 10, 15\%$

(random engagement of change agents) and no planning for communicating change (random selection rule $\mathbb{s} = r$).

Planning the engagement of Change Agents and the Internal Communications can accelerate the Adoption of Change by 80% (Fig. 3), compared to "no planning". More specifically, the best change management strategy is the engagement of change agents based on degree centrality and preferential communication based on the "incoming weight and out-strength" selection rule ($\mathbb{s} = w\mathrm{d}$). The value of planning (decrease of the average CAT) is higher if there are a few change agents in the network.

We compare further the social influence capability of employees of different hierarchical classes. The social influence capability is estimated in terms of centralities. We present (Fig. 4) the results for degree centrality (the best engagement policy) for the four real organizational networks.

We observe that employees in senior positions do not always serve as the best change agents (Fig. 4). For example, all "Executives" in the IT network have a very low degree of centrality compared to other employees, even from the "Administrative Staff". In addition, many "Researchers" in the MC network are more suitable to act as change agents compared to "Project Leaders". On the contrary, several senior employees in both CC and LF networks have also a high degree of centrality.

6 Unreliable Organizational Knowledge Networks

In real knowledge networks, false beliefs spread among agents due to the presence of unreliable channels [19]. The activation $\psi_\kappa(t)$ of each agent κ is its knowledge level, taking values in the interval $[-1, 1]$, with interpretation indicated in Table 2.

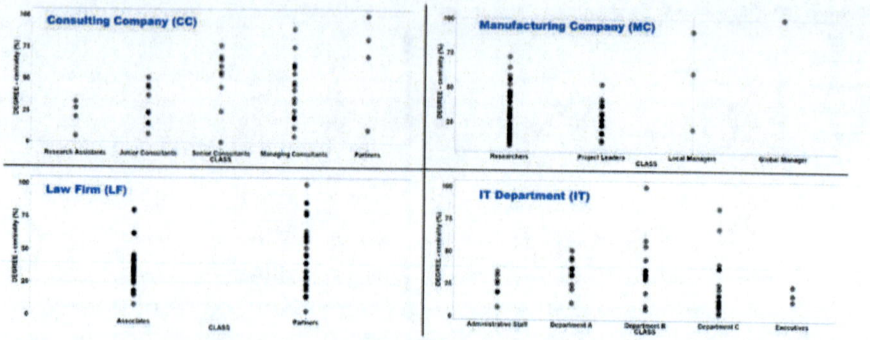

Fig. 4 Employee hierarchical class versus employee centrality in four real organizational networks. Dots represent employees

Table 2 Knowledge levels on a specific knowledge item and interpretation

$\psi_\kappa(t) = 1$	The agent κ has global knowledge
$0 < \psi_\kappa(t) < 1$	The agent κ has partial knowledge
$\psi_\kappa(t) = 0$	The agent κ is ignorant, having no knowledge
$-1 < \psi_\kappa(t) < 0$	The agent κ has partially false beliefs
$\psi_\kappa(t) = -1$	The agent κ has false beliefs in all statements related to global knowledge

The off-diagonal weights $w_{\lambda\kappa}(t)$ (Table 3) estimate the knowledge transfer efficiency through the communication channel $\lambda \rightarrow \kappa$ from agent λ to agent κ. The diagonal weights (Table 4) $w_{\kappa\kappa}(t)$ estimate the knowledge creation efficiency of the agent κ.

We consider randomly evolving weights, according to a uniform distribution. The weight evolution is independent of the knowledge evolution. Knowledge transfer networks include only non-creative agents ($w_{\kappa\kappa} = 0$). Knowledge creation networks include also at least one creative agent ($w_{\kappa\kappa} \neq 0$). The selections of agents are

Table 3 Off-diagonal weight values and interpretation

$w_{\lambda\kappa}(t) = 1$	The communication channel $\lambda \rightarrow \kappa$ does not introduce any limitations or errors
$0 < w_{\lambda\kappa}(t) < 1$	The communication channel $\lambda \rightarrow \kappa$ involves limitations or/and errors which reduce the transmission and/or absorption of knowledge
$w_{\lambda\kappa}(t) = 0$	There is no communication channel from agent λ to agent κ
$-1 < w_{\lambda\kappa}(t) < 0$	The communication channel $\lambda \rightarrow \kappa$ involves limitations or/and errors which result in changes from “justified-true-beliefs” to “false-beliefs”
$w_{\lambda\kappa}(t) = -1$	The communication channel $\lambda \rightarrow \kappa$ changes all “justified-true-beliefs” to “false-beliefs”

Table 4 Diagonal weight values and interpretation

$w_{\kappa\kappa}(t) = 0$ Non-creative agent	Agent κ cannot increase or decrease his/her knowledge $\psi_\kappa(t)$, from time t to $t+1$, due to his/her own creative work only	
$w_{\kappa\kappa}(t) \neq 0$ Creative agent	$0 < w_{\kappa\kappa}(t) \leq 1$ Innovative agent	The knowledge of agent κ is increasing from time t to $t+1$, due to his/her own innovative work only. Agent κ is operating as a "source" of knowledge for the system
	$-1 \leq w_{\kappa\kappa}(t) < 0$ Self-destructive agent	The knowledge of agent κ is decreasing from time t to $t+1$, due to his/her own self-destructive thinking only. Agent κ is operating as a "sink" of knowledge for the system

assumed random with the conventional prioritization (filtering after selection) $\mathbb{f}(\mathbb{s} = r)$.

The **off-diagonal** influence function (2) of the spread dynamics equation (1) takes the form:

$$\Phi_{\nu\kappa}^{\mathbb{f}(\mathbb{s}=r)}(t) = [\![\psi_\nu(t) - \psi_\kappa(t) > 0]\!] \cdot [\![\mathfrak{D}_\kappa^{\mathbb{s}=r}(t) = \nu]\!] \cdot w_{\nu\kappa} \cdot [\psi_\nu(t) - \psi_\kappa(t)] \quad (9)$$

The knowledge creation (**diagonal** influence function) by agent κ during the time interval $(t, t+1]$ is

$$\Phi_{\kappa\kappa}(t) = w_{\kappa\kappa} \cdot [1 - \psi_\kappa(t)] \quad (10)$$

Knowledge creation may be modeled as knowledge transfer from a "knowledgeable self", possessing the globally available knowledge $\psi_\lambda(t) = 1$. Therefore, the knowledge dynamics equation for each agent κ, during the time interval $(t, t+1]$, is

$$\psi_\kappa(t+1) = \psi_\kappa(t) + max\left\{\sum_{\lambda=1}^{N} \Phi_{\lambda\kappa}(t), -1 - \psi_\kappa(t)\right\} \quad (11)$$

The term $max\left\{\sum_{\lambda=1}^{N} \Phi_{\lambda\kappa}(t), -1 - \psi_\kappa(t)\right\}$ in the RHS is introduced in order to keep the knowledge levels bounded from below, in the interval $[-1, 1]$.

We investigate representative spread scenarios for a 5-agent network with (A) non-negative random weights (Fig. 5), (B) non-positive random weights (Fig. 6), and (C) random weights with no sign restriction (Fig. 7).

Knowledge evolution for non-negative weights has been illustrated in two real organizational knowledge networks, namely a "Consulting Company" (CC) of size 46 and a "Research Team of a Manufacturing Company" (MC) of size 77 [19].

We further examine (Figs. 8 and 9) how attacks on links ("breaks" or "infections" of certain "healthy" communication channels) influence knowledge dynamics, in the following two cases:

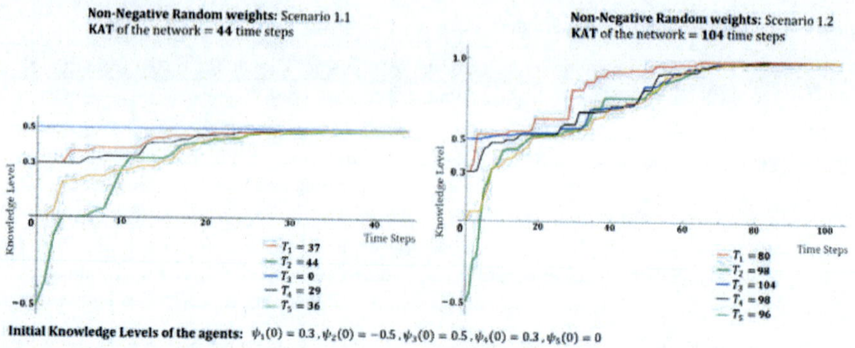

Fig. 5 Knowledge evolution for non-negative random weights. Left: Knowledge transfer may occur, without knowledge creation. Initial false beliefs do not influence significantly the knowledge dynamics of the agents. Right: Only one agent ($\kappa = 1$, red) is creative, whose initial knowledge is not maximal. The non-creative initial "experts" or "gurus" ($\kappa = 3$, blue) will "follow" the creative agents ($\kappa = 1$, red), who eventually become the new "experts"

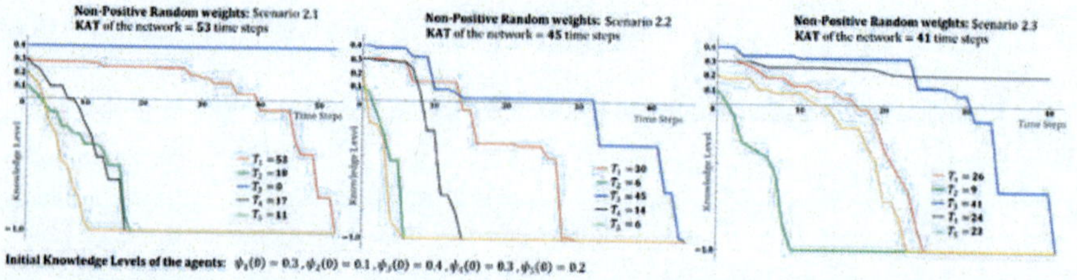

Fig. 6 Knowledge evolution for non-positive random weights. Left: Knowledge transfer may occur, without knowledge creation. False beliefs emerge via unreliable communication channels, even if no false belief is present initially. Middle: All agents are creative. False beliefs may "infect" all agents, resulting in global "knowledge death". In such a case, "resurrection" is possible only by reliable knowledge creation. Right: Only one agent ($\kappa = 3$, blue) is creative, whose initial knowledge is maximal

- **Intervention by Breaking Links** (elimination of certain links)
- **Intervention by Infecting Links** (reversing the sign of certain links).

We explore three attack strategies for the selection of links to attack, namely

- **Attacking Links with High Weights (W-Strategy)**

  Select the links $\lambda \rightarrow \kappa$ with the highest weights $w_{\lambda\kappa}$.

Only the network structure is considered.

- **Attacking Links with High Weights capable to Transfer Knowledge (TW-Strategy)**

  Select the links $\lambda \rightarrow \kappa$ with the highest weights that satisfy the condition at $t = 0$: $\psi_\lambda(t = 0) - \psi_\kappa(t = 0) > 0$.

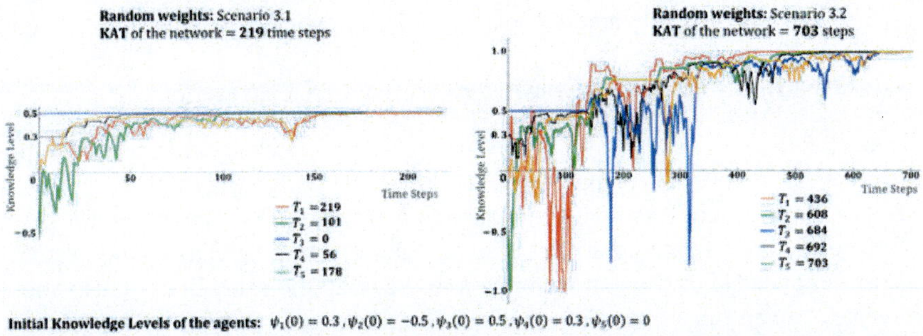

Fig. 7 Knowledge evolution for random weights with no sign restriction. We re-examine the scenarios with non-negative random weights (Fig. 5), with no sign restriction on the weights. Left: Although initial false beliefs have no significant influence on knowledge dynamics (Fig. 5), unreliable communication channels result in significant increase in the knowledge attainment times accompanied by "violent" knowledge fluctuations. Right: One creative agent is sufficient to "enlighten" all agents toward attaining global knowledge, even in the presence of agents with false beliefs and unreliable channels. This is true in general (theorems I and II in [19])

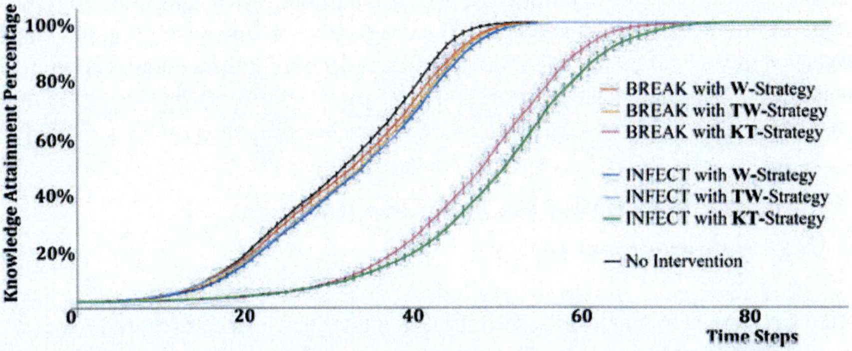

Fig. 8 Knowledge Attainment Diagrams (KAD) of the CC network, after attacking 2% links

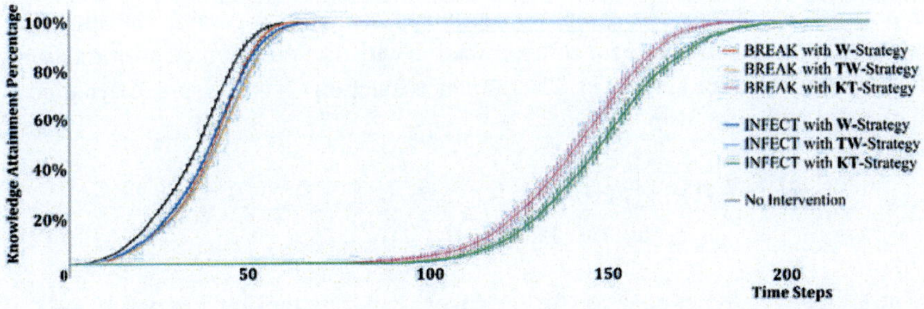

Fig. 9 Knowledge Attainment Diagrams (KAD) of the MC network, after attacking 2% links

The network structure and the possibility for knowledge transfer at $t = 0$ are considered.

- **Attacking Links with High Knowledge Transfer (KT-Strategy)**

  Select the links $\lambda \rightarrow \kappa$ with the highest value of the knowledge transfer at $t = 0$: $[\psi_\lambda(t = 0) - \psi_\kappa(t = 0)] \cdot [\![\psi_\lambda(t = 0) - \psi_\kappa(t = 0) > 0]\!] \cdot w_{\lambda\kappa}$.
  The network structure and the knowledge transfer at $t = 0$ are considered.

"Infecting" links is more harmful than "breaking" links. Infecting links results in slower knowledge attainment, compared to breaking the same links. This is not unexpected, given our observation that infected links give rise to "wild" fluctuations (Fig. 7). Attacking links with high weights, regardless of their capability for knowledge transfer (W-Strategy or TW-Strategy), has a small impact on knowledge attainment (Figs. 8 and 9). The reason for this fact is that the links selected with the W-Strategy or the TW-Strategy may transfer little or even no knowledge. On the contrary, knowledge attainment (Figs. 8 and 9) is significantly influenced by attacks on links with high knowledge transfer (KT-Strategy). Most significantly, attacks based on the KT-Strategy cause delays in the start of knowledge attainment. For example, for the 2% attack, knowledge attainment effectively begins after 20 steps for the CC network (Fig. 8) and after 80 steps for the MC network (Fig. 9). This means that the network remains "non-operational" for a significant duration after the attack.

7 Conflict and Polarization in Co-evolutionary Organizational Networks

In real communication networks, there is strong interdependence of attitudes $\psi_\kappa(t)$ and influence weights $w_{\lambda\kappa}(t)$. The attitudes take values in the interval $[-1, 1]$ assessing the degree of positiveness or negativeness (Sect. 5). The weights may also take values in the interval $[-1, 1]$, interpreted as "attractive" or "repulsive" interpersonal social ties. We discuss evolutionary scenarios assuming the simplest random ($\mathbb{s} = r$) selection rule for communication with the conventional prioritization (filtering after selection) $\mathbb{f}(\mathbb{s} = r)$. The influence function (2) of the spread dynamics equation (1) takes the form:

$$\Phi_{\nu\kappa}^{\mathbb{f}(\mathbb{s}=r)}(t) = [\![|\psi_\nu(t) - \psi_\kappa(t)| \leq \varepsilon_{\nu\kappa}(t)]\!] \cdot [\![\mathfrak{D}_\kappa^{\mathbb{s}=r}(t) = \nu]\!] \cdot w_{\nu\kappa}(t) \cdot [\psi_\nu(t) - \psi_\kappa(t)] \tag{12}$$

The attitude dynamics equation for each agent κ, during the time interval $(t, t+1]$, is

$$\psi_\kappa(t+1) = \psi_\kappa(t) + [\![\psi_\kappa(t) < 0]\!] \cdot max\left\{\sum_{\lambda=1}^{N} \Phi_{\lambda\kappa}(t), -1 - \psi_\kappa(t)\right\}$$

$$+ [\![\psi_\kappa(t) \geq 0]\!] \cdot min\left\{\sum_{\lambda=1}^{N} \Phi_{\lambda\kappa}(t), 1 - \psi_\kappa(t)\right\} \tag{13}$$

The terms $max\left\{\sum_{\lambda=1}^{N} \Phi_{\lambda\kappa}(t), -1 - \psi_\kappa(t)\right\}$ and $min\left\{\sum_{\lambda=1}^{N} \Phi_{\lambda\kappa}(t), 1 - \psi_\kappa(t)\right\}$ are introduced in order to keep the attitude values bounded within the interval $[-1, 1]$.

The evolution of weights depends on both attitudes and weights (co-evolutionary network) (Chap. 7 of [8]). This is in fact a local learning rule which differs for selected and non-selected agents. If the in-neighbor agent λ is selected by agent ($\mathfrak{D}_\kappa(t) = \lambda$), the corresponding weight $w_{\lambda\kappa}(t)$ may change due to Learning, depending on the Relevance Feedback. We consider both positive (amplifying) and negative (damping) feedbacks. The *Relevance Feedback* of agent κ from agent λ is assessed by the function $F_{\lambda\kappa}(t) = 1 - |\psi_\lambda(t) - \psi_\kappa(t)|$, taking values $-1 \leq F_{\lambda\kappa}(t) \leq 1$, depending on the difference of attitudes $|\psi_\lambda(t) - \psi_\kappa(t)|$, with $0 \leq |\psi_\lambda(t) - \psi_\kappa(t)| \leq 2$. The *Learning* of agent κ from agent λ is assessed by the function $L_{\lambda\kappa}(t) = \ell \cdot [\![F_{\lambda\kappa}(t) \geq 0]\!] + (1 - \ell) \cdot [\![F_{\lambda\kappa}(t) < 0]\!]$, where $\ell \in (0, 0.5)$ is a constant, implying the "*learning*" *rate* of the interpersonal ties. The learning function $L_{\lambda\kappa}(t)$ takes values $0 \leq L_{\lambda\kappa}(t) \leq 1$, according to the sign of relevance feedback $F_{\lambda\kappa}(t)$. In this way, the weights dynamics is conditioned by a "*slow-positive*" and "*fast-negative*" relevance feedback $F_{\lambda\kappa}(t)$. The reason underlying this assumption is that trust, and therefore influence, *builds up slowly but may be torn down quickly* [20]. The weight update for the selected agents is

$$w_{\lambda\kappa}(t+1) = (1 - L_{\lambda\kappa}(t)) \cdot w_{\lambda\kappa}(t) + L_{\lambda\kappa}(t) \cdot F_{\lambda\kappa}(t) \tag{14}$$

If the in-neighbor agent λ is not selected by agent ($\mathfrak{D}_\kappa(t) \neq \lambda$), the corresponding weight $w_{\lambda\kappa}(t)$ is subjected to decay with parameter $\zeta \in (0, 1)$:

$$w_{\lambda\kappa}(t+1) = w_{\lambda\kappa}(t) \cdot \zeta \tag{15}$$

Combining Eqs. (14) and (15), we have the *Weights Evolution Formula*:

$$\begin{aligned} w_{\lambda\kappa}(t+1) = {} & [\![\mathfrak{D}_\kappa(t) = \lambda]\!] \cdot [(1 - L_{\lambda\kappa}(t)) \cdot w_{\lambda\kappa}(t) + L_{\lambda\kappa}(t) \cdot F_{\lambda\kappa}(t)] \\ & + [\![\mathfrak{D}_\kappa(t) \neq \lambda]\!] \cdot w_{\lambda\kappa}(t) \cdot \zeta \end{aligned} \tag{16}$$

We investigate the case where two competing groups of influential agents ("promoters" versus "adversaries" of change) operate concurrently within the same organizational network. The social outcome of the co-existence of promoters and adversaries in the same organizational network is characterized by conflicting "pulls" toward the two extremes of attitude $+1$ and -1. We assume that initially ($t = 0$) all agents, except for the promoters and adversaries, are

Same-minded: all agents share the same "neutral" attitude $\psi_\kappa(t)$, taking random values in a narrow interval, as for instance $[-0.05, +0.05]$.

Positively linked: all weights $w_{\lambda\kappa}(t)$ are non-negative.

Due to the above two initial conditions, only "attractive" interactions are possible in the network, according to the Attitude Change Dynamics Equation (13). Therefore, if there is no external intervention, the final distribution of attitudes $\psi_\kappa(t)$ is expected to lie within the same interval [−0.05, +0.05]. However, we assume that some of the agents act as "promoters" or "adversaries":

Promoters influence individuals toward the *promoting attitude* $\psi^+ = 1$, while *at the same time*,

Adversaries influence individuals toward the *adverse attitude* $\psi^- = -1$.

Promoters and adversaries are purposely disseminating targeted messages, beliefs, values, and behaviors, influencing the individuals of the network so that their attitude (of the individuals) to become eventually close to the "target" attitude ψ^{target}, which is equal to $\psi^+ = 1$ (if the influential agent is a "promoter") or equal to $\psi^- = -1$ (if the influential agent is an "adversary"). We assume that "promoters" ($\psi^+ = 1$) and "adversaries" ($\psi^- = -1$) have always the same attitude, like "informed" or "stubborn" agents. The analysis is performed for the 4 real organizational networks of Sect. 3 and presented in Figs. 10, 11, and 12.

From Figs. 10 and 11, we observe that the co-existence of promoters and adversaries may drive the network to the extreme polarization of attitudes: *all* agents eventually adopt *extreme* attitudes $\psi^+ = 1$, $\psi^- = -1$.

Although initially all agents are positively linked, negative links are emerging, reaching eventually a maximum value (blue horizontal lines in Fig. 12). The emergence of negative weights is due to the *feedbacks on weight updates* (network co-evolution). The dynamics of both positive and negative links are symmetric and monotonic.

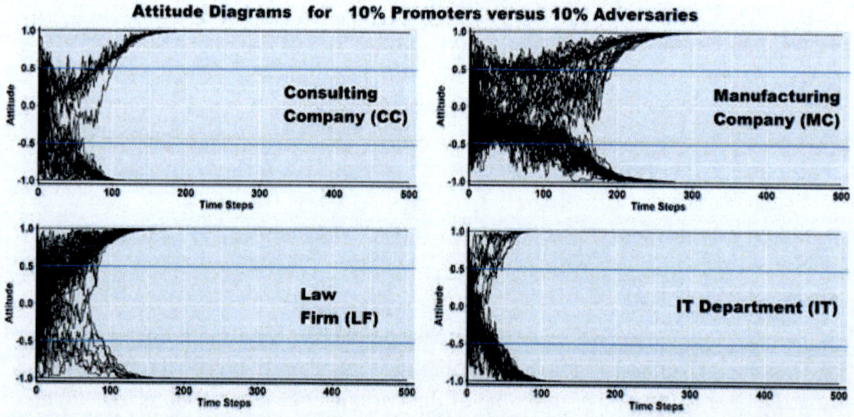

Fig. 10 Attitude diagrams for the 4 real organizational networks. Both promoters and adversaries have high centrality and the learning rate ℓ is 40%. The supporters, neutrals, and opponents are represented in the upper, middle, and lower lanes, defined by the accepted deviation $\delta = 0.5$ which is indicated by the two blue horizontal lines. The number of promoters versus adversaries is 10% versus 10%

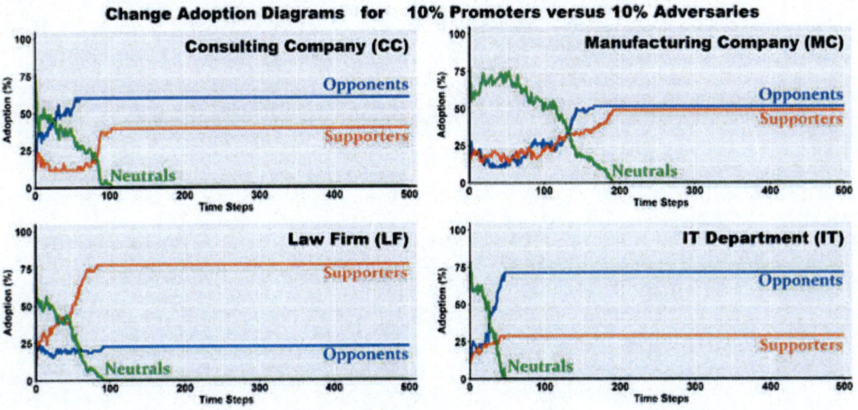

Fig. 11 Change adoption diagrams for the 4 real organizational networks, corresponding to Fig. 10

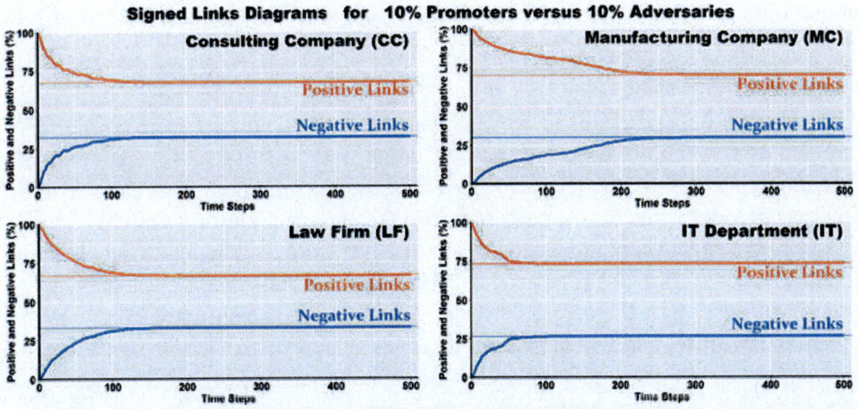

Fig. 12 Signed links diagrams for the 4 real organizational networks. Both promoters and adversaries have high centrality and the learning rate ℓ is 40%. The number of promoters versus adversaries is 10% versus 10%. The blue horizontal line indicates the percentage of *Final Positive Links (FPL)*. The red horizontal line indicates the percentage of *Final Negative Links (FNL)*

In the theory of structural balance, Rapoport noted on page 541 of [21] that the Structure Theorem generates 4 aphorisms corresponding to a balanced graph:

My ***friend's friend*** is my ***friend*** (triangle category **I**)

My ***friend's enemy*** is my ***enemy*** (triangle category **II**)

My ***enemy's friend*** is my ***enemy*** (triangle category **III**)

My ***enemy's enemy*** is my ***friend*** (triangle category **IV**).

Social influence, trust, and friendship are *directed (asymmetric)* relations, and structural balance was formulated in terms of undirected triangles (symmetric relations). We extended structural balance for *directed* networks, considering signed *directed* triangles [22]. The evolution of balance is presented in Fig. 13.

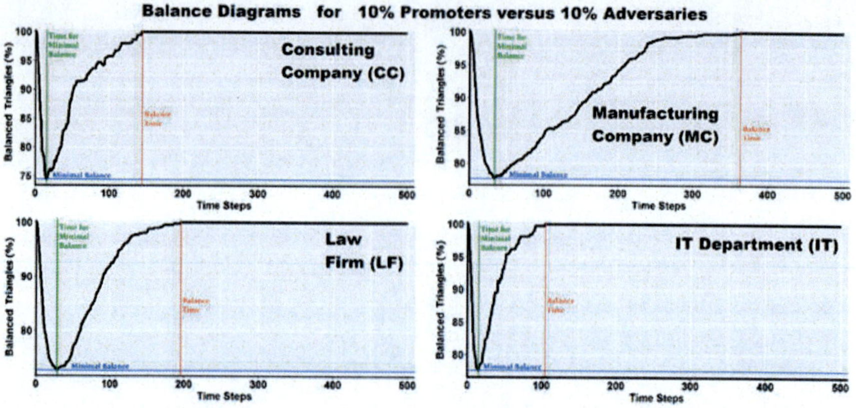

Fig. 13 Balance diagrams for the 4 real organizational networks, corresponding to Figs. 11 and 12. The blue line indicates the minimal value of balance. This minimal value is called *Minimal Balance (MB)*. The green line indicates the time required for balance to take this minimal value. This time is called *Minimal Balance Time (MBT)*. The red line indicates the time required for the re-establishment of balance. This time is called *Balance Time (BT)*

As negative links emerge, network balance decreases rapidly up to a minimal value and afterwards, a "new" balance including both positive and negative links is eventually re-established (Fig. 13). The macroscopic emergence of balance due to microscopic local agent-to-agent interactions is an example of a "micro–macro link". Self-organization manifests because negative links emerge, until the balance of signed directed triangles is re-established. This re-establishment of balance may be seen as *learning* of the social network. The initial balanced network of *same-minded* and *positively linked* agents evolves into a "new" balanced network where *same-minded* agents are *positively linked* and *different-minded* agents are *negatively linked* [22]. This "new" balance emerges concurrently with a polarization of attitudes (Figs. 11, 12, and 13).

8 Intelligence Agents and Centralization in Co-evolutionary Knowledge Networks

We consider knowledge networks with non-negative activations $\psi_\kappa(t) \geq 0$ and weights taking values $0 \leq w_{\lambda\kappa}(t) \leq 1$. The diagonal weight (self-weight) $w_{\kappa\kappa}(t)$ incorporates the capability of agent κ for innovation. We investigate several knowledge dynamics scenarios, with the conventional fs prioritization (selection before filtering) with 4 main selection rules, presented in Table 5.

The **off-diagonal** influence function (2) of the spread dynamics equation (1) takes the form of Eq. (3) in Sect. 3:

Table 5 Probabilistic selection rules for knowledge acquisition

Selection rule $\mathfrak{s}$	Selection probability $p^{\mathfrak{s}}_{\lambda\kappa}(t)$	Bound of rationality
Random $\mathfrak{s} = r$	$p^{r}_{\lambda\kappa}(t) = \frac{[\![w_{\lambda\kappa}(t)\neq 0]\!]}{\sum_{\nu=1,\ \nu\neq\kappa}^{N}[\![w_{\nu\kappa}(t)\neq 0]\!]}$	All incoming links have equal selection probability There is no rationality, as communication is random The agent κ has no intelligence
Knowledge $\mathfrak{s} = \mathfrak{k}$	$p^{\mathfrak{k}}_{\lambda\kappa}(t) = \frac{[\![w_{\lambda\kappa}(t)\neq 0]\!]\cdot\psi_{\lambda}(t)}{\sum_{\nu=1,\ \nu\neq\kappa}^{N}([\![w_{\nu\kappa}(t)\neq 0]\!]\cdot\psi_{\nu}(t))}$	More knowledgeable in-neighbors have higher selection probability Rationality is limited to the awareness of the knowledge of the in-neighbors The agent κ has knowledge-based intelligence
Weight $\mathfrak{s} = w$	$p^{w}_{\lambda\kappa}(t) = \frac{w_{\lambda\kappa}(t)}{\sum_{\nu=1,\ \nu\neq\kappa}^{N} w_{\nu\kappa}(t)}$	Higher incoming weights have higher selection probability Rationality is limited to the awareness of the incoming weights The agent κ has weight-based intelligence
Knowledge–weight $\mathfrak{s} = w\mathfrak{k}$	$p^{w\mathfrak{k}}_{\lambda\kappa}(t) = \frac{w_{\lambda\kappa}(t)\cdot\psi_{\lambda}(t)}{\sum_{\nu=1,\ \nu\neq\kappa}^{N}(w_{\nu\kappa}(t)\cdot\psi_{\nu}(t))}$	Higher products of *weight* · *knowledge* have higher selection probability Rationality is limited to the awareness of the incoming weights and the knowledge of the in-neighbors The agent κ has both knowledge- and weight-based intelligence The agent κ has finer discrimination capability, lower bound of rationality, and higher intelligence, compared to the other cases

$$\Phi^{\mathrm{fs}}_{\nu\kappa}(t) = [\![\psi_{\nu}(t) - \psi_{\kappa}(t) > 0]\!] \cdot [\![\mathfrak{D}^{\mathfrak{s}}_{\kappa}(\psi(t), w_{\mu\kappa}(t)) = \nu]\!] \cdot w_{\nu\kappa}(t) \cdot [\psi_{\nu}(t) - \psi_{\kappa}(t)] \tag{3}$$

The evolution of weights depends on both knowledge levels and weights (co-evolutionary network) (Chap. 7 of [8]). This is in fact a local learning rule which differs for selected and non-selected agents. If the in-neighbor agent λ is selected by agent ($\mathfrak{D}_{\kappa}(t) = \lambda$), the corresponding weight $w_{\lambda\kappa}(t)$ may change due to Learning, depending on the Relevance Feedback. The *Relevance Feedback* of agent κ from agent λ is assessed by the function$F_{\lambda\kappa}(t) = [\![\psi_{\lambda}(t) - \psi_{\kappa}(t) > 0]\!]$ taking binary values, 0 or 1, depending on whether some knowledge is transferred via the channel $\lambda \to \kappa$. If some knowledge is transferred, the channel efficiency (weight) is enhanced, otherwise it weakens. The *Learning* of agent κ from agent λ is assessed by the function $L_{\lambda\kappa}(t) = \ell\cdot[\![F_{\lambda\kappa}(t) = 1]\!] + (1 - \ell)\cdot[\![F_{\lambda\kappa}(t) = 0]\!]$, where $0 < \ell < 0.5$ is a parameter implying the learning rate of the weights. The learning function $L_{\lambda\kappa}(t)$ takes values ℓ or $1-\ell$, *distinguished according to the value of the relevance feedback* $F_{\lambda\kappa}(t)$. That is because we want to incorporate a "*slow-positive*" ($F_{\lambda\kappa}(t) = 1$) and "*fast-negative*" ($F_{\lambda\kappa}(t) = 0$) effect of the relevance feedback $F_{\lambda\kappa}(t)$ on the evolutionary dynamics of the weights. The weight update for the selected agents is

$$w_{\lambda\kappa}(t+1) = (1 - L_{\lambda\kappa}(t)) \cdot w_{\lambda\kappa}(t) + L_{\lambda\kappa}(t) \cdot F_{\lambda\kappa}(t) \tag{17}$$

If the in-neighbor agent λ is not selected by agent ($\mathfrak{D}_\kappa(t) \neq \lambda$), the corresponding weight $w_{\lambda\kappa}(t)$ is subjected to decay with parameter $\zeta \in (0, 1)$ and the weight evolution is given by Eq. (15):

$$w_{\lambda\kappa}(t+1) = w_{\lambda\kappa}(t) \cdot \zeta (15)$$

Combining Eqs. (17) and (15), we have the *Weights Evolution Formula*:

$$\begin{aligned} w_{\lambda\kappa}(t+1) = {} & [\![\mathfrak{D}_\kappa^{\mathfrak{s}}(t) = \boldsymbol{\lambda}]\!] \cdot [(1 - L_{\lambda\kappa}(t)) \cdot w_{\lambda\kappa}(t) + L_{\lambda\kappa}(t) \cdot F_{\lambda\kappa}(t)] \\ & + [\![\mathfrak{D}_\kappa^{\mathfrak{s}}(t) \neq \boldsymbol{\lambda}]\!] \cdot w_{\lambda\kappa}(t) \cdot \zeta \end{aligned} \tag{18}$$

All agents create new knowledge ($w_{\kappa\kappa}(t) > 0$ for $\kappa = 1, 2, \ldots, N$) periodically with period T. The period T is the innovation rate of the system. The selection function $\mathfrak{D}_\kappa(t)$ incorporates (a) the selection of distinct agents ν (with $\nu \neq \kappa$) for knowledge acquisition, based on some selection rule $\mathfrak{s}\big(\mathfrak{D}_\kappa(t) = \mathfrak{D}_\kappa^{\mathfrak{s}}(t) = \nu\big)$, as well as (b) the self-selection of agent κ ($\mathfrak{D}_\kappa(t) = \kappa$) for innovation production with period T. Therefore, the selection function $\mathfrak{D}_\kappa(t)$ is defined as follows:

$$\mathfrak{D}_\kappa(t) = [\![\frac{t}{T} = \left\lfloor \frac{t}{T} \right\rfloor]\!] \cdot \kappa + [\![\frac{t}{T} \neq \left\lfloor \frac{t}{T} \right\rfloor]\!] \cdot \mathfrak{D}_\kappa^{\mathfrak{s}}(t) \tag{19}$$

where $\lfloor x \rfloor$ is the floor function which returns the largest integer not greater than x. For example: $\lfloor 2.2 \rfloor = 2$ and $\lfloor 2.9 \rfloor = 2$.

Therefore, the **diagonal** influence function (2) of the spread dynamics equation (1), representing the Innovation Production from agent κ, takes the form:

$$\Phi_{\kappa\kappa}(t) = [\![\mathfrak{D}_\kappa(t) = \kappa]\!] \cdot w_{\kappa\kappa}(t) \cdot \psi_\kappa(t) \tag{20}$$

The emerging dynamics of network co-evolution is assessed [23] in terms of

- the *average knowledge* of the agents and the "*knowledge inequality*" (standard deviation of knowledge of the agents)
- the *average selection entropy* of the agents
- the *out-degree centralization* of the network.

We explore the impact of 4 significant factors on the emerging co-evolution of the network, namely

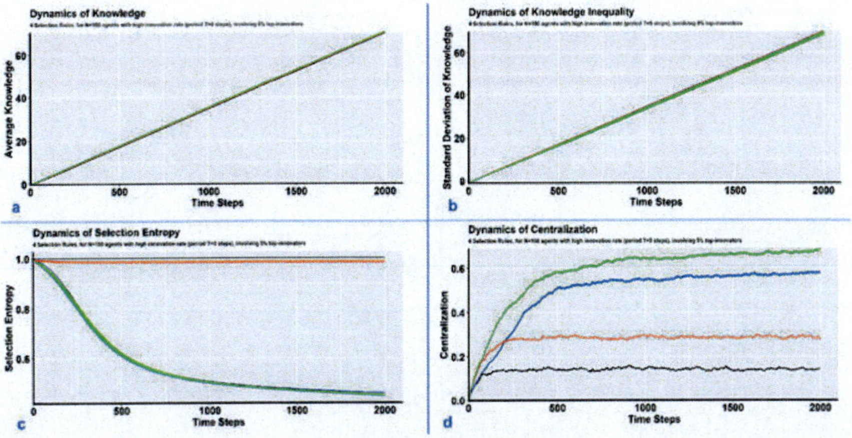

Fig. 14 Simulations for the 4 different probabilistic selection rules: (1) random $s = r$ (black), (2) knowledge $s = k$ (red), (3) weight $s = w$ (blue), and (4) knowledge-weight $s = wk$ (green). Results are illustrated for **a** the average knowledge (up left), **b** the standard deviation of knowledge (up right), **c** the selection entropy (down left), and **d** the centralization (down right). The y axis for the average knowledge (**a**, up left) and the standard deviation of knowledge (**b**, up right) is on the log-10 scale, indicating the exponential increase of knowledge dynamics. Results are illustrated for $N = 100$ agents with high innovation rate (period $T = 5$ steps), involving 5% top-innovators

- the *intelligence* of the agents, expressed by the selection-decision rule for knowledge acquisition (Fig. 14)
- the *innovation rate* of the agents (Fig. 15)
- the *number of "top-innovators"* (Fig. 16)
- the *network size* (Fig. 17).

We observe (Fig. 14) that average knowledge and "knowledge inequality" are both increasing exponentially in time. Intelligent agents reduce selection entropy and centralize the network.

We observe (Fig. 15) that higher innovation rates result in (a) higher average knowledge, (b) higher knowledge inequality, (c) stronger selection entropy decrease, and (d) stronger network centralization increase.

We observe (Fig. 16) that fewer "top-innovators" result in a stronger selection entropy decrease and a stronger network centralization increase.

We observe (Fig. 17) that smaller networks give rise to a stronger centralization increase.

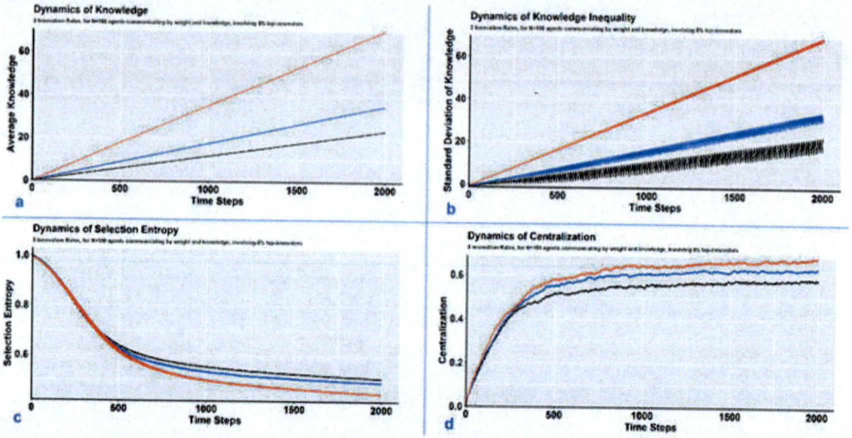

Fig. 15 Simulations for the 3 different innovation rates: (1) low rate with period $T = 15$ steps (black), (2) medium rate with period $T = 10$ steps (blue), and (3) high rate with period $T = 5$ steps (red). Results are illustrated for: **a** the average knowledge (up left), **b** the standard deviation of knowledge (up right), **c** the selection entropy (down left), and **d** the centralization (down right). The y axis for the average knowledge (**a**, up left) and the standard deviation of knowledge (**b**, up right) is on the log-10 scale, indicating the exponential increase of knowledge dynamics. Results are illustrated for $N = 100$ agents communicating by weight and knowledge ($\mathbb{s} = w\mathcal{k}$), involving 5% top-innovators

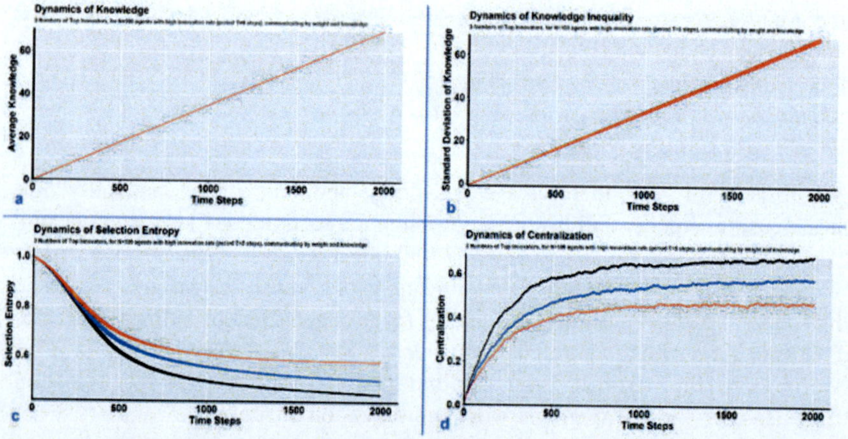

Fig. 16 Simulations for the 3 different numbers of top-innovators: 5% (black), 10% (blue), and 15% (red). Results are illustrated for **a** the average knowledge (up left), **b** the standard deviation of knowledge (up right), **c** the selection entropy (down left), and **d** the centralization (down right). The y axis for the average knowledge (**a**, up left) and the standard deviation of knowledge (**b**, up right) is on the log-10 scale, indicating the exponential increase of knowledge dynamics. Results are illustrated for $N = 100$ agents with high innovation rate (period $T = 5$ steps), communicating by weight and knowledge ($\mathbb{s} = w\mathcal{k}$)

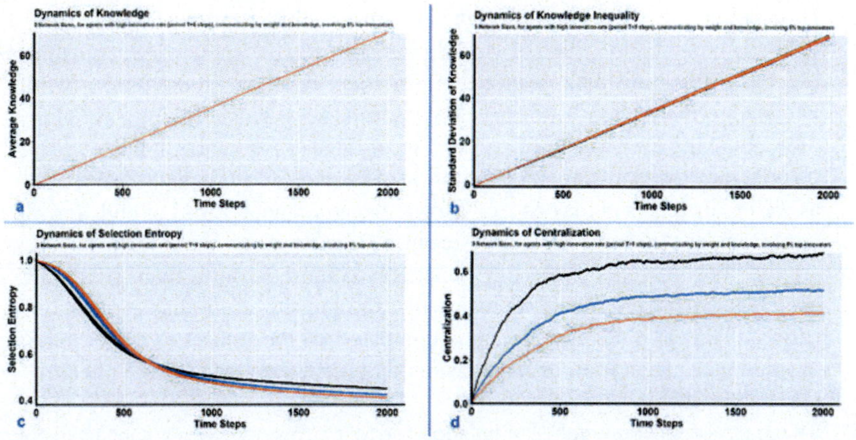

Fig. 17 Simulations for the 3 different network sizes: N = 100 agents (black), N = 200 agents (blue), and N = 300 agents (red). Results are illustrated for **a** the average knowledge (up left), **b** the standard deviation of knowledge (up right), **c** the selection entropy (down left), and **d** the centralization (down right). The y axis for the average knowledge (**a**, up left) and the standard deviation of knowledge (**b**, up right) is on the log-10 scale, indicating the exponential increase of knowledge dynamics. Results are illustrated for agents with high innovation rate (period $T = 5$ steps), communicating by weight and knowledge ($\mathbb{s} = w k$), involving 5% top-innovators

9 Discussion

(A) The risky or costly structural interventions to organizational networks may be avoided, simply by raising the awareness of the agents, so that the "fruitful" selections for knowledge upgrade are more likely to be realized. The impact of the order of implementation of "Filtering" ($\mathbb{f}$) and "Selection" ($\mathbb{s}$) on knowledge attainment (Sects. 3 and 4) reveals the non-commutativity of these two network operations in the context of knowledge networks, indicating a "Non-Boolean Network Logic" [24].

(B) Raising the awareness of agents to implement intelligent selections is a much more effective Knowledge Management Strategy, compared to changing the positions of experts within the network from random to central nodes (Sect. 4).

(C) Senior employees are not always the most suitable to act as change agents. Communicating with "local hubs" results in faster change adoption. Change agents feed the members of the network with "thin slices" of influence, in order to avoid crossing the "confidence bound". Network-based "Planning" reduces the change adoption time by 80% (Sect. 5). As highly influential employees are not detectable from the formal organizational chart alone, Organizational Network Analysis (ONA) is useful to identify the most suitable change agents, and moreover to uncover the communication patterns resisting the adoption of change.

(D) False beliefs are transmitted through unreliable communication channels. Unreliable communication channels undermine seriously the knowledge acquisition (Sect. 6). "Infecting" links is more harmful than "breaking" links, due to "wild" knowledge fluctuations and the elongation of knowledge attainment. Moreover, attacking even a "small" percentage of links ($\leq 5\%$) with high knowledge transfer may result in dramatic elongation of knowledge attainment (over 100%) as well as in delays in the onset of knowledge attainment. Links of high knowledge transfer should be protected, because in Information Warfare, these links are the "best targets". "Infecting" channels is more harmful than "breaking" channels. Managers should be aware that the impact of unreliable communication channels is far more serious compared to the impact of false beliefs or ignorance. Therefore, managers should be timely aware of the "unhealthy" unreliable relationships.

(E) The macroscopic emergence of balance due to microscopic local agent-to-agent interactions is an example of a "micro–macro link" (Sect. 7). Self-organization manifests because negative links emerge, until the balance of signed directed triangles is re-established. This re-establishment of balance may be seen as learning of the social network. The proposed network co-evolution model allows designing control scenarios involving conflict, for example, the planning and simulation of slow or fast re-establishment of balance, or the intentional emergence of structural split and polarization. The action of promoters and adversaries resembles the action of driver nodes [25] for network control.

(F) Rational intelligent agents transform the network into a "centralized world", reducing the entropy of their selections-decisions for knowledge acquisition. The average knowledge, as well as the "knowledge inequality", grows exponentially (Sect. 8). Our work offers some insight into the dynamics of networked societies. The World Wide Web resulted in global interconnectedness. We are interconnected through online social networks, platforms, and communities (Facebook, LinkedIn, Instagram, Twitter, Airbnb, ResearchGate, TripAdvisor, Booking, and YouTube). We are influencing each other, directly or indirectly, intentionally or unintentionally, on a global scale. We share knowledge and we rank individuals, services, or products. This "ranking" is visible, for example, the Followers on Instagram or Twitter, the Endorsements on LinkedIn, the Likes on Facebook, the 5-star experience at Airbnb or Booking or TripAdvisor, the RG Score at ResearchGate, and the Views and Likes on YouTube. Therefore, intelligent decision makers (selecting agents) being aware of this ranking decide with higher rationality, lowing the selection entropy. Intelligent selections are effectively a mechanism of "preferential attachment" (the newcomer is aware of the degree connectivity of all other nodes, and attaches to nodes with higher degrees). This behavior increases the centralization of the network society because more and more individuals tend to select the "best option" (high-ranked agents) and the link weights with the low-ranked agents are weakening. Of course, the correlation between the rank/popularity and the quality of the "best option" is not always positive.

(G) Network structure and dynamics specify the possibilities. The actualization of the possibilities is the implementation of the policy for "Selection" and "Filtering". The utilization of the available possibilities depends on individual choices, including randomness [26]. Each choice (policy adoption) is in fact a passing from "Potentiality" (δυνάμει) to "Actuality" (ενεργεία), realizing the "Entelechy" (Εντελέχεια), as Aristotle defines the terms in Metaphysics (Books 7, 8, 9). The transition from "Net Potentiality" to "Actual Communication" is the Entelechy. "We come much nearer to Aristotle's 'biological' view of space–time" [9]. "Life's history on our planet (is) the single actualized version among millions of plausible alternatives that happened not to occur" [27].

References

1. V.V. Belyi, Fluctuations out of equilibrium. Philos. Trans. A **376** (2018)
2. V.V. Belyi, Thomson scattering in inhomogeneous plasmas: the role of the fluctuation-dissipation theorem. Sci. Rep. **8**(7946) (2018)
3. T.-N. Beuermann, R. Redmer, T. Bornath, Thomson scattering from dense inhomogeneous plasmas. Phys. Rev. E **99**(5) (2019)
4. G. Oster, A. Perelson, A. Katchalsky, Network thermodynamics. Nature **237**, 332–333 (1972)
5. E.R. Lewis, Network thermodynamics revisited. Biosystems **34**(1–3), 47–63 (1995)
6. D.C. Mikulecky, Network thermodynamics and complexity: a transition to relational systems theory. Comput. Chem. **25**(4), 369–391 (2001)
7. M. Newman, *Networks*, 2nd edn. (Oxford University Press, Oxford, 2018)
8. S. Thurner, R. Hanel, P. Klimek, *Introduction to the Theory of Complex Systems* (Oxford University Press, 2018)
9. I. Prigogine, *From Being to Becoming: Time and Complexity in the Physical Sciences* (W H Freeman & Co, San Francisco, 1980)
10. I. Prigogine, R. Herman, *Kinetic Theory of Vehicular Traffic* (American Elsevier Pub. Co., 1971)
11. R. Herman, I. Prigogine, A two-fluid approach to town traffic. Science **4389**(204), 148–151 (1979)
12. M. Browne, *Stuck in Traffic? Consult a Physicist* (1997)
13. F. Schweitzer, An agent-based framework of active matter with applications in biological and social systems. Eur. J. Phys. **40**(014003) (2018)
14. E. Ioannidis, N. Varsakelis, I. Antoniou, Communication policies in knowledge networks. Phys. A **492**, 360–374 (2018)
15. E. Ioannidis, N. Varsakelis, I. Antoniou, Change agents and internal communications in organizational networks. Phys. A: Stat. Mech. Appl. **528** (2019)
16. E. Ioannidis, *Modeling Knowledge Networks*. Ph.D. thesis (School of Mathematics, Faculty of Exact Sciences, Aristotle University of Thessaloniki, 2018)
17. E. Ioannidis, N. Varsakelis, I. Antoniou, Experts in knowledge networks: central positioning and intelligent selections. Phys. A **509**, 890–905 (2018)
18. J. Battilana, T. Casciaro, Change agents, networks, and institutions: a contingency theory of organizational change. Acad. Manag. J. **55**(2), 381–398 (2012)
19. E. Ioannidis, N. Varsakelis, I. Antoniou, False beliefs in unreliable knowledge networks. Phys. A **470**, 275–295 (2017)
20. F. Walter, S. Battiston, F. Schweitzer, A model of a trust-based recommendation system on a social network. Auton. Agent. Multi-Agent Syst. **16**(1), 57–74 (2008)

21. A. Rapoport, Mathematical models of social interaction, in *Handbook of Mathematical Psychology* (Wiley, Oxford, 1963), pp. 493–580
22. E. Ioannidis, N. Varsakelis, I. Antoniou, Promoters versus adversaries of change: agent-based modeling of organizational conflict in co-evolving networks. Mathematics **8**(12), 1–25 (2020)
23. E. Ioannidis, N. Varsakelis, I. Antoniou, Intelligent agents in co-evolving knowledge networks. Mathematics **9**(1) (2021)
24. I. Antoniou, I. Gialampoukidis, E. Ioannidis, Age and time operator of evolutionary processes, in *Quantum Interaction* (Springer, 2016), pp. 51–75
25. Y.-Y. Liu, J.-J. Slotine, A.-L. Barabási, Controllability of complex networks. Nature **473**, 167–173 (2011)
26. H. Atmanspacher, R. Bishop, *Between Chance and Choice: Interdisciplinary Perspectives on Determinism* (Imprint Academic, 2007)
27. I. Prigogine, *The End of Certainty* (Free Press, 1997)

Thermal Fluctuations Induced Emergence of Umbilical Defects in Nematic Liquid Crystal Cells

Esteban Aguilera, Marcel G. Clerc, David Pinto-Ramos, and Valeska Zambra

Abstract Optical vortices are equally relevant for their fundamental features as beams with topological properties and applications in image processing, telecommunications, optical tweezers, and quantum information. The interaction of light beams with umbilical defects in liquid crystal cells is a natural source of optical vortices. Here we investigate, experimentally and theoretically, the mechanisms of the matter vortices that appear in liquid crystal cells and establish statistical laws that govern them. Based on an adequate stochastic equation, the law for the number of nucleated vortices as a function of anisotropy, bifurcation parameter, and noise level intensity is set. Experimental results show a fair agreement with the theoretical findings.

1 Introduction

In the last decades, a great effort has been developed to understand spiral output light beams about their axis of propagation, *orbital angular momentum of light or optical vortex* [1–5]. These beams have a donut-like structure, that is, the beam intensity cancels out into the center, generating a phase singularity into the envelope. Around the point of zero intensity, the phase distribution forms an N-armed spiral, with N being the topological charge [2–6]. These optical vortices have aroused interest from both the fundamental and applied point of view. The photonic applications ranging from optical tweezers [7–9], enhancement of astronomical images [10], quantum computation [11], wavefront sensors [12], and data transmission [13]. From a fundamental point of view, the interchange of angular momentum between light and matter has attracted attention (see the collected articles [5] and references therein).

E. Aguilera · M. G. Clerc (✉) · D. Pinto-Ramos · V. Zambra
Departamento de Física and Millenium Institute for Research in Optics, Facultad de Ciencias Físicas y Matemáticas, Universidad de Chile, Casilla 487-3, Santiago, Chile
e-mail: marcel@dfi.uchile.cl

D. Pinto-Ramos
e-mail: david.pinto@ug.uchile.cl

L. Brenig et al. (eds.), *Nonequilibrium Thermodynamics and Fluctuation Kinetics*, Fundamental Theories of Physics 208,
https://doi.org/10.1007/978-3-031-04458-8_15

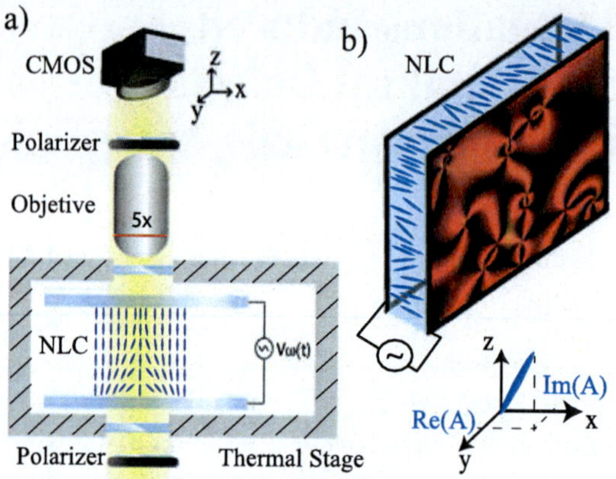

Fig. 1 Vortex nucleation in a nematic liquid crystal cell with homeotropic anchoring and negative dielectric anisotropic constant. **a** Schematic representation of the experimental setup. The rods are a schematic representation of the average orientations of molecules. **b** Snapshot of a vortex gas obtained in the nematic liquid crystal (NLC) cell. The lower inset is a schematic representation of the director and complex amplitude

Different methods have been used to generate optical vortices based on diffractive elements [14], deformable mirrors [15], holograms [16], spiral phase plates [17], nanostructured glass plates [18], and helical structures of liquid crystals [19–22]. In most of these methods, the light beam interacts with a material object which has a helicity. Hence, to control the optical vortex, it is important to have an adequate alignment between the light beam, the target, and the geometry of it. In the case of liquid crystals with photosensitive walls, the light induces a vortex in the matter (umbilical defect), with which interacts, generating an optical vortex [22–24]. These matter vortices are described by a nonlinear amplitude equation, the Ginzburg-Landau equation with real coefficients [24–26]. When a sufficiently large electric field is applied to a nematic liquid crystal cell with homeotropic anchoring and negative anisotropic dielectric constant, a gas of umbilical defects emerge (see Fig. 1). These defects later begin to be annihilated by pairs with opposite charges [27]. The emergence process and statistical rules that this phase singularity gas followed have not been established.

Using a liquid crystal cell with homeotropic anchoring and negative dielectric anisotropic constant under the effect of a transverse voltage allows us to study the statistical laws that govern the nucleation of vortices. Theoretically, based on a stochastic amplitude equation, the Ginzburg Landau equation with additive noise, we establish the law for the number of nucleated vortices as a function of anisotropy, bifurcation parameter, and intensity of the noise level. Experimental results show a qualitative agreement with the theoretical findings.

2 Experimental Setup

Nematic liquid crystals are nonlinear optical media [27, 28], composed by rod-like molecules that have a preferential orientational order but not a positional one. This state of matter shares features of solids and liquids, such as fluidity and birefringence. Introducing a liquid crystal inside a cell, that is, it is sandwiched between two confining layers, the molecules oriented according to anchoring conditions. Homeotropic anchoring is characterized by molecules that are oriented orthogonal to cell walls, as illustrated in Fig. 1. If the dielectric anisotropic constant of the liquid crystal is negative, when applying a vertical electric field, the molecules tend to orient orthogonal to it. This generates different domains connected by orientation defects or phase singularities, *matter vortices* [27]. Let us consider a 15 μm thick cell, (SB100A150uT180 manufactured by Instec), filled with nematic liquid crystal LC BYVA- 01 (Instec) with dielectric anisotropy $\epsilon_a = -4.89$, birefringence $\Delta n = n_e - n_o = 0.1$, rotation viscosity $\gamma = 204$ mPas, splay and bend elastic constant, respectively, $K_1 = 17.65$ pN and $K_3 = 21.39$ pN. This sample is placed inside a thermal control chamber (Linkam LTS420), which in turn is inserted inside a microscope (Leica DM2700P), in between the crossed linear polarizers. The thermal control chamber allows precise control of the temperature of the liquid crystal cell. Likewise, cross-polarizer microscopy enables an efficient vortex detection methodology. Figure 1a shows a schematic representation of the experimental setup. To monitor the images a CMOS camera is connected to the microscope. A sinusoidal voltage with a frequency 100 Hz is applied to the sample.

Maintaining the temperature at 26 °C, the voltage is turned on, the dynamics of vortex nucleation and annihilation are recorded. Figure 2a depicts the temporal evolution of the observed umbilical defects. To figure out vortex evolution, we have considered a voltage sweep between 9.0 Vpp and 30.0 Vpp. Likewise, keeping the voltage at 15 Vpp it is switched on and sweeping the temperature between 25 and 80 °C, the dynamics of vortex nucleation is analyzed. From the chart in Fig. 2a, we infer that there is an abrupt process of vortex nucleation. The vortices are subsequently annihilated by pairs of opposite charges, generating a coarsening process characterized by a power law [29].

3 Theoretical Description

To shed light on the vortex nucleation mechanisms, theoretically, we consider the liquid crystal cell close to orientational instability, which is described by the stochastic Ginzburg-Landau equation [24–26]

$$\partial_t A = \mu A - |A|^2 A + \nabla^2 A + \delta \partial_{\eta,\eta} \bar{A} + \sqrt{T} \zeta(\mathbf{r}, t), \tag{1}$$

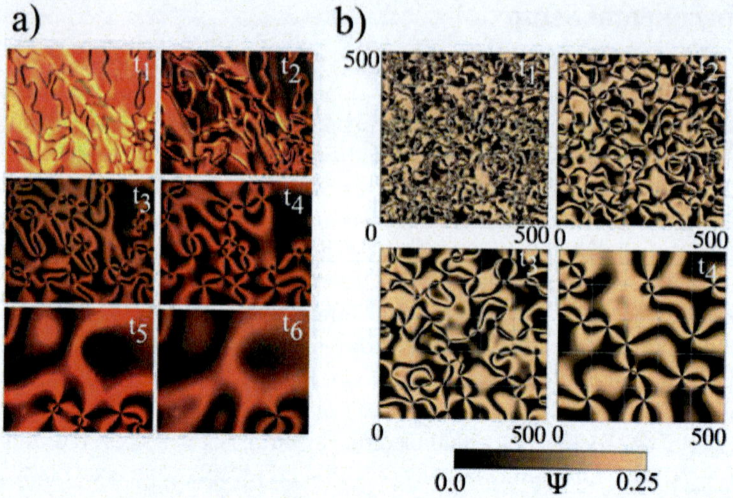

Fig. 2 Nucleation and evolution of umbilical defects in a liquid crystal cell driven by an electric field. **a** Experimental temporal evolution of the number of vortices as a function of time at fixed driven voltage of 15 Vpp, 100 Hz, and 26 °C. Right panel account for a temporal sequence of snapshots of the liquid crystal cell driven by an electric field and constant temperature ($t_1 < t_2 < t_3 < t_4 < t_5 < t_6$). **b** Numerical temporal sequence of polarized field $\Psi(\mathbf{r}, t) = Re(A)Im(A)$ obtained by the numerical simulations of the stochastic Ginzburg-Landau equation (1) with $\mu = 1.0$, $\delta = 0.0$, and $T = 0.01$

where the complex field $A(\mathbf{r}, t)$ accounts for the amplitude of the critical mode that describes the deviation of the molecular director with respect to the vertical direction. $\bar{A}$ accounts for the complex conjugate of A. μ is the bifurcation parameter that is proportional to the voltage minus the critical Fréederickz voltage [24–26]. For a planar anchoring cell, the Fréederickz voltage has the analytical expression $V_{FT} \equiv 2\pi\sqrt{K_3/\epsilon_a}$. $\delta = K_1 - K_2/(K_1 + K_2)$ is the parameter that accounts for the anisotropy of the liquid crystal elastic constants. $\partial_\eta \equiv \partial_x + i\partial_y$ is a differential operator, note that the Laplacian operator satisfies $\nabla^2 = \partial_{\eta,\bar{\eta}}$. $\zeta(\mathbf{r}, t)$ is a Gaussian white noise with zero mean value $\langle\zeta\rangle = 0$ and correlation $\langle\zeta(\mathbf{r}, t)\bar{\zeta}(\mathbf{r}', t')\rangle = \delta(t - t')\delta(\mathbf{r} - \mathbf{r}')$ and T accounts for the noise intensity level. The main sources of noise are inherent thermal fluctuations and electrical fluctuations on the applied voltage.

For $\mu \leqslant 0$, the Ginzburg-Landau Eq. (1) has a null solution $A = 0$ as a stable equilibrium, which corresponds to molecules that are not reoriented, *homeotropic state*. For $\mu > 0$, this state becomes unstable by means of a degenerate pitchfork bifurcation, giving rise to the appearance of vortices [26]. Figure 2b illustrates the emergence of vortices in model Eq. (1) as a result of stochastic fluctuations. As in the experiment, the uncontrollable fluctuations, noise, nucleates a large number of vortices that are subsequently annihilated by opposite pairs. Numerical simulations were implemented using a finite differences scheme in space that uses a centered stencil of five grid points with Runge-Kutta order-4 algorithm, with a 500×500 points grid temporal step $dt = 0.0004$, and Neumann boundary conditions.

4 Vortices Nucleation Law

Numerically, we have monitored the number of vortices at a given instant as a function of the bifurcation parameter μ. Figure 3a summarizes the results found. From these charts, we infer that the number of vortices grows linearly with the bifurcation parameter. Likewise, we note that this behavior is not modified when we change the anisotropy δ. Experimentally, we have studied the number of umbilical defects in a given instant as a function of the voltage applied to the sample. We found that the number of defects grows with the voltage, which shows a qualitative agreement with the numerical results (cf. Fig. 3b). The defects emerge from the homeotropic state, due to the inherent fluctuations of the system. Hence, to understand the nucleation process, we approximate Eq. (1) by its deterministic linear part and consider the Fourier mode decomposition $A = A_{\vec{k}} e^{\sigma t + i(k_x x + k_y y)}$, after straightforward calculations we get

$$\sigma = \mu - k_x^2(1+\delta) - k_y^2(1-\delta) \pm 2i\delta k_x k_y, \tag{2}$$

where $Re(\sigma)$ is the growth rate mode, k_x and k_y are wavenumber modes in the horizontal directions. $\sigma(k_x, k_y) > 0$ corresponds to unstable modes. Notice that white noise is characterized by excited in the same manner all modes. The boundary conditions and geometric dimensions of the system determine the wavenumbers of modes. For simplicity, if we consider periodic boundary conditions and a square domain wavenumbers take the form $k_x = 2\pi n/L$ and $k_y = 2\pi m/L$, where L is the size of the box and $\{n, m\}$ are integer numbers. The nodes of the spatial modes correspond to zeros of the amplitude; that is, these nodes correspond to phase singularities (vortex germ) for the modes. The mode with the maximum number of vortices (nodes) corresponds to $\sigma = 0$. To calculate this maximum number of vortices, we proceed by calculating the number of modes in one direction $[\sigma(n^c, k_y = 0) = 0]$, then in the other $[\sigma(k_x = 0, m^c) = 0]$, and finally we determine the maximum number of vortices (nodes) by

$$N = n^c m^c = \left(\frac{L}{2\pi}\right)^2 \frac{\mu}{\sqrt{(1-\delta^2)}}. \tag{3}$$

Note that all other unstable modes have a similar expression (3) multiplied by a proper fraction. Hence, the number of vortices is proportional to the previous expression, in particular to the bifurcation parameter, which is consistent with what is observed numerically and experimentally (see Fig. 3). Likewise, we note that this result predicts that the number of vortices diverges when δ^2 tends to 1. This limit physically corresponds when one of the elastic constants diverges. This phenomenon is observed when there is a nematic-smectic transition ($K_2 \to \infty$) when the temper-

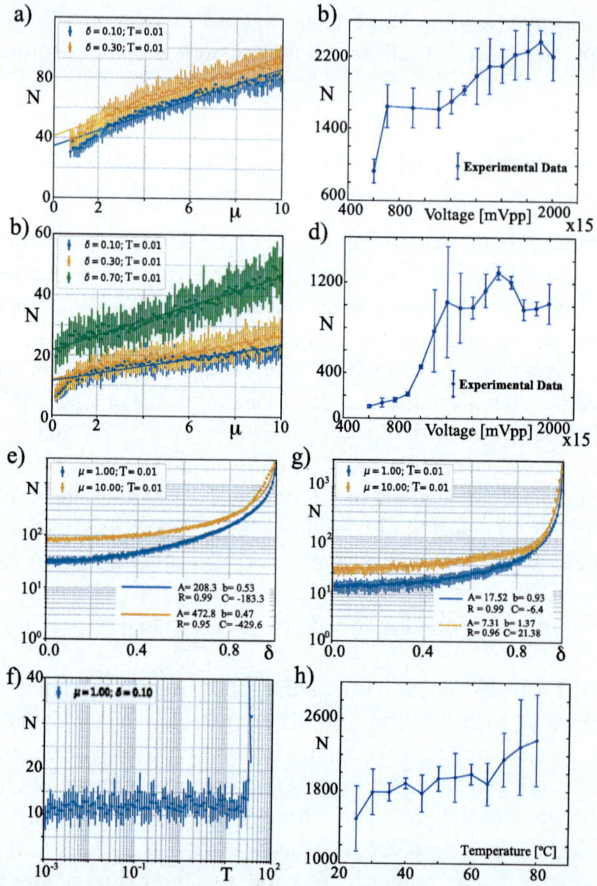

Fig. 3 The number of defects in a given instant as a function of the bifurcation parameter and noise level intensity. The number of defects obtained from numerical simulations of Eq. (1) with $\mu = 1.0$ at $t = 12$ (**a**) and $t = 60$ (**b**). The points with a bar account for mean value and standard deviation obtained after carrying out for each parameter 30 realizations. Number of umbilical defects as a function of the driven voltage at $t = 0.5$ s (**c**) and $t = 1.0$ s (**d**). The points with a bar account for mean value and standard deviation obtained after five experimental realizations. The number of defects in a given instant as a function of the anisotropic parameter δ. obtained from numerical simulations of Eq. (1) at $t = 12$ (**e**) and $t = 60$ (**g**). The points with a bar account for mean value and standard deviation obtained after carrying out for each parameter 20 realizations. The continuous curves were obtained using the fitting function $N = A/(1-\delta^2)^b + C$. The simulations and fitting parameters are specified in insets. (**f**) The number of defects in a given moment as a function of the noise intensity level T. (**h**) Umbilical defects number as a function of the temperature after 1 s of applying voltage 15 Vpp. The points with a bar account for mean value and standard deviation obtained after carrying out 5 experimental realizations

ature is modified [30]. However, the liquid crystal under our study does not present this type of transition. The study of vortex nucleation near the nematic-smectic transition is in progress. Figure 3 shows the number of vortices at a given moment as a function of the anisotropy parameter δ. This result shows an excellent agreement with expression (3). To study its trend, we have used a more general fitting function of the form $N = A/(1-\delta^2)^b + C$, which can take into account the nonlinear effects and errors of the vortex measurement method. From charts, Fig. 3e and g, note that the critical exponent b evaluated at higher times is dissimilar that predicted theoretically. This effect is due to the fact that nonlinear terms begin to play a non-negligible role. Experimentally, we cannot carry out a similar analysis since elastic anisotropy δ is determined by intermolecular interactions that we cannot control.

Formula (3) does not depend on the noise intensity level T. Indeed, the number of vortices (nodes) does not depend on the intensity of the noise, however, their presence is essential to stimulate unstable modes. Figure 3f shows that effectively the noise intensity level does not affect the number of vortices created. When the noise intensity is very large the linear theory is no longer valid and the vortices are no longer related to the linear modes (see Fig. 3f). To investigate experimentally fluctuations that are inherent to our system, we have estimated the number of vortices in a given moment as a function of temperature. Figure 3h summarizes the results found. We deduce that there is a tendency to increase the number of vortices with temperature. The increase in temperature has a double effect; on the one hand, it increases the thermal fluctuations and, in turn, modifies the elastic constants [30]. This last effect is responsible for the increase found in the number of vortices.

5 Noise Induced Emergence of One-Dimensional Defect

The above vortex creation mechanism should be a general defect creation mechanism not only valid in two dimensions. A simple dimensionless model of one-dimensional topological defects is the dissipative bistable model, which has the form [31, 32]

$$\partial_t u = \epsilon u - u^3 + \partial_{xx} u + \sqrt{T}\chi(t,x), \tag{4}$$

where $u(x,t)$ is a one-dimensional order parameter, ϵ is a bifurcation parameter, $\chi(x,t)$ is a Gaussian white noise with zero mean value $\langle\chi\rangle = 0$ and correlation $\langle\chi(x,t)\chi(x',t')\rangle = \delta(t-t')\delta(x-x')$ and T accounts for the noise intensity level.

The bistable model Eq. (4) is monostable for negative ϵ, where the equilibrium is $u = 0$. When the bifurcation parameter changes sign and becomes positive, the system exhibits a pitchfork bifurcation, giving rise to two equilibria $u = \pm\sqrt{\epsilon}$, bistable regime. In the latter regime, when inhomogeneous initial conditions are considered, it exhibits domain walls between these two equilibria. These domain walls are usually called kinks (u_k) or anti-kinks. Analytically these solutions have the form

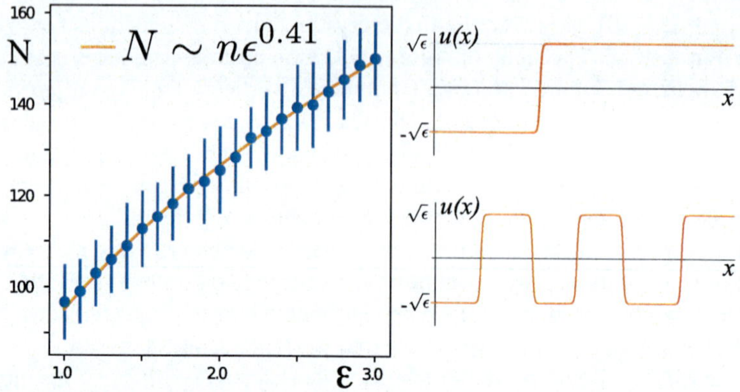

Fig. 4 Number of kinks in a given instant as a function of the bifurcation parameter ϵ obtained from numerical simulations of Eq. (5) with $T = 0.1$ at $t = 20$. The points with a bar account for mean value and standard deviation obtained after carrying out for each parameter 80 realizations. The continuous curves were obtained using the fitting function $N = n\epsilon^{0.41}$. The right panels account for a single kink and a kink gas, respectively

$$u_k(x) = \pm\sqrt{\epsilon}\tanh\left(\frac{\sqrt{\epsilon}}{2}(x - x_0)\right), \tag{5}$$

where x_0 accounts for the position of the wall, that is, $u_k(x = x_0) = 0$. Figure 4 shows a single kink and several kink-antikink interacting solutions obtained from numerical simulations of Eq. (4). These solutions are topological since the destruction of a kink employing continuous deformations can only be carried out through collision with an anti-kink.

When one starts from the null state, the fluctuations induce the emergence of kink and anti-kink solutions. Figure 4 shows how the number of kinks changes as a function of the bifurcation parameter. This chart shows that the number of kinks grows with the bifurcation parameter and has a power law. To understand this law, we use the same strategy used to understand vortex emergence. Then we linearize the dynamics around the zero value $u = 0$, which takes the form

$$\partial_t u = \epsilon u + \partial_{xx} u + \sqrt{T}\chi(t, x) \tag{6}$$

Introducing the ansatz $u = u_0 e^{ikx+\sigma t}$, in the above equation, one get $\sigma = \epsilon - k^2$. Hence, all unstable modes are in range $\{-\sqrt{\epsilon}, \sqrt{\epsilon}\}$. The unstable mode with the most spatial oscillations corresponds to $k = \sqrt{\epsilon} = 2\pi N/L$, where N is the number of domains or zeros of the critical mode. Then the number of domains satisfies the relation

$$N = \frac{L\sqrt{\epsilon}}{2\pi}. \tag{7}$$

Therefore, the number of defects or kinks grows with the square root of the bifurcation parameter. Figure 4 shows a good qualitative agreement. However, the exponent is different. The possible origin of the difference between the exponent of analytical expression (7) and the numerical one (see Fig. 4) are nonlinearities and interactions of kinks and anti-kink that are neglected in the analytical expressions.

6 Conclusion and Remarks

Despite many studies of vortices for their fundamental properties associated with particles with topological properties and their interest in various technological applications such as telecommunications, image processing, and optical tweezers, the study of vortex nucleation in real physical systems had not been addressed to our knowledge. Based on linear theory and stochastic fluctuations, we can establish that the matter vortices are a consequence of the different excited unstable spatial modes. The above is summarized by formula (3) multiplied by a constant that accounts for the effect of all unstable modes. Therefore, we can establish that the number of vortices grows proportionally to the bifurcation parameter; it is inverse to the square of the elastic anisotropy and does not depend on the level of the noise intensity. Experimental observations show a qualitative agreement with theoretical findings. Vortices are an intrinsically nonlinear nature phenomenon; however, we show that the generic mechanism for creating vortices in nature is based on a simple linear theory of critical spatial modes.

Experimental imperfections, which give rise to heterogeneous parameters, can be a source of vortices. To understand the effect of heterogeneities is by modifying the linear problem to non-constant coefficients, which generates that the modes depend on these coefficients. The inclusion of this type of effect is in progress. Likewise, in the developed theory, we have considered no spatial and temporal correlation, white noise; however, the system may exhibit spatial correlations, which may be a consequence of the anisotropic elastic coupling of the liquid crystal. This type of correlation may be responsible for exciting some privileged unstable nodes, *stochastic resonance*. The inclusion of these phenomena can improve the simplified description presented.

Acknowledgements The authors thank Enrique Calisto, Michal Kowalczyk, and Michel Ferre for fructified discussions. MGC acknowledges financial support from ANID–Millennium Science Initiative Program–ICN17_012, Millennium Institute for Research in Optics and FONDECYT 1210353 project.

References

1. A. Sommerfeld, Lectures on theoretical physics, in *Optics*, vol. IV (Academic Press, New York, 1954)
2. J. Nye, M. Berry, Dislocations in wave trains. Proc. R. Soc. Lond. A. **336**, 6–31 (1974)
3. L. Allen, M.W. Beijersbergen, R.J.C. Spreeuw, J.P. Woerdman, Orbital angular momentum of light and the transformation of Laguerre-Gaussian laser modes. Phys. Rev. A **45**, 8185 (1992)
4. M.S. Soskin, M.V. Vasnetov, in *Progress in Optics*, vol. 42, ed. by E. Wolf, (Elsevier, 2001), p. 219

5. L. Allen, S.M. Barnett, M.J. Padgett, *Optical Angular Momentum* (CRC Press, 2003)
6. L.M. Pismen, *Vortices in Nonlinear Fields* (Oxford Science, 1999)
7. D.G. Grier, A revolution in optical manipulation. Nature **424**, 810–816 (2003)
8. V.G. Shvedov, A.V. Rode, Y.V. Izdebskaya, A.S. Desyatnikov, W. Krolikowski, Y.S. Kivshar, Giant optical manipulation. Phys. Rev. Lett. **105**, 118103 (2010)
9. M. Padgett, R. Bowman, Tweezers with a twist. Nat. Photonics **5**, 343–348 (2011)
10. F. Tamburini, G. Anzolin, G. Umbriaco, A. Bianchini, C. Barbieri, Overcoming the Rayleigh criterion limit with optical vortices. Phys. Rev. Lett. **97**, 163903 (2006)
11. H.H. Arnaut, G.A. Barbosa, Orbital and intrinsic angular momentum of single photons and entangled pairs of photons generated by parametric down-conversion. Phys. Rev. Lett. **85**, 286–289 (2000)
12. K. Murphy, C. Dainty, Comparison of optical vortex detection methods for use with a Shack-Hartmann wavefront sensor. Opt. Express **20**, 4988–5002 (2012)
13. J. Wang, J.-Y. Yang, I.M. Fazal, N. Ahmed, Y. Yan, H. Huang, Y. Ren, Y. Yue, S. Dolinar, M. Tur, A.E. Willner, Terabit free-space data transmission employing orbital angular momentum multiplexing. Nat. Photonics **6**, 488–496 (2012)
14. V.Y. Bazhenov, M.V. Vasnetsov, M.S. Soskin, Laser beams with screw dislocations in their wavefronts. JETP Lett. **52**, 429–431 (1990)
15. R.K. Tyson, M. Scipioni, J. Viegas, Generation of an optical vortex with a segmented deformable mirror. Appl. Opt. **47**, 6300–6306 (2008)
16. J. Arlt, K. Dholakia, L. Allen, M.J. Padgett, The production of multiringed Laguerre?Gaussian modes by computer-generated holograms. J. Mod. Opt. **45**, 1231–1237 (1998)
17. M.W. Beijersbergen, L. Allen, H.E.L.O. van der Veen, J.P. Woerdman, Astigmatic laser mode converters and transfer of orbital angular momentum. Opt. Commun. **96**, 123–132 (1993)
18. M. Beresna, M. Gecevicius, P.G. Kazansky, T. Gertus, Radially polarized optical vortex converter created by femtosecond laser nanostructuring of glass. Appl. Phys. Lett. **98**, 201101 (2011)
19. D. Voloschenko, O.D. Lavrentovich, Optical vortices generated by dislocations in a cholesteric liquid crystal. Opt. Lett. **25**, 317319 (2000)
20. L. Marrucci, C. Manzo, D. Paparo, Optical spin-to-orbital angular momentum conversion in inhomogeneous anisotropic media. Phys. Rev. Lett.**96**, 163905 (2006)
21. E. Brasselet, N. Murazawa, H. Misawa, S. Juodkazis, Optical vortices from liquid crystal droplets. Phys. Rev. Lett. **103**, 103903 (2009)
22. R. Barboza, U. Bortolozzo, G. Assanto, E. Vidal-Henriquez, M.G. Clerc, S. Residori, Vortex induction via anisotropy stabilized light-matter interaction. Phys. Rev. Lett. **109**, 143901 (2012)
23. R. Barboza, U. Bortolozzo, G. Assanto, E. Vidal-Henriquez, M.G. Clerc, S. Residori, Harnessing optical vortex lattices in nematic liquid crystals. Phys. Rev. Lett. **111**, 093902 (2013)
24. R. Barboza, U. Bortolozzo, M.G. Clerc, S. Residori, E. Vidal-Henriquez, Adv. Opt. Photonics **7**, 635 (2015)
25. T. Frisch, S. Rica, P. Coullet, J.M. Gilli, Spiral waves in liquid crystal. Phys. Rev. Lett. **72**, 1471–1464 (1994)
26. M.G. Clerc, E. Vidal-Henriquez, J.D. Davila, M. Kowalczyk, Symmetry breaking of nematic umbilical defects through an amplitude equation. Phys. Rev. E **90**, 012507 (2014)
27. S. Chandrasekhar, *Liquid Crystals* (Cambridge University, New York, 1977)
28. L.M. Blinov, *Structure and Properties of Liquid Crystals* (Springer, New York, 2011)
29. V. Zambra, M.G. Clerc, R. Barboza, U. Bortolozzo, S. Residori, Umbilical defect dynamics in an inhomogeneous nematic liquid crystal layer. Phys. Rev. E **101**, 062704 (2020)
30. C. Chevallard, M.C. Clerc, Inhomogeneous Fréedericksz transition in nematic liquid crystals. Phys. Rev. E **65**, 011708 (2001)
31. J. Garcia-Ojalvo, J.M. Sancho, *Noise in Spatially Extended Systems* (Springer, New York, 1999)
32. D.A. Contreras, M.G. Clerc, Internal noise and system size effects induce nondiffusive kink dynamics. Phys. Rev. E **91**, 032922 (2015)

Thermodynamic Instability of the Atmospheric Boundary Layer as a Precursor of an Earthquake

Sergey Pulinets

1 The Thermal Anomalies Before Earthquakes

The history of short-term earthquake forecast is the most complex and intricate in modern seismology research. Starting from approval of concepts of the physical precursors [22] through its complete neglection [5] to reasonable understanding that there is some light at the end of the tunnel [11]. Inside this mess, there is a small spot that probably will lead us to the solution of the problem. It is the thermal anomalies registered before earthquakes. They were known for many years [13] but remained behind the main scene. Even important results [21], unfortunately, published and limited and margin scientific sources were not known until they attracted our attention.

According to [13], the increase in the air temperature and drop of relative humidity were observed during season/month when the strong $M > 5.4$ earthquakes happened in Central Asia (Turkmenia and Uzbekistan) [21] were surprised by excessive drop of the air pressure few days before the earthquakes $M > 4.5$ in Tajikistan (Fig. 1). But this effect gives clue to understanding of the observed phenomena.

Confirmation of the [13] results from Central Asia was obtained in Mexico while registering the atmospheric conditions around the time of M7.8 Colima earthquake on 21 January 2003 [16] when the synchronous variations of the air temperature (increase) and relative humidity (decrease) were registered 1 week before the main shock of Colima earthquake. These results were laid in foundation of the physical mechanism of the observed thermal anomalies [17] and to further development of the Lithosphere-Atmosphere-Ionosphere Coupling (LAIC) model [18].

It should be noted that the pre-earthquake thermal anomalies characteristic to practically 100% of earthquakes and are used now as reliable earthquake precursors.

S. Pulinets (✉)
Space Research Institute Russian Academy of Sciences Profsoyuznaya 84/32,
Moscow 117997, Russia
e-mail: pulse1549@gmail.com

L. Brenig et al. (eds.), *Nonequilibrium Thermodynamics and Fluctuation Kinetics*,
Fundamental Theories of Physics 208,
https://doi.org/10.1007/978-3-031-04458-8_16

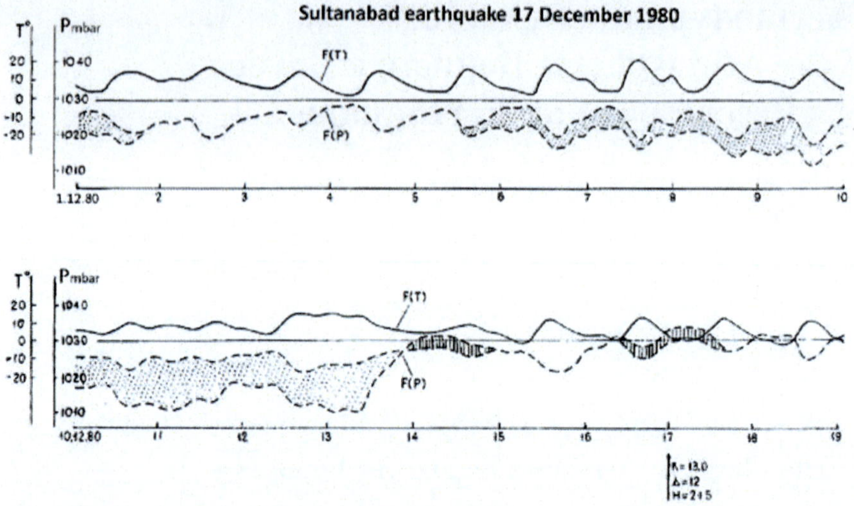

Fig. 1 Variations of the air temperature and pressure around the time of $M4.5$ earthquake near Sultanabad registered at Dushanbe, 14 km distance from the epicenter. Dot shadowed areas demonstrate the anomalous excessive drop of air pressure, vertical line shadowed areas—periods of excessive increase of the air pressure

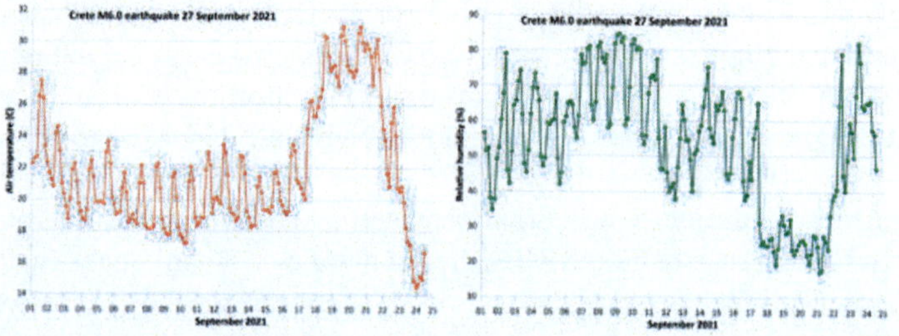

Fig. 2 Left panel—the air temperature variations over the epicenter of Crete $M6$ earthquake at altitude 100 m. Right panel—the relative humidity variations for the same case

So as not to be unfounded, we demonstrate in Fig. 2 the variations of air temperature and the relative humidity before the Crete M6 earthquake on 27 September 2021.

One can see that anomalies start on 17 September, 10 d before the main shock. It should be noted that there were no mesoscale atmospheric phenomena in this region, absolutely quiet weather, and if we use the official weather forecast, we would never expect what is happened with the temperature and humidity after 17 September.

2 Air Ionization as a Main Source of Pre-earthquake Atmospheric Anomalies

It was established that sharp increase of radon emanation before earthquake within the zone of earthquake preparation [26] in which size is determined by estimation made by [3]. Up to the altitudes near 1 km radon is the main source of the air ionization [7]. In general, enhanced radon emissions have been observed prior to earthquakes and this has been recorded all over the world [15, 20]. Let us consider what effect radon can provide in the near-ground layer of atmosphere [18].

Thus, the decay of radon ^{222}Rn releases α-particles with energy $E\alpha = -5.46$ MeV. Since the energy of atmospheric gas ionization is in the range from 10 to 30 eV, each α-particle released by radon can produce on the average $\sim 3 \times 10^5$ electron-ion pairs. It is known from literature data [8, 23, 25] that the average radon activity near the Earth's surface is ~2000 Bq Bq/m^3. Here, it should be noted that the measurements were performed at a depth of around 70 cm from the Earth's surface in three quite different geological regions (Mexico, Turkey, and Russia). These were not periods of increased seismic activity. Since radon is known to be six times heavier than air, its accumulation occurs in lowlands (ravines, gorges, and folds); this fact explains the choice of the height at which the measurements were conducted. Also, it is known that during the preparation of earthquakes, the radon activity can increase by an order of magnitude; however, the reference level for calculation was taken of order of ~2000 Bq Bq/m^3. Then, in view of the ionization capacity of radon α-particles of $\sim 3 \times 10^5$ electron–ion pairs, the ion formation rate is $\sim 6 \times 10^8$ m^{-3} s^{-1}.

As it was described in [17] after series of plasma-chemical reactions, the final ions undergo the hydration process leading to formation of condensation nucleus. This process is called the Ion Induced Nucleation (IIN) [10]. A hydrated ion of a size of around 1–3 μm (particles of this size range are observed by the AERONET network a few days before the earthquake) contains around 0.4×10^{12} water molecules. The latent heat constant U_0 is:

$$U_0 \sim \times 10^3 \mathrm{J/mole} \qquad 1 \text{ mole} = 6.022 \times 10^{23}. \tag{1}$$

For a given radon activity and the formation of hydrated ions of a size of ~1 μm, the release of latent heat is ~16 W/m^2, which is consistent with experimentally recorded fluxes of infrared outgoing longwave radiation (OLR) [14]. Since 1 eV = 1.6×10^{-19} J, a given radon activity of 2000 Bq Bq/m^3 leads to an expenditure on ionization of 1.7×10^{-9} J/m^3 s^{-1}. The ratio of IIN-induced heat to the energy spent on the ionization of atmospheric gases is then 16/(1.7 × 10–9) ~ 10^{10}, which is evidence that the energetic efficiency of the process is extremely high.

3 Autocatalytic Reaction Leading to Thermal Instability

We should take into account that we deal with the essentially non-equilibrium system which the zone of earthquake preparation represents. According to [9], "Non-equilibrium in such systems can be associated with the inhomogeneity of the distribution of temperature and composition of the substance over the volume, with the flows of matter within the system, and with more complex types of non-equilibrium caused by propagation of electromagnetic waves, flow of electric current, etc. Systems been far from thermodynamic equilibrium can exchange with the environment not only energy, but also the mass of a substance. Such systems, in contrast to closed ones, are called open. The energy in them can dissipate and irreversibly transform into other types of energies, for example, into the energy of vibrational or thermal motion of atoms. Therefore, sometimes such systems are called dissipative."

One more factor of instability that should be mentioned – is the presence of water within our volume of analysis in two phases (gaseous and liquid) with active phase transition due to ion's hydration. Such state is also the source of system instability [6]. The hydration is essentially differing from usual condensation that is traditionally considered in water phase transitions in atmosphere. It gives opportunity to extract the additional amount of thermal energy from the latent state what will be considered further in paragraph devoted to the Atmospheric Chemical Potential (ACP).

According to [12] for processes development in dissipative media, we should consider the following factors (schematically presented in Fig. 3).

1. The threshold sensitivity.
   Radon is emanating from the Earth's crust continuously regardless the presence of seismic activity, but as it was mentioned earlier, its flux essentially growth at the last stage of the seismic cycle of earthquake preparation, and after reaching some threshold starts to generate the chain of processes leading to generation of thermal and electromagnetic precursors of earthquakes.
2. The process becomes the so-called exacerbation process [12] leading to hyperbolic growth of some parameters.
   We can see in the Fig. 3 the curve of radon variation before the Kobe earthquake in Japan in January 1995 described in [27]. Another picture from [24] shows the avalanche increase of the number of ions produced by corona discharge with increasing the voltage on the needle-producing discharge. The maximum value of the voltage in experiment is 3.5 kV while the energy of α-particles emitted by radon during decay is 5.46 MeV.
3. The exacerbation processes are usually characterized by auto-catalytic reactions [12].
   In Plasma-chemical reaction described in [17] hydronium $H_3O^+(H_2O)_n$ is one of the final products, and it supports the increase of water molecules attached to the formed cluster ions, and launches the reaction shown in the top of the Fig. 3 again and again (positive feedback) playing the role of catalyzer [1, 4]).

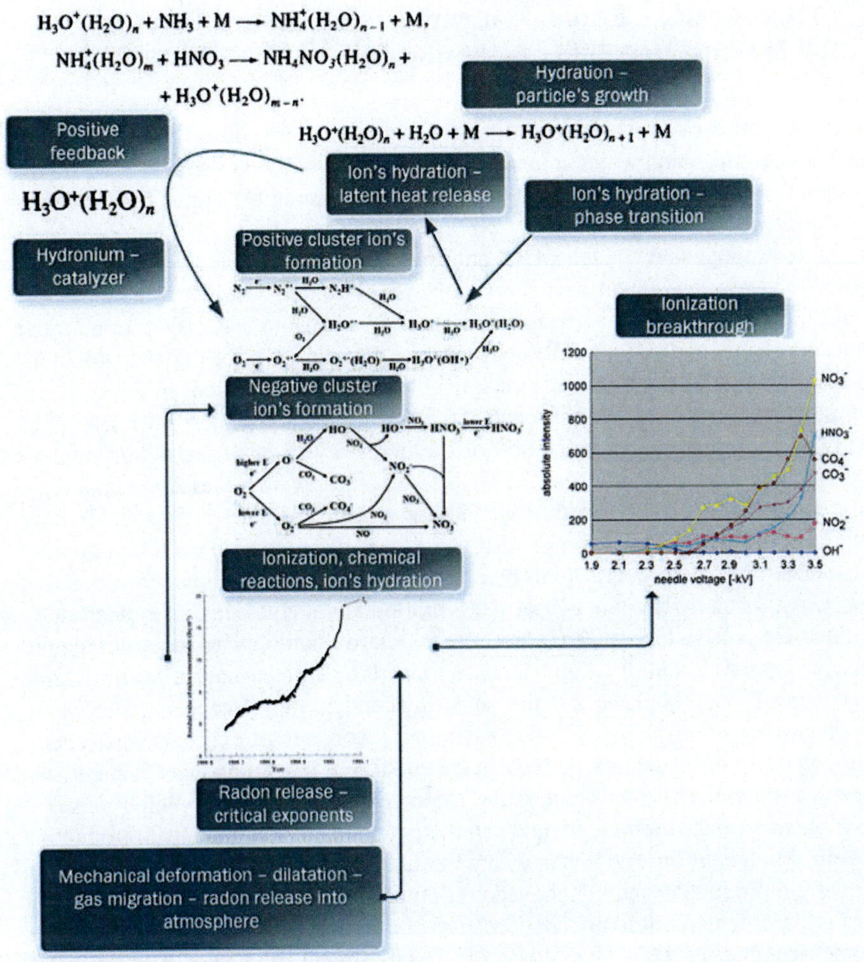

Fig. 3 Consequences of increased radon emanation due to dilatation process initiating the exacerbation regime with instability stimulated by autocatalytic exothermal reaction. Letter *M* in formulas means the presence of any neutral gas molecule

The larger number of water molecules attached to ions formed after ionization and formation of terminal ions, the more essential heat release we observe. As indicator of the process intensity, the size of formed cluster ions could serve. But there exists more reliable measure of the thermal instability intensity called the correction of chemical potential of water vapor in atmosphere [18]. To be short, we call it atmospheric chemical potential (ACP).

4 Atmospheric Chemical Potential as a Measure of Thermal Instability in Atmosphere

Here we will explain why the result of pressure excessive drop (Fig. 1) surprised the paper authors, and why the meteorological agencies are not able to forecast the observed variations of air temperature and relative humidity (Fig. 2) before strong earthquakes.

All thermodynamic models of the atmosphere, while considering the phase transitions of water and latent heat fluxes, operate with latent heat (per mole, or per molecule) as a constant at a given temperature. It is equal to 0.422 eV per one water molecule. But if one will carefully deal with real data, he can often find that violations of the gas equation are observed (inexplicable additional variations in temperature, humidity, and pressure). We approached this phenomenon from the other side when studying the processes of ionization of the boundary layer of the atmosphere by radon, the release of which from the earth's crust sharply increases at the final stage of preparation for a strong earthquake. We gave some estimates in Sect. 3. Here we will go straight to the final result. As it turned out, atmospheric ions formed during the ionization of atmospheric gases by energetic alpha particles emitted by radon during decay, instantly become hydrated. Hydration is not equivalent to condensation because the process takes place at any level of relative humidity and does not require saturated vapor. But all the same, a phase transition of water molecules from a free to a bound state takes place, and, just as during condensation, latent heat is released. The phase transition of the first order (evaporation/condensation) occurs when chemical potentials of interacting molecules are equal. For one-component systems, the chemical potential is equivalent to the thermodynamic potential, which makes it possible to estimate thermodynamic parameters, for example, humidity, through the chemical potential U. And this is where miracles begin. It turned out that at a high rate of ion production and high concentration of ions, the released amount of latent heat is larger than for ordinary condensation of the same number of water molecules condensed/attached. How much larger? If we talk about earthquakes, then this value (the difference between the released latent heat during hydration and during normal condensation) ranges from 0.01 to 0.1 eV, i.e. in extreme cases, it can reach about 25% of the latent heat constant. This difference we call the correction of the chemical potential of the water vapor in the atmosphere.

After radon activity exceeds the corresponding threshold, the hydration takes explosive character, what leads to sharp decrease of the relative humidity which cannot be predicted by meteorological models, because they do not see the saturation to predict the humidity drop. That's why the forecast of relative humidity variation shown in Fig. 2 is impossible by the ordinary meteorological means. We can say that from the point of view of traditional meteorology the observed phenomena is anomaly.

Now we can explain Fig. 1. According to Dalton law, the total atmospheric pressure is the sum of the gas constituents' pressures. Here we can consider the sum of dry air pressure and the water vapor pressure. Because the authors of the paper did not take into account the drop of relative humidity due to ionization process, they couldn't explain the excessive drop of the total air pressure few days before the earthquake.

What is advantage of the ACP and how it is calculated? According to [2], the relative humidity H(t) can be expressed as:

$$H(t) = \frac{\exp(-U(t)/k_B T)}{\exp(-U_0/k_B T)} = \exp\left(\frac{U_0 - U(t)}{k_B T}\right) = \exp\left(-\frac{0.032\varDelta U\ \cos^2 t}{(k_B T)^2}\right) \tag{2}$$

where $U(t)$ is the current latent heat per one molecule, U_0 – the latent heat of evaporation, and $\varDelta U$ – is the required value of the chemical potential correction ACP. Using (2), it could be expressed through the air temperature T and relative humidity H as:

$$\varDelta U = 5.8 \times 10^{-10}\,(20\,T\,g + 5463)^2\,\ln(100/H). \tag{3}$$

The advantage of these parameters is that it expresses intensity of the ionization process taking into account both the changes of air temperature and relative humidity and can be used to monitor the global radon activity without direct radon measurements, which is especially valuable in inaccessible regions of the planet.

What is also very useful – it is the self-similarity of ACP variations before earthquakes for the same area. In the different regions of the globe, these variations could differ what is quite understandable. The radon activity could vary not only before earthquakes but for any cases of activation of tectonic faults what makes possible to map the active tectonic faults without geologic prospection. Another advantage of ACP is that it reacts also on the volcanic activity and its increase is observed before the volcano eruption what provides one more application of ACP in real forecast of the seismic and volcanic activity.

The practical applications are not the purpose of the current paper, so just for understanding of beauty of this parameter and the scale of atmospheric anomalies before the strong earthquakes I would like to demonstrate the spatial distribution of the ACP before the strong M7.7 Jamaica earthquake on 28 January 2020. This distribution was registered 5 d before the earthquake, 23 January (Fig. 4).

5 Conclusion

We considered the thermal instability in the boundary layer of atmosphere, which appears as a result of development, the critical processes characteristic to the open nonequilibrium systems with dissipation within the period of the strong earthquake approaching. It is initiated by the sharp increase of the radon release within the earthquake preparation zone, which initiates the strong modification of the boundary layer of atmosphere. The thermal anomalies and especially the derived atmospheric

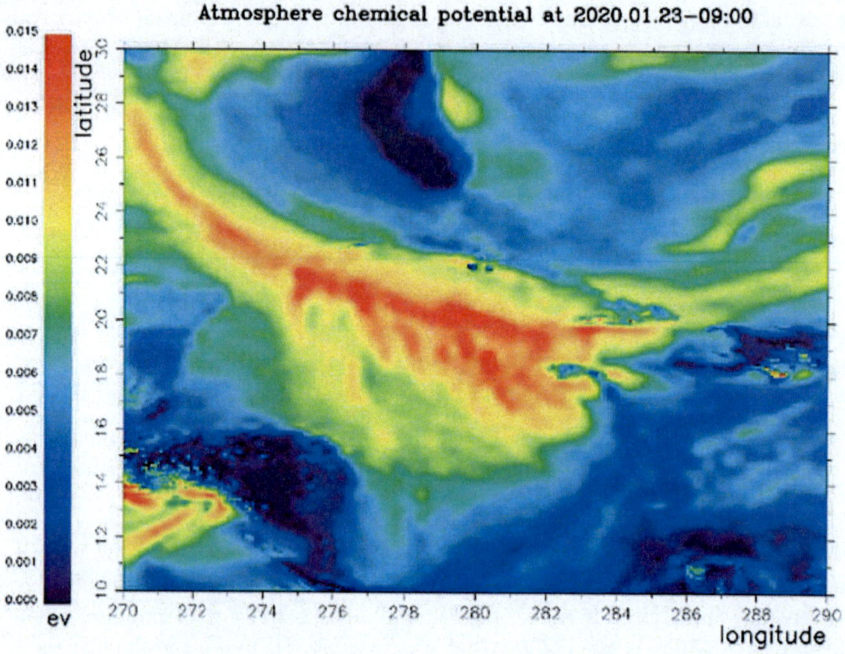

Fig. 4 Scorpio-shaped spatial distribution of the ACP 5 d before the M7.7 Jamaica earthquake

chemical potential could be used in practical application for the short-term volcano and earthquakes forecast [19].

6 Four Dimensions of My Interaction with Slava Belyi

We are brothers. We have the common *Alma Mater* – Lomonosov Moscow State University, even more, the common Faculty of Physics, even deeper, the common division – Radiophysics, and finally, the common cathedra with the strange name: "General Physics for Mekhmat and Wave Processes". This name does not reflect the scientific themes and directions, but only the fact that professors from our cathedra were lecturers at the Mechanico-Mathematical faculty of Moscow State University. In reality, the professor's staff was a kind of embryo of completely new direction in Radiophysics. Just during period of our University studies, the fundamental transformation of classical oscillation theory happened with its conversion into nonlinear oscillations, formation of statistical radiophysics, non-equilibrium thermodynamics, nonlinear open systems and synergetics. Professors Stratonovich, Klimontovich, Romanovsky, Shmalgausen were the real founders of these directions, and communications with them at their lectures and seminars formed the axis of our movement into the science. Professor Klimontovich became a scientific advisor of Slava for many years.

Slava was 4 years older than me, so our contacts at the university were not very tight, but the first and main dimension of our interaction was the spirit of our cathedra which we saved for all our life and could distinguish ourselves due to this spirit being present inside us.

Our second encounter happened in small scientific town Troitsk located South from Moscow. Again, we found ourselves in a common cradle for many years, now for professional work in the Pushkov Institute of Terrestrial Magnetism, Ionosphere and Radiowave Propagation of the Russian Academy of sciences (IZMIRAN). Slava appeared there in 1971, and me in 1972. There was the unique situation at IZMIRAN where co-existed two ionospheric departments simultaneously because of incompatibility of their heads Dr. Lev Lobachevsky and Prof. Natalia Ben'kova. There was the official explanation that the first department provided the application research, and the second—the fundamental research of the ionosphere. In relation to us it was quite opposite. Slava was working on fundamental problems of plasma physics, and I was involved in the space instrumentation development. In mutual communications, Slava impressed with his absolute calm, but this tranquility was based on an inner sense of self-righteousness. By the end of my carrier at IZMIRAN, I became the Deputy Director of the Institute and formally Slava was my subordinate. I checked the employee yearly reports and sometimes wondered that some of them had several papers published, and Slava only one. But in reality, it could be publication in *Nature* accompanied by the State Prize of the Soviet Union.

Troitsk was the small city and in those times the majority were the scientists greedy for contacts and social activity. With transformation of the country, we searched for new forms of such activity and one of ideas was the organization of the Rotary club in Troitsk. As a Chairman of the Club of Scientists council with group of like-minded scientists we became the Charter members of the Troitsk Rotary club. Slava was also between the first members of the club. He shared his experience on contacts with Rotarians in Belgium and France and proposed some interesting ideas for our activity and he did this with calmness and confidence as always.

In casual conversations in club with Slava, I learned about his contacts with Professor Prigogine, one of the founders of Synergetics, and this became the fourth dimension of our personal contacts. At those times I started to work on the physics of earthquake precursors and realized that our geospheres all together compose a complex open system in which the critical process development could be explained with the help of synergetic approach. The thermal instability of atmospheric boundary layer that is discussed in this article is a bright example of synergy ideas application, and I want to dedicate this paper to Slava.

References

1. M.A. Biondi, Atmospheric electron-ion and ion-ion recombination processes. Canad. J. Chemistry. **47**, 1711–1719 (1969)
2. K.A. Boyarchuk, A.V. Karelin, R.V. Shirokov, *The basic model of the ionized atmosphere kinetics* (VNIIEM Publ, Moscow, 2006), p. 320
3. I.R. Dobrovolsky, S.I. Zubkov, V.I. Myachkin, Estimation of the size of earthquake preparation zones. Pageoph. **117**, 1025–1044 (1979)
4. Castleman A.W. (1973) Aerosol chemistry physics and chemistry of upper atmosphere. Ed. BM McComac. Dordrecht, Holland. p. 143-157
5. R.J. Geller, D.D. Jackson, Y.Y. Kagan, F. Mulargia, Earthquake cannot be predicted. Science, New Series **275**(5306), 1616–1617 (1997)
6. H. Haken, *Synergetics* (Springer-Verlag, Berlin Heidelberg, 1983), p. 390
7. W.A. Hoppel, R.V. Anderson, J.C. Willett, Atmospheric Electricity in the Planetary Boundary Layer, - in Studies in Geophysics, in *The Eart's Electrical Envirinment*. (National Academy Press, Washington. D.C, 1986), pp. 149–165
8. S. Inan, T. Akgül, C. Seyis, R. Saatçilar, S. Baykut, S. Ergintav, M. Baş, Geochemical monitoring in the Marmara region (NW Turkey): A search for precursors of seismic activity. J. Geophys. Res. **113**, B03401 (2008). https://doi.org/10.1029/2007JB005206
9. Kadomtsev B.B., Ryazanov A.I. (1983) What the synergetics is? Priroda No8: 2-11
10. Kathmann S.M., Schenter G.K., and Garrett B.C. (2005) Ion-Induced Nucleation: The Importance of Chemistry. Physical Rev. Lett. 94: 116104. https://doi.org/10.1103/PhysRevLett.94.116104
11. V. Keilis-Borok, (2002) EARTHQUAKE PREDICTION: State-of-the-Art and Emerging Possibilities. Annu. Rev. Earth Planet. Sci. **30**, 1–33 (2002). https://doi.org/10.1146/annurev.earth.30.100301.083856
12. Knyazeva E.N., Kurduimov S.P. (2003) Peculiar properties of nonequilibrium processes in open dissipative media. In: The Problems of XXI Century Geophysics. Nauka Publ., Moscow, p. 37-65 (in Russian)
13. M.R. Mil'kis, Meteorological Precursors of Earthquakes. Izvestiya, Earth Physics. **22**, 195–204 (1986)
14. D. Ouzounov, D. Liu, K. Chunli, G. Cervone, M. Kafatos, P. Taylor, Outgoing longwave radiation variability from IR satellite data prior to major earthquakes. Tectonophysics. **431**, 211–220 (2007)
15. E. Petraki, D. Nikolopoulos, D. Panagiotaras, D. Cantzos, P. Yannakopoulos et al., Radon-222: A Potential Short-Term Earthquake Precursor. J Earth Sci Clim Change **6**, 282 (2015). https://doi.org/10.4172/2157-7617.1000282
16. S.A. Pulinets, D. Ouzounov, L. Ciraolo, R. Singh, G. Cervone, A. Leyva, M. Dunajecka, A.V. Karelin, K.A. Boyarchuk, A. Kotsarenko, Thermal, atmospheric and ionospheric anomalies around the time of the Colima M7.8 earthquake of 21 January 2003. Annales Geophysicae. **24**, 835–849 (2006)
17. S.A. Pulinets, D. Ouzounov, A.V. Karelin, K.A. Boyarchuk, L.A. Pokhmelnykh, The physical nature of the thermal anomalies observed before strong earthquakes. Physics and Chemistry of the Earth. **31**, 143–153 (2006)
18. S.A. Pulinets, D.P. Ouzounov, A.V. Karelin, D.V. Davidenko, Physical Bases of the Generation of Short-Term Earthquake Precursors: A Complex Model of Ionization-Induced Geophysical Processes in the Lithosphere-Atmosphere-Ionosphere-Magnetosphere System. Geomagnetism and Aeronomy. **55**, 540–558 (2015). https://doi.org/10.1134/S0016793215040131
19. Pulinets S. and Ouzounov D. (2018) The possibility of earthquake forecasting. Learning from Nature. IOP Publishing, Bristol, December 2018. 167 p., https://iopscience.iop.org/book/978-0-7503-1248-6
20. A. Riggio, M. Santulin, Earthquake forecasting: a review of radon as seismic precursor. Bulletin of Geophysics and Oceanography. **56**, 95–114 (2015)

21. Sadovsky M.A., Bannov Yu.A., Mirzoev K.M., Negmatullaev S.Kh. (1995) The Phenomenon of Violation of an Equilibrium State of the Temperature and Pressure Functions in Atmosphere in Enclosed Volumes before Earthquakes. Earthquakes Forecast No 6: 242-266. Donish Publishing, Tajik Acad. Sci
22. C.H. Scholz, L.R. Sykes, Y.P. Aggarwal, Earthquake prediction: A physical basis. Science. **181**, 803–809 (1973)
23. N. Segovia, S.A. Pulinets, A. Leyva, M. Mena, M. Monnin, M.E. Camacho, M.G. Ponciano, V. Fernandez, Ground radon exhalation, an electrostatic contribution for upper atmospheric layers processes. Radiation Measurements. **40**, 670–672 (2005)
24. K. Sekimoto, M. Takayama, Influence of needle voltage on the formation of negative core ions using atmospheric pressure corona discharge in air. International Journal of Mass Spectrometry. **261**, 38–44 (2007)
25. A.A. Spivak, Volumetric activity of soil radon within the zones of tectonic faults, in *Geophysics of inter-geospheres interaction*. ed. by V.V. Adushkin (GEOS Publ, Moscow, 2008), pp. 235–246
26. J.-P. Toutain, J.-C. Baubron, Gas geochemistry and seismotectonics: a review. Tectonophysics **304**, 1–27 (1998)
27. Y. Yasuoka, G. Igarashi, T. Ishikawa, S. Tokonami, M. Shinogi, Evidence of precursor phenomena in the Kobe earthquake obtained from atmospheric radon concentration. Applied Geochemistry. **21**, 1064–1072 (2006)

Multistability of Vector Solitons in High-Q Resonators

B. Kostet, Y. Soupart, E. Averlant, K. Panajotov, and M. Tlidi

Abstract We investigate the formation of vector solitons in weakly birefringent high-Q resonators. The presence of nonlinear polarization mode coupling in optical resonators subject to a coherent optical injection allows stabilizing up to two families of bright or dark vector dissipative solitons, depending on the dispersion properties of the system. We use coupled Lugiato–Lefever equations to investigate the dynamical properties of interacting laser fields confined in the Kerr optical resonators. Anomalous and normal dispersion regimes are considered, and it is shown that in both cases two branches of dissipative solitons coexist and exhibit different peak powers and different polarization properties. In these regimes, the input–output characteristics possess either a bistable or a tristable homogeneous response. The coexistence of two branches of localized states is not possible without taking into account the polarization degrees of freedom. The stabilization mechanism of these localized states is attributed to a subcritical modulational instability in the case of anomalous dispersion and to a front-locking mechanism in the normal dispersion regime. Their bifurcation diagrams exhibit either a homoclinic or a heteroclinic snaking type of instability, depending on the dispersion properties.

B. Kostet (✉) · Y. Soupart · E. Averlant · M. Tlidi
Département de Physique, Faculté des Sciences, Université Libre de Bruxelles (U.L.B.), CP 231, Campus Plaine, 1050 Bruxelles, Belgium
e-mail: Bilal.Kostet@ulb.be

Y. Soupart
e-mail: Youri.Soupart@ulb.be

E. Averlant
e-mail: e.averlant@ernestrichard.edu.brussels

M. Tlidi
e-mail: Mustapha.Tlidi@ulb.be

K. Panajotov
Department of Applied Physics and Photonics (IR-TONA), Vrije Universiteit Brussels, Pleinlaan 2, 1050 Brussels, Belgium
e-mail: kpanajot@b-phot.org

Institute of Solid State Physics, Bulgarian Academy of Sciences, 72 Tzarigradsko Chaussee Boulevard, 1784 Sofia, Bulgaria

L. Brenig et al. (eds.), *Nonequilibrium Thermodynamics and Fluctuation Kinetics*, Fundamental Theories of Physics 208,
https://doi.org/10.1007/978-3-031-04458-8_17

1 Introduction

Dissipative structures are inherent to out-of-equilibrium systems that are subject to mechanisms that tend to restore uniformity (a transport process such as diffusion, dispersion, diffraction, or thermal conductivity) and which compete with nonlinear interactions that tend to locally amplify the field intensity (chemical reactions or matter–light interaction), while dissipation of energy competes with injection [1, 2]. These properties can be found in many real-life systems, e.g. in the fields of chemistry [3, 4], biology [5], ecology [6–9], and nonlinear optics [10–12]. Dissipative structures can be spatially extended patterns and/or localized in time. In the first case, the organization of matter or energy in the bulk of the material often originates from a modulational instability (MI). Some of the most common spatial patterns are stripes, hexagons, and honeycombs [13]. Furthermore, when the modulational instability appears subcritically, there can exist a pinning region where isolated spots of the pattern are embedded on a homogeneous background which are often called dissipative solitons (DSs) [14]. Another type of dissipative solitons that can appear in the presence of bistability, without any pattern or any specific wavelength emerging, is due to the front-locking mechanism [15, 16]. In this case, the interaction of two fronts, i.e. heteroclinic connections between the two stable homogeneous steady states (HSSs), (also called continuous wave, CW solutions in optics), is responsible for the appearance of the DSs. For both of these formation mechanisms, the two balances between the nonlinear effect and the transport process on the one hand, and between pumping and dissipation on the other hand, make the DSs robust structures with an intrinsic size defined by the dynamical properties of the system only. This leads to very interesting prospects in the field of optics for possible applications such as information processing and optical storage [17–19].

Staying in the field of optics, considering the polarization degrees of freedom leads to richer dynamics. In free propagation, i.e. in the absence of an optical resonator, it has been shown that new modulational instabilities can appear, and with them, symmetry breaking, domain wall vector solitons, rotating vector soliton bound states [20], dark–bright vector solitons [21] and vector flat-top solitons [22]. In the presence of a Kerr resonator, the third-order dependency of the polarization on the electric field implies that the medium becomes birefringent. In particular, when the polarization state of the DSs evolves in time while the group velocities of the two different polarization components are locked, they are called group-velocity-locked vector solitons (GVLVSs) [23, 24]. Another common case corresponds to the locking of the polarization states of the two components, called polarization-locked vector solitons (PLVSs) [25, 26]. Other situations exist, such as vector solitons with locked and precessing states of polarization [27] or group-velocity-locked vector soliton molecules [28].

Most of the above-mentioned physical systems are spatially extended, where DSs correspond to spots in the two- or three-dimensional bulk of the material. However, in the field of nonlinear optics, they can be obtained in small area waveguides where diffraction can be neglected, so that the transport phenomenon role is carried out by the chromatic dispersion of light. In this case, the DSs are called temporal DSs and present themselves as pulses propagating indefinitely in the cavity without any spreading of their temporal profile. This makes the temporal DSs very attractive for information processing as they could serve as bits in an all-optical buffer [29–31]. Another property of temporal DSs gathering ever more increasing interest is the optical spectrum built over many roundtrips in the cavity, which is made of equally spaced lines, also called optical frequency combs, allowing a myriad of applications in spectroscopy, metrology, and photonics [32–35]. When the generation of these optical frequency combs is due to the nonlinearity of a Kerr medium, they are called Kerr combs. In particular, a lot of this interest has been focused on high-Q resonators which have seen rapid development in recent years. This is partly due to their ability to host the Kerr combs in small devices with low pump power and various other important properties such as on-chip integration [36], octave-spanning spectra [37], or tuning of the central frequency [38].

In this chapter, we investigate numerically the formation of vector temporal solitons with different polarization states and intensities in optical resonators for different cases. This approach is valid for both all-fiber macroscopic resonators and microscopic resonators. First, we consider the anomalous dispersion regime in which we investigate bright DSs generated by the patterning phenomenon of modulational instability. We pursue with the normal dispersion regime where we place ourselves far from any MI, with low detunings and bistability between stable CW solutions, and consider dark DSs generated by the front-locking mechanism. In the case of higher detunings, we show that the polarization degrees of freedom lead to the appearance of tristability of stable CWs, bringing a high degree of multistability of DSs with very different polarization states.

2 The Vectorial Lugiato–Lefever Model

In 1987, L. A. Lugiato and R. Lefever theoretically evidenced the possibility of a spontaneous emergence of spatial stationary dissipative structures in the transverse plane of a laser beam circulating in a passive optical resonator filled with a nonlinear Kerr medium [10]. They provided a model allowing for the description of the spatio-temporal evolution of the electric field envelope E in such a system and that is since known as the Lugiato–Lefever equation (LLE) :

$$\frac{\partial E}{\partial t} = E_i - (1 + i\theta)E + i|E|^2 E + i\frac{\partial^2 E}{\partial x^2}. \tag{1}$$

Here, t is the normalized time, x is the normalized coordinate along the cavity, and θ is the frequency detuning of the injected field to the cavity resonance frequency. The self-organization of the electric field within this out-of-equilibrium system and the robustness of the resulting structures are explained by the appropriate balances of, on the one hand, diffraction with nonlinearity and, on the other hand, of the internal losses with the pump E_i. This breakthrough in the field of nonlinear optics aroused a great interest among the scientific community, and the Lugiato–Lefever (LL) model was (and still is) widely used and developed.

Our purpose here is to adapt the LLE in order to take into account the polarization degrees of freedom of the electric field propagating inside a passive optical resonator. Moreover, when the transverse dimensions can be neglected (e.g. by using waveguides), so can the diffraction. The transport phenomenon playing its role is the chromatic dispersion and it is possible to find stable structures that are now localized in time instead of space. Such *temporal* dissipative solitons form a pulse train in the output branch of the resonator whose power spectral distribution is made out of evenly spaced teeth, forming so-called optical frequency combs. Adding a polarization degree of freedom will result in a vectorial LL model made out of two coupled LLEs, one for each polarization component of the electric field. Let us first briefly recall basic notions about polarization and how to characterize it. The reader familiar with the Stokes parameters and the polarization ellipse can skip the next section.

Characterizing the Polarization

In this section, we recall the fundamental notions and quantities necessary to study light polarization following the presentation made in [39]. From a classical point of view, the propagation of light corresponds to the spatio-temporal evolution of an electromagnetic wave constituted of an electric and a magnetic field oscillating perpendicularly relative to one another and both lying in the plane normal to the direction of propagation. Transverse waves such as electromagnetic waves possess a fundamental property called *polarization* describing the oscillatory behavior of the electric field vector in the transverse plane. Polarization of incoming light can be constantly evolving with time in a random fashion. Such light is said to be *unpolarized.* In that case, during the spatio-temporal evolution of the electric field, the successive orientations of the vector are uncorrelated and no dominant polarization state emerges. However, polarization is a central concept in nonlinear optics and laser physics since lasers generally produce quasi-monochromatic light which is always at least partially polarized. Different quantities can be used to characterize the polarization properties of an electric field $\mathbf{E}$ based on its components in the transverse plane (E_x, E_y). Among them, widespread are the *Stokes parameters*, defined as

$$\mathbf{S} = \begin{bmatrix} S_0 \\ S_1 \\ S_2 \\ S_3 \end{bmatrix} = \begin{bmatrix} |E_x|^2 + |E_y|^2 \\ |E_x|^2 - |E_y|^2 \\ E_x^* E_y + E_x E_y^* \\ i(E_x E_y^* - E_x^* E_y) \end{bmatrix} = \begin{bmatrix} P_h + P_v \\ P_h - P_v \\ P_{\pi/4} - P_{3\pi/4} \\ P_l - P_r \end{bmatrix} \in \mathbb{R}^4, \tag{2}$$

where * stands for the complex conjugate.

While S_0 is the total intensity, S_1 and S_2 give information about the fluxes of light polarized *linearly* along the horizontal x-direction (P_h) and the vertical y-direction (P_v) and along the directions at angles 45° $(P_{\pi/4})$ and 135° $(P_{3\pi/4})$ with respect to the x-direction, respectively. The sign of S_1 and S_2 indicates which flux overcomes

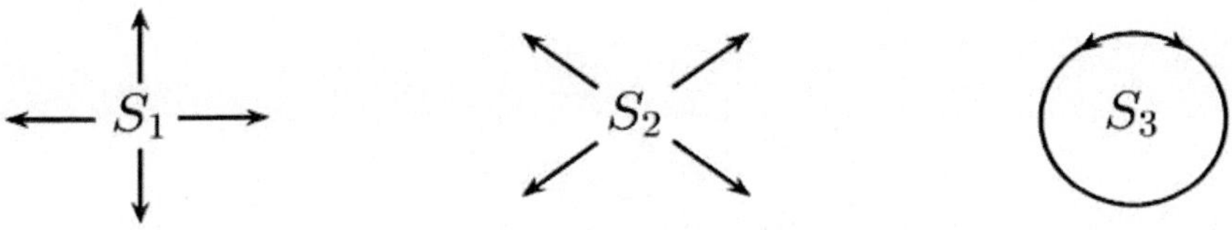

Fig. 1 Schematic representation of the Stokes parameters. The oscillations of the electric field in the transverse plane along the horizontal x-direction and the vertical y-direction are captured by S_1 while S_2 gives informations about the directions at angles $\pi/4$ and $3\pi/4$ with respect to the horizontal direction. Finally, S_3 quantifies the amount of right ($S_3 < 0$) or left ($S_3 > 0$) circularly polarized light

the other while their magnitude quantifies that excess. Finally, the *circular* polarization is described by parameter S_3, whose sign tells us if it is left-handed ($S_3 > 0$) or right-handed ($S_3 < 0$). Considering a light beam propagating in the z-direction (normal to the page), Fig. 1 gives a schematic representation of the Stokes parameters, sometimes seen as the components of a quantity **S**, the Stokes (pseudo-)vector (as it does not transform as a vector), and fully describing the polarization state of light. Note that we generally work with the normalized quantities $s_i = S_i/S_0 \in [-1, 1]$.

Pure polarization states as presented in Fig. 1 are given by $\mathbf{s}_1 = [1, \pm 1, 0, 0]^T$, $\mathbf{s}_2 = [1, 0, \pm 1, 0]^T$, and $\mathbf{s}_3 = [1, 0, 0, \pm 1]^T$. However, in general, the tip of the electric field vector draws an ellipse in the course of its evolution, combining non-zero linear and circular components of polarization.

Based on the Stokes parameters and in order to complete our characterization of the polarization state of the output light, we can define several other quantities. The *degree of polarization* (DoP) takes a value between 0 and 1 and is the ratio of the total polarized flux and the total flux:

$$\mathrm{DoP}(\mathbf{S}) = \sqrt{s_1^2 + s_2^2 + s_3^2}. \tag{3}$$

Unpolarized light exhibits a zero DoP whereas light in a single state of polarization is fully polarized and the corresponding DoP is equal to 1. That general feature can be refined into its constitutive parts, namely the *degree of linear polarization* (DoLP) and the *degree of circular polarization* (DoCP), given by

$$\mathrm{DoLP}(\mathbf{S}) = \sqrt{s_1^2 + s_2^2}, \quad \text{and} \quad \mathrm{DoCP}(\mathbf{S}) = s_3. \tag{4}$$

DoLP varies from 0 (for circularly polarized or unpolarized light) to 1 (for linearly polarized light) while DoCP goes from -1 (for left circular polarization) to 1 (for right circular polarization). Light that is either unpolarized or linearly polarized has a DoCP of 0.

In the general case, we can define the ellipticity ϵ of the polarization ellipse

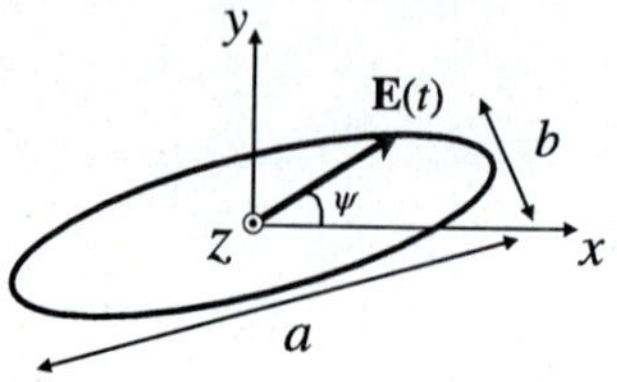

Fig. 2 Polarization ellipse in the (x, y)-transverse plane for light propagating in the z-direction. The major and minor axes of the ellipse are denoted by a and b, respectively

$$\epsilon = \frac{|s_3|}{\sqrt{s_1^2 + s_2^2 + s_3^2} + \sqrt{s_1^2 + s_2^2}} = \frac{b}{a}, \tag{5}$$

which is the ratio of the minor axis to the major axis of the ellipse describing the trajectory followed by the tip of the electric field vector (see Fig. 2).

Vectorial LL model

We consider the propagation of light inside a Kerr resonator submitted to coherent optical pumping (see Fig. 3). The ring cavity is filled with a birefringent medium, and its slow and fast axes are oriented along the x- and y-directions, respectively. This model is suitable to describe macroscopic fiber cavities as well as microscopic resonators. We take into account the polarization degrees of freedom and we neglect the phenomenon of diffraction so that the nonlinearity of the medium balances with the chromatic dispersion only, and the field only varies along the longitudinal direction. The Brillouin and Raman scatterings are not considered so that, over one roundtrip, the linear polarization components $\hat{E}_{x,y}(z, \tau)$ of the slowly varying electric field envelope obey the nonlinear Schrödinger equations that take the general forms [40]:

$$\frac{\partial \hat{E}_x}{\partial z} + \hat{\beta}_{1,x}\frac{\partial \hat{E}_x}{\partial \tau} + i\hat{\beta}_{2,x}\frac{\partial^2 \hat{E}_x}{\partial \tau^2} + \frac{\alpha_{i,x}}{2}\hat{E}_x = i\gamma\left(|\hat{E}_x|^2 + \frac{2|\hat{E}_y|^2}{3}\right)\hat{E}_x + \frac{i\gamma}{3}\hat{E}_x^*\hat{E}_y^2 e^{-2i\Delta\hat{\beta}z}, \tag{6a}$$

$$\frac{\partial \hat{E}_y}{\partial z} + \hat{\beta}_{1,y}\frac{\partial \hat{E}_y}{\partial \tau} + i\hat{\beta}_{2,y}\frac{\partial^2 \hat{E}_y}{\partial \tau^2} + \frac{\alpha_{i,y}}{2}\hat{E}_y = i\gamma\left(|\hat{E}_y|^2 + \frac{2|\hat{E}_x|^2}{3}\right)\hat{E}_y + \frac{i\gamma}{3}\hat{E}_y^*\hat{E}_x^2 e^{2i\Delta\hat{\beta}z}, \tag{6b}$$

where * stands for the complex conjugate. The spatial variable z represents the longitudinal coordinate while the temporal variable $\tau = t - \hat{\beta}_1 z$ is expressed in a reference frame moving at the mean group-velocity : $\hat{\beta}_1 = |\hat{\beta}_{1,x} + \hat{\beta}_{1,y}|/2 = v_g^{-1}$. Indeed, the first-order dispersion coefficient $\hat{\beta}_{1,j}$ represents the inverse speed of

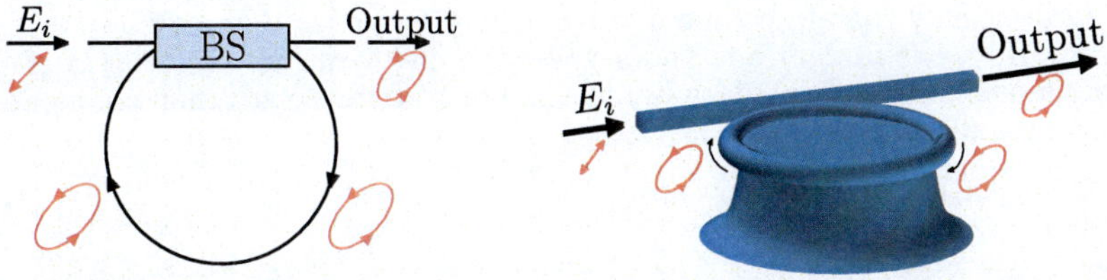

Fig. 3 (Left) Schematics of a typical optical macroscopic fiber resonator pumped with a linearly polarized field coherently coupled to the intracavity field at the beam splitter (BS). (Right) Schematics of a microscopic toroidal resonator or microring cavity. We will see in the next sections that the electric field adopts a elliptical polarization along its path in the cavity. Reproduced from [41]

the envelope of the electric field component $\hat{E}_j$ in the material. The second-order dispersion coefficient $\hat{\beta}_{2,j}$ characterizes the group-velocity dispersion of component $\hat{E}_j$ and we neglect higher order terms in the expansion of the propagation constant around the carrier frequency ω_0:

$$\hat{\beta}_j(\omega) = \sum_{k=0}^{\infty} \frac{\hat{\beta}_{k,j}}{k!}(\omega - \omega_0)^k, \quad \hat{\beta}_{k,j} = \left. \frac{d^k \hat{\beta}_j}{d\omega^k} \right|_{\omega_0}. \tag{7}$$

The zeroth-order coefficient $\Delta\hat{\beta} = \hat{\beta}_{0,x} - \hat{\beta}_{0,y} = 2\pi|n_x - n_y|/\lambda$ represents the difference between wavenumbers corresponding to the polarization components and therefore characterizes the birefringence of the material. The parameters $\alpha_{i,j}$ stands for the internal linear losses in each direction of polarization. Finally, the nonlinear coefficient γ accounts for the Kerr nonlinearity.

In order to consider the superposition of the intracavity field with the linearly polarized input beam, we impose the following boundary conditions coupling the fields at roundtrips m and $m+1$:

$$\hat{E}^{m+1}_{x,y}(z = 0, \tau) = \sqrt{T}\hat{E}_{ix,iy} + \sqrt{R}e^{-i\delta_{x,y}}\hat{E}^{m}_{x,y}(L, \tau), \tag{8}$$

where T and R are, respectively, the intensity transmission and reflection coefficients at the beam splitter, $\hat{E}_{ix,iy}$ are the polarization components of the source field, and $\delta_{x,y}$ are the phase differences between the injected field and the circulating field after having traveled a cavity length L. To derive the mean-field Lugiato–Lefever model, we integrate the field equations (6) over one roundtrip and apply the boundary conditions (8). Under the assumptions of high-finesse ($T \ll 1$), and of detuning and coupling coefficient of order 1, we introduce the continuous limit by $t = mt_R$ and

$$\frac{\partial \hat{E}_{x,y}(t, \tau)}{\partial t} = \frac{\hat{E}^{m+1}_{x,y}(z = 0, \tau) - \hat{E}^{m}_{x,y}(z = 0, \tau)}{t_R}, \tag{9}$$

with t_R the cavity roundtrip time and t a slow time variable describing the field evolution from one roundtrip to another. We can now write the vectorial Lugiato–Lefever model as two coupled evolution equations for the polarization components of the renormalized electric field envelopes $E_{x,y}$ [42, 43],

$$\frac{\partial E_x}{\partial t} = E_i \cos(\psi) - (1 + i\theta_x)E_x + i\left(|E_x|^2 + \frac{2}{3}|E_y|^2\right) E_x + \Delta\beta_1 \frac{\partial E_x}{\partial \tau} + i\eta \frac{\partial^2 E_x}{\partial \tau^2}, \tag{10a}$$

$$\frac{\partial E_y}{\partial t} = E_i \sin(\psi) - (1 + i\theta_y)E_y + i\left(|E_y|^2 + \frac{2}{3}|E_x|^2\right) E_y - \Delta\beta_1 \frac{\partial E_y}{\partial \tau} + i\eta \frac{\partial^2 E_y}{\partial \tau^2}. \tag{10b}$$

It has to be noted that, in what follows, the fast time τ will be interpreted as a spatial variable. Indeed, knowing the group-velocity of light within the cavity of fixed length L, we can identify a position corresponding to each time τ. Consequently, for a given roundtrip (meaning that we fix the value of t), the observation of $E(t,\tau)$ over the interval going from $\tau = 0$ to $\tau = t_R = L\frac{n}{c}$ precisely gives the evolution of the electric field along the cavity for the selected round trip, as τ is expressed in a reference frame moving at the group-velocity of light in the cavity. The following renormalization factors were used in order to obtain the non-dimensional system of equations (10) :

$$E_{x,y} = \sqrt{\frac{\gamma L}{\alpha}}\hat{E}_{x,y}, \qquad E_{ix,iy} = \sqrt{\frac{\gamma L T}{\alpha^3}}\hat{E}_{ix,iy}, \tag{11}$$

$$t = \frac{\alpha}{t_R}t', \qquad \tau = \sqrt{\frac{2\alpha}{|\hat{\beta}_2|L}}\tau', \tag{12}$$

$$\theta_{x,y} = \frac{1}{\alpha}\delta_{x,y}, \qquad \beta_{1,x,y} = \sqrt{\frac{2\alpha L}{|\hat{\beta}_2|}}\hat{\beta}_{1,x,y}. \tag{13}$$

We consider that the input field E_i is linearly polarized in a direction oriented with an angle ψ with respect to the slow axis and that the total losses $\alpha = (\alpha_i L + T)/2$ are the same for each direction of polarization and renormalized to 1. The frequency detunings between the injected field components and the closest corresponding cavity resonance are given by $\theta_{x,y}$ and we placed ourselves in a reference frame moving at a speed being the mean group-velocity between the polarization components, hence $\Delta\beta_1 = (\beta_{1,x} - \beta_{1,y})/2$. The parameter $\Delta\beta_1$ is known as the group-velocity mismatch (GVM) parameter; however, in what follows, we will assume that no GVM affects the propagation of light inside the cavity which is a reasonable approximation for practical applications [44]. The group-velocity dispersion (GVD) coefficient is taken to be the same in both directions. In other words, it is here assumed that the second-order dispersion acts identically on both components of polarization of the electric field. Parameter $\eta = \pm 1$ stands for the sign of the GVD. Finally, the factor 2/3 is the cross-phase modulation (XPM) coefficient through which occurs the coupling between the field components. We complete our discussion of the model by giving some physical values of the parameters that one can encounter in physical applications [29] and [45]. While using silica fibers for optical transmission of information, it is suitable to use a carrier wave of wavelength in the vicinity of $\lambda = 1.5\,\mu$m. Indeed, in this spectral region, internal losses due to absorption and scattering can go down to $\alpha_i \approx 0.2$ dB/km. In this conditions, a typical value for the group-velocity dispersion coefficient is $\hat{\beta}_2 \approx -20$ ps^2/km. The nonlinear coefficient γ characterizes the nonlinear response of the material to the presence of the electric field and is of the order of 2 (W km)$^{-1}$. A typical value for the refractive index for silica optical fibers is $n \approx 1.467$. When considering optical frequency combs, a fundamental property is the free spectral range (FSR), corresponding to the constant spacing between spectral lines. This FSR is equal to the repetition rate of the propagating pulse in the cavity, $\dfrac{1}{t_R}$.

Linear stability analysis

The homogeneous steady state solutions (HSSs) of system (10) satisfy the conditions $\partial E^s_{x,y}/\partial\tau = \partial E^s_{x,y}/\partial t = 0$ and therefore obey the equations

$$I_{ix,iy} = \left[1 + \left(\theta_{x,y} - I_{x,y} - \frac{2}{3} I_{y,x}\right)^2\right] I_{x,y}, \tag{14}$$

with $I_{ix,iy} = E^2_{ix,iy}$ and $I_{x,y} = |E^s_{x,y}|^2$. For fixed values of the control parameters that are the frequency detunings $\theta_{x,y}$ and the injected field intensities $E_{ix,iy}$, Eq. (14) possesses up to five physical solutions. A linear stability analysis of the HSSs brings insights concerning the dynamics of the system as a function of the control parameters and allows one to identify regimes that are interesting to investigate. In the case of zero GVM ($\Delta\beta_1 = 0$), we will perform it by adding a small perturbation to the HSSs ($E_{x,y} = E^s_{x,y} + E'_{x,y}$) and splitting the fields components into their real and imaginary parts. To the first order in the perturbations $E'_{x,y}$, system (10) rewrites

$$\begin{aligned}\frac{\partial E'_{x,r}}{\partial t} &= -E'_{x,r} + \theta_x E'_{x,i} - 2\left[E^s_{x,r}E'_{x,r} + E^s_{x,i}E'_{x,i} + \frac{2}{3}\left(E^s_{y,r}E'_{y,r} + E^s_{y,i}E'_{y,i}\right)\right] E^s_{x,i} \\ &\quad - \left(|E^s_x|^2 + \frac{2|E^s_y|^2}{3}\right) E'_{x,i} - \beta_2 \frac{\partial^2 E'_{x,i}}{\partial\tau^2},\end{aligned} \tag{15a}$$

$$\begin{aligned}\frac{\partial E'_{x,i}}{\partial t} &= -E'_{x,i} - \theta_x E'_{x,r} + 2\left[E^s_{x,r}E'_{x,r} + E^s_{x,i}E'_{x,i} + \frac{2}{3}\left(E^s_{y,r}E'_{y,r} + E^s_{y,i}E'_{y,i}\right)\right] E^s_{x,r} \\ &\quad + \left(|E^s_x|^2 + \frac{2|E^s_y|^2}{3}\right) E'_{x,r} + \beta_2 \frac{\partial^2 E'_{x,r}}{\partial\tau^2},\end{aligned} \tag{15b}$$

$$\begin{aligned}\frac{\partial E'_{y,r}}{\partial t} &= -E'_{y,r} + \theta_y E'_{y,i} - 2\left[E^s_{y,r}E'_{y,r} + E^s_{y,i}E'_{y,i} + \frac{2}{3}\left(E^s_{x,r}E'_{x,r} + E^s_{x,i}E'_{x,i}\right)\right] E^s_{y,i} \\ &\quad - \left(|E^s_y|^2 + \frac{2|E^s_x|^2}{3}\right) E'_{y,i} - \beta_2 \frac{\partial^2 E'_{y,i}}{\partial\tau^2},\end{aligned} \tag{15c}$$

$$\frac{\partial E'_{y,i}}{\partial t} = -E'_{y,i} - \theta_y E'_{y,r} + 2\left[E^s_{y,r}E'_{y,r} + E^s_{y,i}E'_{y,i} + \frac{2}{3}\left(E^s_{x,r}E'_{x,r} + E^s_{x,i}E'_{x,i}\right)\right]E^s_{y,r}$$
$$+\left(|E^s_y|^2 + \frac{2|E^s_x|^2}{3}\right)E'_{y,r} + \beta_2\frac{\partial^2 E'_{y,r}}{\partial\tau^2}, \tag{15d}$$

where subscripts r and i denote real and imaginary part, respectively.
We now specify the shape of the perturbations $\mathbf{E}'$ that are taken under the form of normal modes of constant amplitude $\tilde{\mathbf{E}}$ and of frequency and wavenumber $\lambda, \omega \in \mathbb{C}$, respectively, $\mathbf{E}' = \tilde{\mathbf{E}}\exp(\lambda t + i\omega\tau)$, with

$$\mathbf{E}' = \begin{bmatrix} E'_{x,r} \\ E'_{x,i} \\ E'_{y,r} \\ E'_{y,i} \end{bmatrix}, \quad \text{and} \quad \tilde{\mathbf{E}} = \begin{bmatrix} \tilde{E}_{x,r} \\ \tilde{E}_{x,i} \\ \tilde{E}_{y,r} \\ \tilde{E}_{y,i} \end{bmatrix}. \tag{16}$$

In order to investigate the temporal stability of the homogeneous states with respect to such perturbations, we substitute $\mathbf{E}'$ in Eq. (15) and set the spatial derivatives to zero. This leads us to the following eigenvalue problem:

$$\mathbf{M_T}\tilde{\mathbf{E}} = \lambda\tilde{\mathbf{E}} \tag{17}$$

with

$$\mathbf{M_T} = \begin{bmatrix} -1-2E^s_{x,r}E^s_{x,i} & \mathcal{A}_1 & -\frac{4}{3}E^s_{y,r}E^s_{x,i} & -\frac{4}{3}E^s_{y,i}E^s_{x,i} \\ \mathcal{A}_2 & -1+2E^s_{x,r}E^s_{x,i} & \frac{4}{3}E^s_{y,r}E^s_{x,r} & \frac{4}{3}E^s_{y,i}E^s_{x,r} \\ -\frac{4}{3}E^s_{x,r}E^s_{y,i} & -\frac{4}{3}E^s_{x,i}E^s_{y,i} & -1-2E^s_{y,r}E^s_{y,i} & \mathcal{A}_3 \\ \frac{4}{3}E^s_{x,r}E^s_{y,r} & \frac{4}{3}E^s_{x,i}E^s_{y,r} & \mathcal{A}_4 & -1+2E^s_{y,r}E^s_{y,i} \end{bmatrix}, \tag{18}$$

the temporal evolution matrix of the perturbation, and

$$\mathcal{A}_1 = \theta_x - 2\left(E^s_{x,i}\right)^2 - \left(|E^s_x|^2 + \frac{2|E^s_y|^2}{3}\right),$$
$$\mathcal{A}_2 = -\theta_x + 2\left(E^s_{x,r}\right)^2 + \left(|E^s_x|^2 + \frac{2|E^s_y|^2}{3}\right),$$
$$\mathcal{A}_3 = \theta_y - 2\left(E^s_{y,i}\right)^2 - \left(|E^s_y|^2 + \frac{2|E^s_x|^2}{3}\right),$$
$$\mathcal{A}_4 = -\theta_y + 2\left(E^s_{y,r}\right)^2 + \left(|E^s_y|^2 + \frac{2|E^s_x|^2}{3}\right).$$

For each value of the injection amplitude E_i, there are either one, three, or five physical homogeneous steady state solutions to the system (17) for each of which correspond four eigenvalues of $\mathbf{M_T}$ that characterize the temporal evolution of the perturbation. More precisely, the real part λ_r of the temporal eigenvalue informs about the growth rate of the perturbation. If all λ_r are negative, the perturbation will die out and the system will fall back on the base state $\mathbf{E}^s$, which is subsequently qualified as *stable*. If, on the contrary, the dominant eigenvalue has a positive real part, $\mathbf{E}'$ will grow with time, driving the perturbed state $\mathbf{E}$ away from the base state $\mathbf{E}^s$. In that case, the base state is said to be *unstable* with respect to perturbations of the form $\mathbf{E}'$. A zero dominant eigenvalue prevents us to conclude from the restricted *linear* stability analysis. The imaginary part λ_i of the temporal eigenvalue characterizes modulations and periodic behavior in the temporal evolution of the perturbation.

Considering the stable stationary states, it is worth looking at the spatial evolution of the perturbation or, in other words, their spatial stability. Back to system (15), we now set the temporal derivative to zero, inject the perturbation (16), and isolate the spatially dependent terms. This results in the eigenvalue problem

$$\mathbf{M_S}\tilde{\mathbf{E}} = \Omega\tilde{\mathbf{E}} \tag{19}$$

with $\Omega = \omega^2$,

$$\mathbf{M_S} = -\frac{1}{\beta_2}\begin{bmatrix} \mathcal{B}_1 & 1-2E^s_{x,r}E^s_{x,i} & -\frac{4}{3}E^s_{x,r}E^s_{y,r} & -\frac{4}{3}E^s_{x,r}E^s_{y,i} \\ -1-2E^s_{x,r}E^s_{x,i} & \mathcal{B}_2 & -\frac{4}{3}E^s_{y,r}E^s_{x,i} & -\frac{4}{3}E^s_{y,i}E^s_{x,i} \\ -\frac{4}{3}E^s_{x,r}E^s_{y,r} & -\frac{4}{3}E^s_{x,i}E^s_{y,r} & \mathcal{B}_3 & 1-2E^s_{y,i}E^s_{y,r} \\ -\frac{4}{3}E^s_{x,r}E^s_{y,i} & -\frac{4}{3}E^s_{x,i}E^s_{y,i} & -1-2E^s_{y,r}E^s_{y,i} & \mathcal{B}_4 \end{bmatrix}, \tag{20}$$

the spatial evolution matrix of the perturbation, and

$$\mathcal{B}_1 = \theta_x - 2(E^s_{x,r})^2 - \left(|E^s_x|^2 + \frac{2|E^s_y|^2}{3}\right),$$
$$\mathcal{B}_2 = \theta_x - 2(E^s_{x,i})^2 - \left(|E^s_x|^2 + \frac{2|E^s_y|^2}{3}\right),$$
$$\mathcal{B}_3 = \theta_y - 2(E^s_{y,r})^2 - \left(|E^s_y|^2 + \frac{2|E^s_x|^2}{3}\right),$$
$$\mathcal{B}_4 = \theta_y - 2(E^s_{y,i})^2 - \left(|E^s_y|^2 + \frac{2|E^s_x|^2}{3}\right).$$

Regarding the spatial stability, we focus on the presence of a *modulational instability* which is characterized by a transition in the spectrum of Ω from a pair of complex conjugated eigenvalues to real eigenvalues:

$$\Omega^{\text{init}} = \{a_1 \pm ia_2; b_1 \pm ib_2\} \rightarrow \Omega^{\text{fin}} = \{a_3; a_4; b_3 \pm ib_4\}, \tag{21}$$

with $a_1, a_2, a_3, a_4, b_1, b_2, b_3, b_4 \in \mathbb{R}_*^+$.
Indeed, for stationary states, if the dynamics of the perturbation (16) is characterized by a real wavenumber $\omega \in \mathbb{R}$, then it evolves as $\mathbf{E}' = \tilde{\mathbf{E}} \exp(\tilde{\omega}\tau)$ where $\tilde{\omega} = i\omega$ is purely imaginary. Consequently, the perturbation leads to a modulation of the amplitude of the electric field along the cavity, without global growth or decay, and results in the formation of a periodic pattern (sometimes called wavetrain) filling the cavity. In the fields of nonlinear optics and laser physics, the transition (21) is referred to as modulational instability.

3 Modulational Instability in the Case of Anomalous Dispersion

In this section, we investigate how the polarization of light affects the usual solutions found in the LL model in the case where the second-order dispersion is anomalous, i.e. $\eta = +1$. We begin with a standard linear stability analysis of the continuous wave solutions in this case, with respect to a perturbation of the form $\exp(i\omega\tau + \lambda t)$. We fix the detuning along the fast axis θ_y, while θ_x and E_i are the control parameters. The resulting stability map in the parameter space is shown in Fig. 4. For some representative values of θ_x (cuts (a),(b), and (c)), the total intracavity field intensity $S_0 = |E_x|^2 + |E_y|^2$ is plotted as a function of the fast time τ in Fig. 5. Along these cuts, curves with positive slope ($\partial S_0/\partial E_i > 0$) are stable, while curves with negative slope ($\partial S_0/\partial E_i < 0$) are unstable. For very low values of θ_x, moving along the cut (a), (corresponding to the input–output characteristic curve Fig. 5a), we encounter the threshold from which the single stable CW state (region I) becomes a single modulationally unstable state (region II). For higher injection values, we enter the hysteresis loop (region IV) meaning that a domain of bistability between two modulationally unstable states appears. We can continue past the hysteresis loop, where only the upper modulationally unstable state is left.
As the detuning θ_x is increased (going upwards on the map), we can see that the value of the E_i threshold of the bistability between MI states (the border between regions II and IV) monotonically decreases, while the threshold of the MI bifurcation (the border between regions I and II) decreases before increasing again, at which point it becomes superseded by the birth of a new bistability, between one stable CW state and one modulationally unstable state (region III). Two distinct bistabilities exist for different ranges of the injection power at this point, as shown in the input–output characteristics Fig. 5b. The threshold corresponding to the disappearance of region III and the one corresponding to the appearance of region IV start converging for higher values of θ_x, until both different types of bistabilities coincide which leads to the red region V where one CW state coexists with two different modulationally unstable states. The cut (c) illustrates this region, with the corresponding CW intensity curve shown in Fig. 5c.

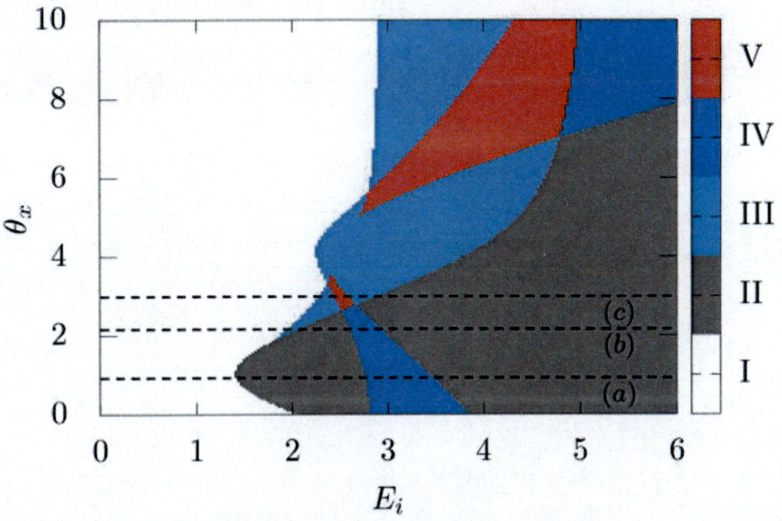

Fig. 4 Stability map of the CW solutions in the parameter plane (E_i, θ_x). Region I corresponds to parameters for which the system only hosts one single stable steady state. Region II corresponds to parameters for which the system only hosts one single modulationally unstable state. Region III corresponds to parameters for which the system hosts bistability between one stable steady state and one modulationally unstable state. Region IV corresponds to parameters for which the system hosts bistability between two modulationally unstable states. Region V corresponds to parameters for which the system hosts tristability between one stable steady state and two modulationally unstable states

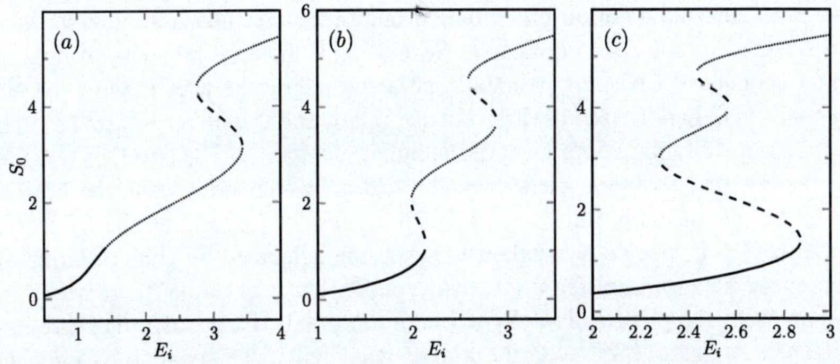

Fig. 5 Steady states corresponding to the cuts taken along the dashed lines in Fig. 4, with the total intensity S_0 given as a function of the injection for fixed values of the x detuning **(a)** $\theta_x = 1$, **(b)** $\theta_x = 2.25$, and **(c)** $\theta_x = 3$. Other parameters are the same as in Fig. 4. Full lines correspond to stable states, dashed lines correspond to unstable states, and dotted lines correspond to modulationally unstable states

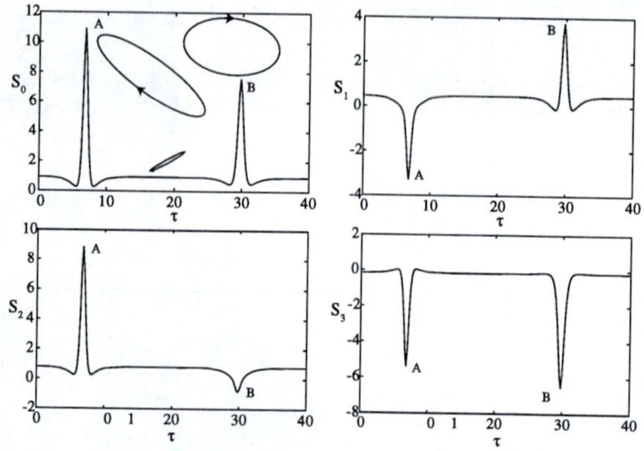

Fig. 6 Profiles of the non-normalized Stokes parameters S_0, S_1, S_2, and S_3 as a function of the fast time τ for two coexisting bright DSs with different properties. Parameters are $E_i = 2.54, \theta_x = 2.75, and\, \theta_y = 4.3$. Reproduced from [46]

The bistability between a stable CW state and a subcritically arising MI pattern is a key ingredient to the generation of DSs [11]. This coexistence leads to a range of the control parameter (where DSs can be stabilized) called the pinning range. Thanks to the polarization properties of our system, two different MI patterns exist, and in particular two different MI patterns can coexist with a stable CW state in the red areas of Fig. 4, meaning that it should be possible to generate two different types of DSs in this cavity for overlapping values of the two pinning ranges associated with each type of DS. This is shown in Fig. 6 with the profiles of two different bright DSs as a function of the fast time τ, obtained by direct numerical simulation of Eq. (10) with periodic boundary conditions. We show their total intensity S_0, and we characterize their polarization properties through the other non-normalized Stokes parameters S_1, S_2, and S_3, as defined in Sect. 2. S_0 is the total intensity of light. S_1 is the component of light that is linearly polarized along the axes x and y. S_2 also corresponds to light polarized linearly, but diagonally at 45° with respect to the x- and y-axes. S_3 is the circular component of the light polarization. The two DSs obtained in the same physical system clearly exhibit different intensities, with the brighter pulse on the left of the profile, since it has a higher peak power on the S_0 profile. The brighter DS is labeled A, while the darker one is labeled B. Their polarization properties are also quite different. The A-type DS has a negative S_1, representing a component of the polarization aligned with the y-axis. Its positive S_2 represents a component aligned at 45° with the x-axis, and finally, the negative S_3 indicates a right-handed circular component. The B-type DS has a positive S_1 (component aligned with the x-axis), negative S_2 (component aligned at 45°C with the x-axis), and negative S_3 (right-handed circular component). The background exhibits a very small positive S_2, a moderate positive S_3, and a very small negative S_3, so that its polarization is mostly aligned at 45° with the x-axis, showing only a very slight ellipticity.

It is well known that DSs generated by subcritical modulational instabilities exhibit bifurcation diagrams in the form of homoclinic snaking, emerging from the modulational instability bifurcation in the input–output characteristic curve [47]. We

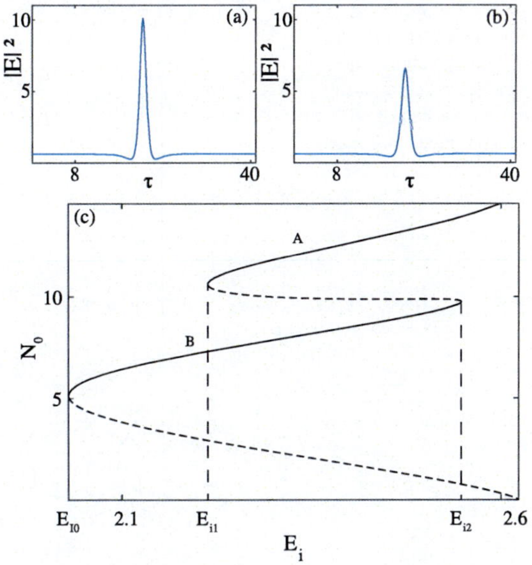

Fig. 7 (a) Profile of an A-type DS and (b) Profile of a B-type DS for fixed parameters of $E_i = 2.54$, $\theta_x = 2.75$, and $\theta_y = 4.3$. (c) Bifurcation diagram showing the L2-norm N_0 as a function of injected field amplitude E_i. Dashed (full) lines are unstable (stable) solutions. Reproduced from [46]

verify how this behavior changes due to the polarization by performing a numerical continuation in the parameter space with a predictor–corrector method initiated by a numerical simulation profile, both using periodic boundary conditions. The detuning parameters are all fixed, while the amplitude of the injection is varied. The results are shown in Fig. 7(c), where the normalized L2-norm $N_0 = \int (S_0 - \tilde{S}_0)/L \, d\tau$ is shown as a function of E_i. $\tilde{S}_0$ corresponds to the CW intensity, so that the contribution of the background is removed from the norm. As usual for a homoclinic snaking, an unstable branch emerges from the MI bifurcation of the HSS, then reaches a turning point where it becomes stable, which corresponds to the B-type DS. The DS then grows for increasing injection amplitude, until it reaches a saddle-node bifurcation where the solution becomes unstable. Then, the unstable branch goes through another turning point where it becomes the stable A-type DS. There is only a finite range of parameters where the two types of DSs coexist, between the values E_{i1} and E_{i2}. The profiles of the two coexisting solitons, types A and B, are shown in Fig. 7a, b, respectively, for fixed parameters of $E_i = 2.54$, $\theta_x = 2.75$, and $\theta_y = 4.3$. Homoclinic snaking bifurcation diagrams usually present themselves in the form of two branches oscillating across the pinning range, one corresponding to an even number of identical peaks, and the other one corresponding to an odd number of identical peaks. These two branches continue until the DSs entirely fill up the cavity, and they connect back to the patterned state. The system is then highly multistable, as these solutions all coexist in the pinning range. In our case, the behavior is more complex, as the system can host different types of vector DSs. The numerical simulations shown in Fig. 8 illustrate what happens in this case. As usual, even or odd numbers of DSs hosted in the intracavity field each correspond to a homoclinic snaking branch. For the regions where both pinning ranges coexist, the main branches divide into

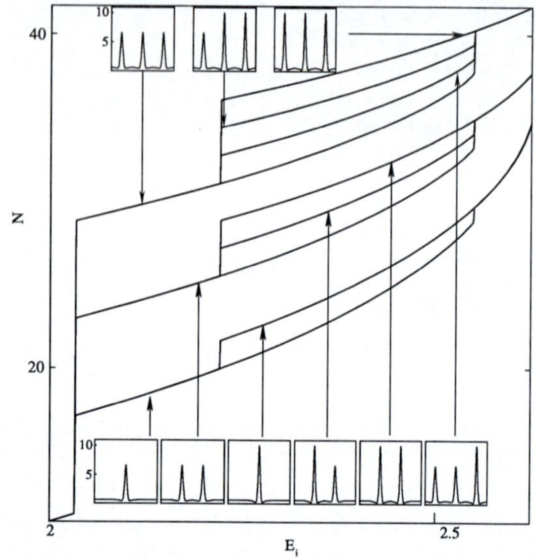

Fig. 8 Evolution of the L2-norm $\mathcal{N}$ as a function of injected field amplitude E_i for numerical simulations taken with a step of 0.001 for E_i. Reproduced from [46]

sub-branches for the different combinations of A-type and B-type DSs (see insets). In this figure, numerical simulations allow to draw the stable branches only.

4 Front-Locking in the Case of Normal Dispersion

In this section, we will be operating in the normal dispersion regime, i.e. $\eta = -1$, as this will allow us to avoid MI and focus on CW solutions only. Fronts, sometimes also called switching waves, are heteroclinic connections between different CWs. Their dynamics, including the generation of dark DSs, has been studied in the scalar case [48], without taking into account the polarization of light.

We start with the linear stability analysis of the CWs. Figure 9(a) shows the parameter space (θ_x, E_i), with a region of bistability between two stable CW states in blue, while white regions correspond to monostability. Figure 9(b) shows the bistable curve corresponding to a section of the map for a fixed value of θ_x. The system can host a mixed state, called front, connecting these two stable CWs as shown in Fig. 9(c). These fronts are generally not stationary. Depending on the value of the pump, one of the states will invade the other. For low values of E_i, the front will move toward regions where S_0 is higher, meaning that the CW corresponding to the lowest total intensity will invade the system (left panel). For high values of E_i, the front moves in the opposite direction and the CW corresponding to the highest total intensity will invade the system (right panel). There exists a point in-between, called the Maxwell point, where the front is stationary as both states are equally stable so that they will not invade each other.

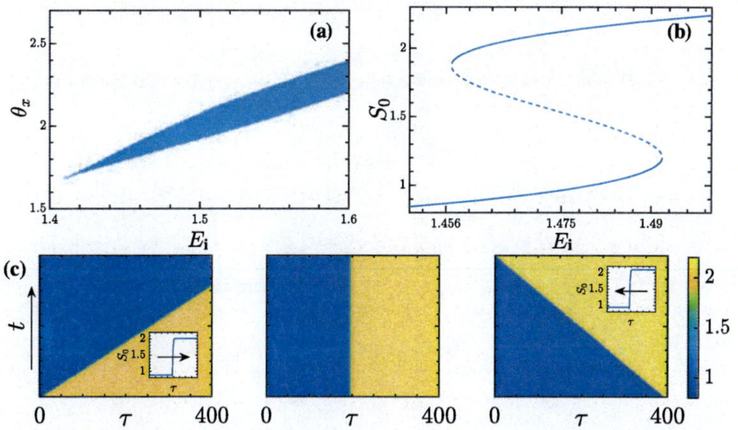

Fig. 9 (a) Stability map in the θ_x-E_i plane, with $\theta_y = 1.95$. The white (blue) region indicates monostability (bistability). (b) Bistable curve obtained for $\theta_x = 1.90$ and $\theta_y = 1.95$. Stable (unstable) CW states are denoted by solid (dashed) lines. (c) Front propagation to the right, below the Maxwell point (left panel: $E_i = 1.462$), at the Maxwell point (middle panel: $E_i = 1.46652$), and above the Maxwell point (right panel: $E_i = 1.471$). The boundary conditions are fixed to the CW values. Reproduced from [49]

We can notice that the fronts possess oscillatory tails on the bottom of their lower branch. This is an essential ingredient for the formation of DSs, as it is through these tails that the interaction between fronts is mediated. When the oscillations are absent, and the front is only exponentially decaying, there is no interaction between fronts so that it is impossible to stabilize the DSs [50–52]. The oscillatory tails however allow for an oscillatory potential of interaction with positive and repulsive interactions, depending on the relative positions of the fronts, and with equilibrium positions where those attractive and repulsive interactions are perfectly balanced. A numerical simulation with periodic boundary conditions of two fronts converging toward each other is summarized in the space–time map at the top of Fig. 10. In this case, the amplitude of the injected light is higher than the Maxwell point, so that the upper state tries to invade the system. The two well-separated fronts (Fig. 10(a)) then reach an equilibrium position as shown in Fig. 10(b), meaning that the repulsion of the fronts counterbalances their attraction as well as the front motion, and the DS is formed. In the case where the amplitude of the injected light is lower than the Maxwell point, the two fronts should start close enough so that the repulsive interaction will beat the invading motion and move them toward their equilibrium position. In our case, only the lower part of the fronts exhibit such oscillatory tails, so that only dark DSs will be formed. Two different dark DSs can be formed depending on the exact initial conditions, Fig. 10(b) or Fig. 10(c). The latter is the simplest solution, with a simple soliton consisting of only a dip in the total light intensity, while the first one is another more complex solution, with a single bump at the bottom of the profile. These two solutions thus can coexist for the same values of the parameters. They also

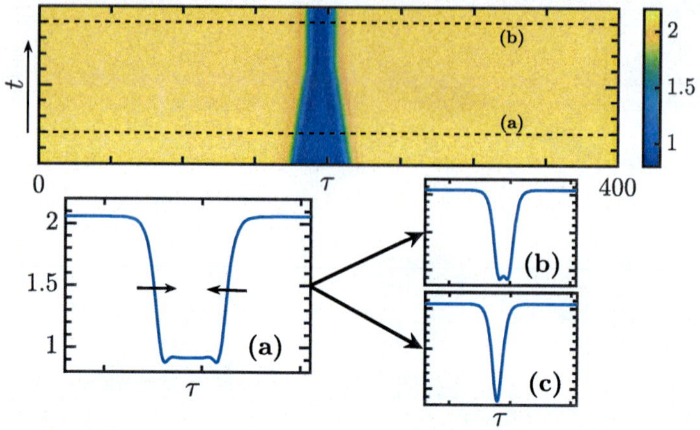

Fig. 10 Two fronts interact in an attractive way, which leads to the formation of vectorial dark dissipative solitons with the total intensity in the τ-t map (top panel). (**a**, **b**) Cross-sections along the dashed lines indicated in the τ-t map. (**c**) stable single dip DVDS. Numerical simulations of Eq. 10 are obtained for the parameters $E_i = 1.4675$, $\theta_x = 1.90$, and $\theta_y = 1.95$. Reproduced from [49]

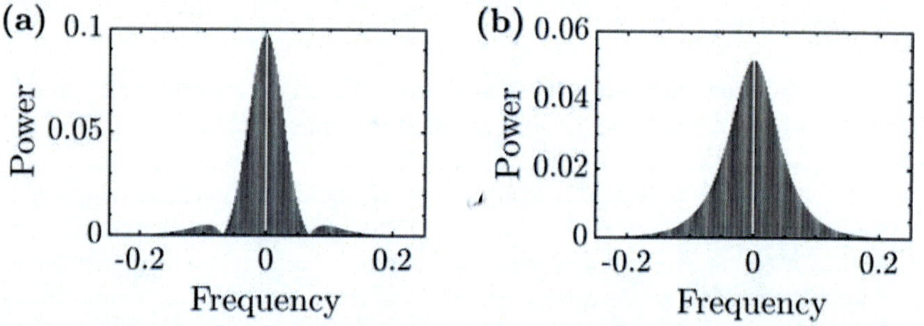

Fig. 11 Optical frequency combs corresponding to the two coexisting profiles shown in Fig. 10. Parameters are $E_i = 1.46655$, $\theta_x = 1.90$, and $\theta_y = 1.95$. Reproduced from [49]

have similar spectral properties, as the spectral contents of DSs are optical frequency combs shown in Fig. 11. The simplest DS shows an envelope in the shape of a sech^2 curve [34], while the solution from Fig. 10b shows additional bumps at each side of the envelope of the comb. The two combs share the same FSR, as this is only determined by the repetition rate which is the same for all of our DSs, since they move in the cavity with the same group velocity.

To investigate this coexistence, we show the bifurcation diagram for these two solutions with the detuning parameters fixed, the injected field amplitude as the control parameter, and we describe the branches through the normalized L2-norm $\mathcal{N} = \int S_0/L d\tau$, where L is the size of the system. The profiles obtained through numerical simulations were used to initiate a predictor–corrector continuation algorithm [53]. The so obtained diagram, shown in Fig. 12, is called a collapsed heteroclinic snaking, because the branch oscillates with an exponentially decreasing amplitude around the Maxwell point until it collapses on it. A closeup of the collapsed snaking curve is shown alongside the bifurcation diagram. Profiles associated with the points (a–h) are shown underneath the bifurcation diagram. Each turning point of the snaking curve corresponds to a change of stability, and a new stable state brings a new solution, with an additional bump at the extremum of the DS each time, as can be seen on profiles (a–d). This branch emerges from the point SN_1 where the CW regains its stability and the higher stable CW appears, and after oscillating and collapsing, it connects to the point SN_2 where the lower stable CW loses its stability. This point actually corresponds to a bifurcation to a MI state, that exists for a range of the injection amplitude that is extremely small before turning unstable. When the branch has collapsed on the Maxwell point and thus stops oscillating, the solution stops changing and only the width of the DS varies as we continue along the curve. Indeed, profiles (e–h) show that the width increases as we go down the branch, meaning that the lower state invades the entire system. At the end of this branch, we connect back to the lower stable CW at the point SN_2 as the lower state has entirely invaded the system.

The DSs formed by this mechanism are robust structures, and they can coexist in the same system as shown in Fig. 13 which depicts the space–time map of a numerical simulation with periodic boundary conditions, initialized with random noise added to a constant value between the two stable CWs. We also show their normalized Stokes parameters, S_0 to s_3, to illustrate that the two different coexisting dark DSs exhibit slightly different peak powers and polarization properties.

5 Tristability in the Case of Normal Dispersion

We will now consider the propagation of light in a pumped Kerr cavity in the normal dispersion regime for higher values of the detuning parameters. As this parameter is increased, the hysteresis loop formed by the CWs widens and undergoes new bifurcations. In particular, a second hysteresis loop can appear leading to the existence of three different solutions for the same value of the detuning, as shown by the linear stability analysis represented in Fig. 14 in the phase space (E_i, θ_x). In this map, light blue regions correspond to two stable solutions coexisting, while light red regions correspond to three stable solutions coexisting. At the very bottom of the figure (low values of θ_x), for a fixed value of the detunings and varying E_i, only one light blue region (II) is encountered, corresponding to a simple hysteresis loop. Higher in the map, two distinct blue regions can be encountered while varying E_i for fixed θ_x as highlighted by the dashed line (a), which is the above-mentioned case of two hysteresis loops. The input–output characteristics of the CWs for this cut are shown in Fig. 15(a) with the output intensity as a function of the injected field amplitude. Because the two hysteresis loops widen as the detuning is increased,

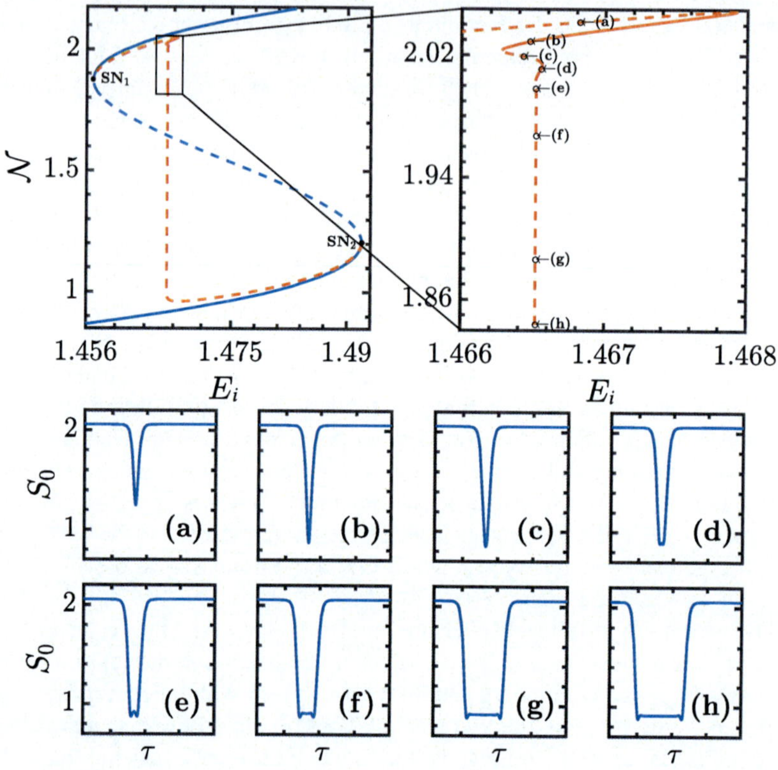

Fig. 12 Bifurcation diagram showing the L2-norm $\mathcal{N}$ as a function of injected field amplitude E_i. The right panel is a zoom around the snaking curve. Solid (dotted) curves correspond to stable (unstable) localized solutions. **(a–h)** Profiles of the total field intensity S_0. Parameters are $\theta_x = 1.90$ and $\theta_y = 1.95$. Periodic boundary conditions were used. Reproduced from [49]

they end up overlapping, leading to the first light red tristable region (IV). As θ_x is increased, this tristable region disappears as the upper hysteresis loop collapses into a single monotonous curve, going back into simple bistability represented by the light blue region II. The light red region IV then reappears for higher θ_x values through the inverse process, as a new hysteresis loop reappears near the upper saddle-node bifurcation. An example of tristable region is highlighted by the dashed line (b), for which the corresponding input–output characteristics of the CWs are shown in Fig. 15(b). This curve shows the coexistence of three stable solutions for the same value of all parameters. We can also notice in Fig. 14 the presence of regions with modulationally unstable states bordering on the right the bistable and tristable regions. These states will not be relevant here as we will operate far from them, and thus far from any branch corresponding to modulationally unstable states.

The polarization properties of each of these CW solutions are already of interest. Despite an injected light with a constant and completely linear polarization, cross-

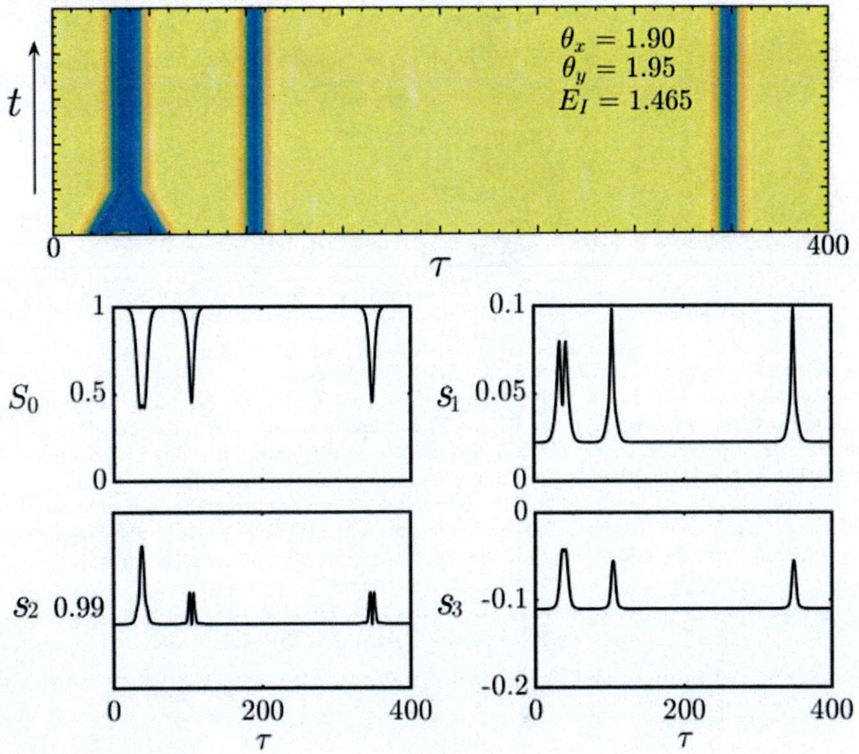

Fig. 13 (Top) (τ, t) map of the total intensity S_0. Parameters are the same as in Fig. 11. (Bottom) The Normalized Stokes parameters S_0, s_1, s_2, and s_3 as functions of the fast time τ. Reproduced from [49]

phase modulation (XPM) allows the intracavity fields to show a rich complexity in how each of the Stokes parameters evolves as shown in the previous sections. To investigate this further, the normalized Stokes parameters for the current set of parameters are shown in Fig. 16. All of the Stokes components undergo hysteresis loops at the same values of the pump, as those hystereses correspond to the ones from Fig. 15, but with very different output values, so that the three CWs exhibit very different polarization properties. The linear components can be in any configuration; however, one common trait between those CWs is that they are elliptically polarized, even if slightly. It can also be noted that the sign of s_3 changes between both curves Fig. 15(a) and Fig. 15(b), indicating that a change in the value of the detunings can also lead to notable variations in the polarization properties of the intracavity light.

Similarly to the previous section, the coexistence of stable CW solutions for fixed values of the parameters allows for fronts connecting them to emerge. The most important feature of these fronts is the presence of oscillatory tails damped around the CW, allowing two fronts evolving toward each other to interlock and subsequently constitute a stable structure embedded in a CW background and with a

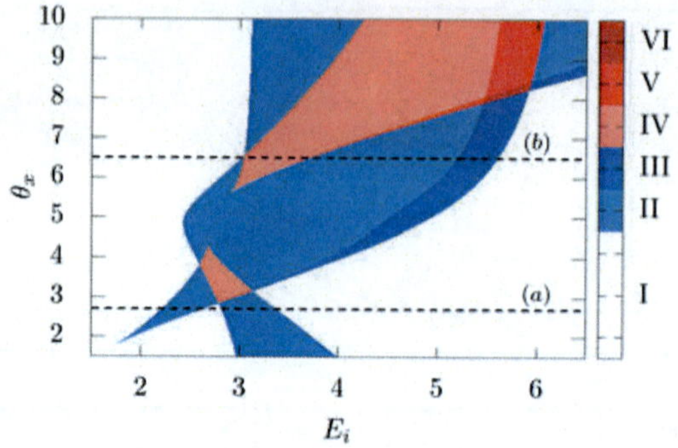

Fig. 14 Stability regions in the parameter space E_i–θ_x. Parameters are: $\theta_y = 5$ and $\eta = -1$. Region I corresponds to monostability, with the presence of only one stable state. Region II corresponds to bistability between two stable states. Region III corresponds to bistability between one stable state and one modulationally unstable state. Region IV corresponds to tristability between three stable states. Region V corresponds to tristability between two stable states and one modulationally unstable state. Finally, region VI corresponds to tristability between one stable state and two modulationally unstable states. Examples of two consecutive bistable curves and of a tristable curve along the dashed lines (**a**) and (**b**) are shown in Fig. 15a, b, respectively. Reproduced from [41]

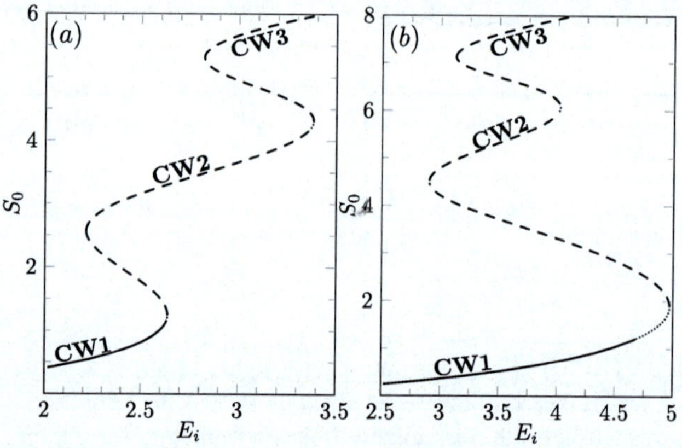

Fig. 15 Bistable curve (**a**) and tristable curve (**b**) obtained for $\theta_y = 5$. These curves are taken along the dashed lines (**a**) and (**b**) in Fig. 14 corresponding to $\theta_x = 2.7$ and $\theta_x = 6.5$, respectively. Full lines correspond to stable states, dashed lines correspond to unstable states, and dotted lines correspond to modulationally unstable states. Reproduced from [41]

finite width. This *front-locking mechanism* arises as a result of the complex balance between attractive and repulsive interactions occurring through the front oscillatory tails, with the stable structures being dissipative solitons.

The bifurcation diagram for these dark DSs, corresponding to the case with two distinct hysteresis loops shown in Fig. 15(a), is presented in Fig. 17. Similarly to the previous bifurcation diagrams, it was obtained through numerical continuation with a predictor–corrector algorithm that was initialized with a profile obtained by

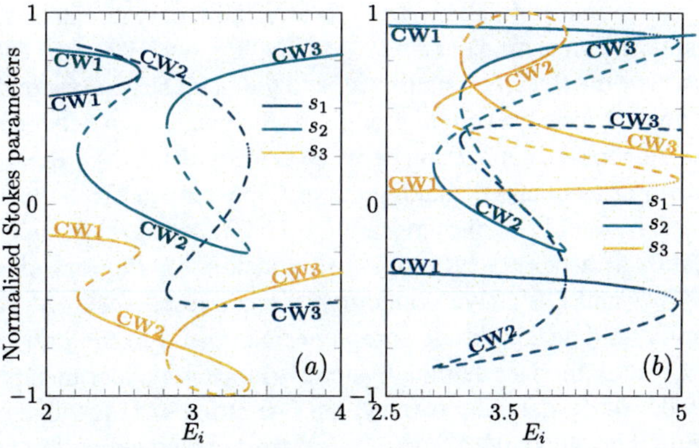

Fig. 16 (a) Stokes parameters of the CW solutions plotted in Fig. 15a and (b) Stokes parameters of the CW solutions plotted in Fig. 15b. The meaning of the dashed, dotted, and solid lines is the same as in Fig. 15. The parameters are the same as in Fig. 15. The injected light is linearly polarized, with $s_1 = s_3 = 0$ and s_2=1. Reproduced from [41]

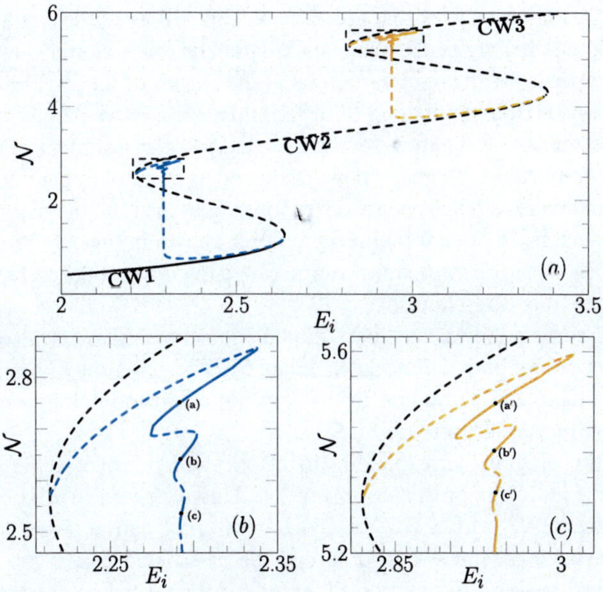

Fig. 17 (a) Bifurcation diagram showing the L2-norm $\mathcal{N}$ as a function of the injected field amplitude E_i. Stable (unstable) states are denoted by solid (dashed) lines. MI states are denoted by dotted lines. Parameters are $\theta_x = 2.7$ and $\theta_y = 5$. (b) Close-up on the snaking curves collapsing onto the Maxwell point of the left bistability. (c) Close-up on the snaking curves collapsing onto the Maxwell point of the right bistability. Reproduced from [41]

time-stepping numerical simulations with periodic boundary conditions. Two types of fronts can appear, connecting either CW3 with CW2, or CW2 with CW1. For each type of front, a type of DS can be stabilized, and the branches corresponding to each of these two types of dark DSs (light orange for the connection between CW3 with CW2 and dark blue for the connection between CW2 with CW1) present themselves again in the form of a collapsed snaking curve. They emerge from the saddle-node bifurcations corresponding to the appearance of CW3 and CW2 and connecting to the modulational instability bifurcation responsible for the loss of stability of, respectively, CW1 and CW2. We notice that the collapsed snakings exhibit more oscillations and that their amplitudes are larger compared to the previous section, which is explained by the fact that the hysteresis loops are wider than previous due to the higher value of the detuning parameters. The profiles of the Stokes parameters corresponding to the points (a)–(c) and (a′)–(c′) are depicted in Fig. 18 and show that again, each turning point leading to a new stable branch of dark DS corresponds to a new type of solution, with an additional bump at the bottom of the profile. As a dark DS corresponds to an excursion from a background corresponding to a CW solution into another CW solution for a short domain of the fast time τ, the Stokes parameters behave in the same way as the ones of the CW solutions. Since the CWs have very different polarization properties that are always elliptical, the DSs exhibit the same properties. DSs that belong to the same snaking curve share more or less the same polarization properties with slight variations as the shapes of the profiles differ. The differences between DSs that belong to different snaking curves are however more pronounced, as the two CWs involved are not the same in each case. This means that the background values, the peak powers as well as the Stokes parameters are all very different in this case. The spectra corresponding to the total intensity profiles of these solutions are again optical frequency combs, shown in Fig. 19. The frequency combs of the DSs share the same free spectral range regardless of their types, as they all share the same repetition rate.

The branch corresponding to modulationally unstable states is also shown in Fig. 20 to evidence that both collapsed snakings occur in a regime that is far enough from it, so that there can be no confusion between our front-locking-induced DSs and modulational instability-induced DSs.

The bifurcation diagram corresponding to the case with two overlapping hysteresis loops shown in Fig. 15b is presented in Fig. 21. This time, the two types of fronts connecting either CW3 with CW2, or CW2 with CW1 can appear for the same values of the system parameters, and so can the associated dark DSs. We obtain again a collapsed snaking bifurcation diagram for the branches corresponding to each of these two types of DSs (light orange for the connection between CW3 with CW2 and dark blue for the connection between CW2 with CW1), except that the two oscillating curves now overlap for a certain range of the injected field amplitude called C. As such, two types of dark DSs exist again, (a)–(c) connecting CW1 to CW2 and (a′)–(c′) connecting CW2 to CW3, shown in Fig. 22. The two types of DSs, similarly to the previous case, have very different peak powers, background values, and polarization properties. In the region C, the system can host dark DSs of the two types, (a)–(c) and (a′)–(c′). However, they cannot be hosted at the same time

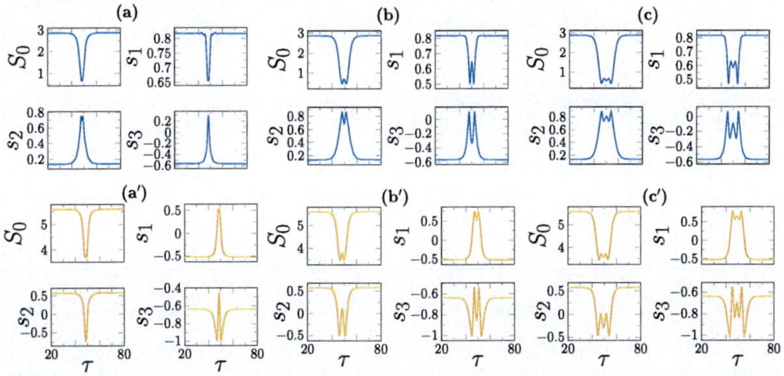

Fig. 18 Profiles of the Stokes parameters S_0, s_1, s_2, and s_3 as a function of the fast time τ for the stable DS solutions indicated in Fig. 17. The size of the system was taken as $L = 100$. Profiles **(a)**–**(c)** correspond to the region highlighted on the left, while (a′)–(c′) correspond to the region highlighted on the right. Injection amplitude values are $E_i =$ (a) 2.2917, (b) 2.2922, (c) 2.2926 (a′) 2.9394, (b′) 2.9406, and (c′) 2.9406. Reproduced from [41]

Fig. 19 Vector Kerr combs corresponding to Fourier transform of the stable DS solutions shown in Fig. 18. Reproduced from [41]

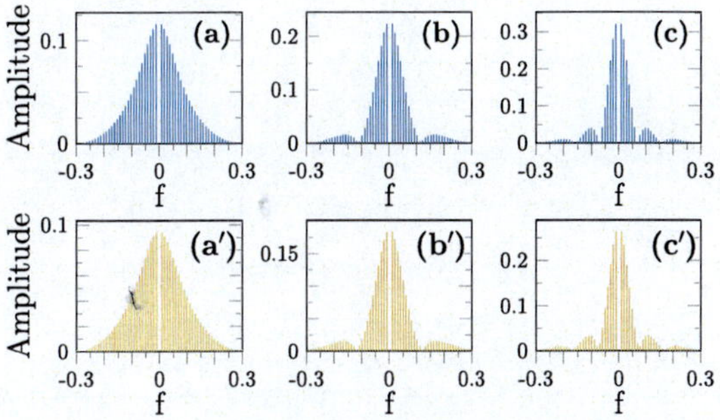

Fig. 20 Close-up showing the MI branch emerging from the end of the upper snaking curve from Fig. 17a. Reproduced from [41]

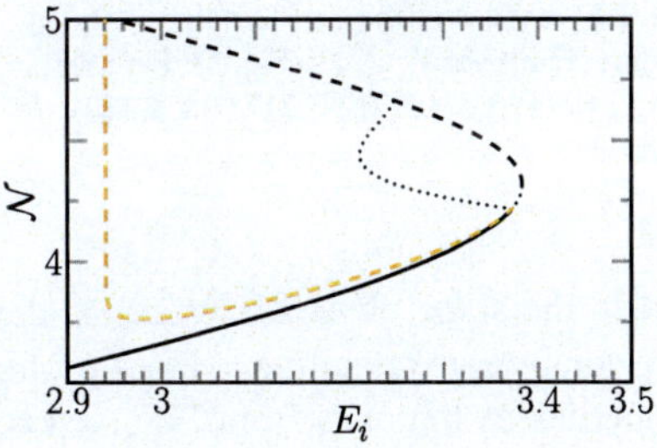

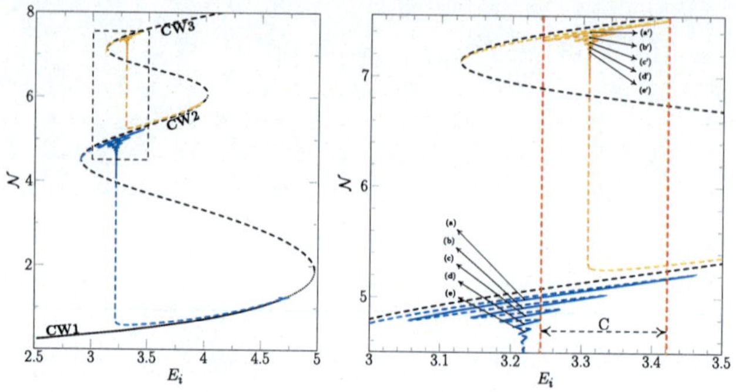

Fig. 21 Tristable heteroclinic snaking. Left panel: Bifurcation diagram showing the L2-norm $\mathcal{N}$ as a function of the injected field amplitude E_i. Stable (unstable) states are denoted by solid (dashed) lines. MI states are denoted with dotted lines. Parameters are $\theta_x = 6.5$ and $\theta_y = 4.5$. The size of the system was taken as $L = 200$. Right panel: Close-up on the snaking curves collapsing onto the Maxwell point of each respective bistability showing the coexistence region C. Reproduced from [41]

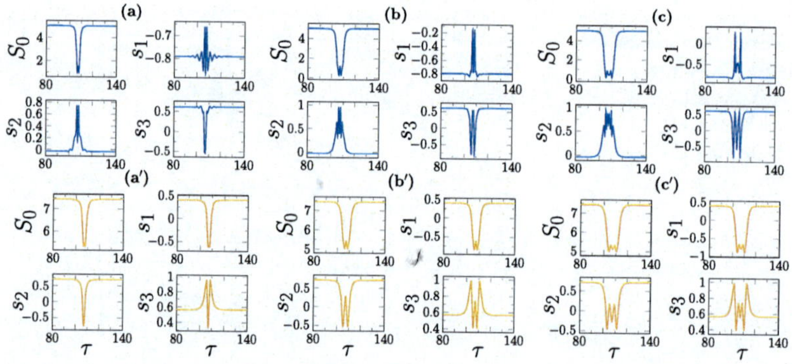

Fig. 22 Profiles of the Stokes parameters S_0, s_1, s_2, and s_3 as a function of the fast time τ for the stable solutions indicated in Fig. 21. Parameters are the same as in Fig. 21. Injection amplitude values are $E_i =$ (a) 3.2195, (b) 3.217, (c) 3.2209, (a′) 3.3094, (b′) 3.3087, and (c′) 3.3085. Reproduced from [41]

and in the same physical system as they require different background intensities. The combs corresponding to these profiles are drawn in Fig. 23. They share the same properties as the ones from Fig. 19, except for a different FSR that is only due to the fact we used a different cavity length in this case, for the sake of numerical convenience.

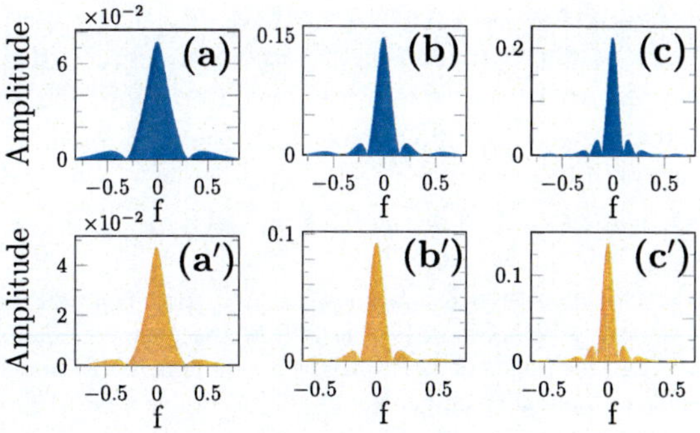

Fig. 23 Vector Kerr combs corresponding to the stable solutions shown in Fig. 22. Reproduced from [41]

6 Conclusions and Perspectives

We have investigated the formation of dissipative solitons in high-Q micro- and macro-resonators driven by a linearly polarized injected light. We have studied the influence of the polarization degrees of freedom in these optical resonators that are described by the vectorial Lugiato–Lefever equation. These additional degrees of freedom impact the homogeneous steady states solutions by creating a new critical second-order transition allowing for the generation of multistability. We have shown that there can be up to three coexisting stable solutions. Two operating regimes were considered: anomalous and normal dispersion.

In the anomalous dispersion regime, the linear stability analysis indicated that one or more homogeneous solutions suffer modulational instability. We have focused on the situation where two modulationally unstable homogeneous solutions coexist with one stable CW solution. Besides periodic solutions, an infinite number of DSs characterized by either a odd or even number of peaks coexist with the periodic and the CW solutions. The system can host two types of DSs, as the stable CW solution provides the continuous background needed to host them, and the system exhibits multistability. We have shown that their bifurcation diagram each follows an homoclinic snaking type of bifurcation. These two types of DSs exhibit different peak powers and polarization properties and can coexist in the same physical system.

In the normal dispersion regime, we first showed that for low values of the detuning, we obtain a bistable regime where DSs can be formed thanks to the interaction between switching waves, or front, connecting pairs of CW solutions. The bifurcation diagram of these DSs obeys collapsed heteroclinic snaking. Then, for higher values of the detuning, there can be up to the three coexisting CW solutions. The system can also host two types of DSs, each exhibiting a separate collapsed snaking. Again, the two types of DSs exhibit different peak powers and polarization properties. For the right values of the detuning parameters, these two collapsed snaking curves can have an overlapping domain of stability and the two types can coexist for the same values of the system parameters, however not in the same physical system.

These results bring new insights into the properties of dissipative structures in driven resonators when the polarization degrees of freedom are taken into account. In order to further complete the description, the influence of the group-velocity mismatch between the two polarization components should be considered. However neglected in the present study, this property can significantly impact the dynamics, as it leads to a drift of the DSs caused by a breaking of the $\tau \rightarrow -\tau$ symmetry.

Acknowledgements K.P. acknowledges support from the Fonds Wetenschappelijk Onderzoek-Vlaanderen FWO (G0E5819N) and the Methusalem Foundation. M.T. acknowledges financial support from the Fonds de la Recherche Scientifique FNRS under Grant CDR no. 35333527 "Semiconductor optical comb generator". A part of this work is supported by the "Laboratoire Associé International" University of Lille—ULB on "Self-organization of light and extreme events" (LAI-ALLURE).

References

1. I. Prigogine, R. Lefever, Symmetry breaking instabilities in dissipative systems. ii. J. Chem. Phys. **48**(4), 1695–1700 (1968)
2. P. Glansdorff, I. Prigogine, *Thermodynamic Theory of Structure, Stability and Fluctuations* (Willey, 1971)
3. V. Castets, E. Dulos, J. Boissonade, P. De Kepper, Experimental evidence of a sustained standing turing-type nonequilibrium chemical pattern. Phys. Rev. Lett. **64**(24), 2953–2957 (1990)
4. Q. Ouyang, H.L. Swinney, Transition from a uniform state to hexagonal and striped turing patterns. Nature **352**, 610–612 (1991)
5. J.D. Murray, *Mathematical Biology I (An Introduction* (Springer, Berlin, 2001)
6. R. Lefever, O. Lejeune, On the origin of tiger bush. Bull. Math. Biol. **59**(2), 263–294 (1997)
7. M. Rietkerk, S.C. Dekker, P.C. De Ruiter, J. van de Koppel, Self-organized patchiness and catastrophic shifts in ecosystems. Science **305**(5692), 1926–1929 (2004)
8. M. Tlidi, R. Lefever, A. Vladimirov, On vegetation clustering, localized bare soil spots and fairy circles, in *Dissipative Solitons: From Optics to Biology and Medicine* (Springer, 2008), pp. 1–22
9. M. Tlidi, K. Staliunas, K. Panajotov, A.G. Vladimirov, M.G. Clerc, Localized structures in dissipative media: from optics to plant ecology (2014)
10. L.A. Lugiato, R. Lefever, Spatial dissipative structures in passive optical systems. Phys. Rev. Lett. **58**(21), 2209 (1987)
11. A.J. Scroggie, W.J. Firth, G.S. McDonald, M. Tlidi, R. Lefever, L.A. Lugiato, Pattern formation in a passive kerr cavity. Chaos, Solitons & Fractals **4**(8–9), 1323–1354 (1994)
12. M. Tlidi, K. Panajotov, Cavity solitons: dissipative structures in nonlinear photonics. Rom. Rep. Phys **70**, 406 (2018)
13. G. Kozyreff, M. Tlidi, Optical patterns with different wavelengths. Phys. Rev. E **69**, 066202 (2004). (Jun)
14. M. Tlidi, P. Mandel, R. Lefever, Localized structures and localized patterns in optical bistability. Phys. Rev. Lett. **73**(5), 640 (1994)
15. W. van Saarloos, P.C. Hohenberg, Fronts, pulses, sources and sinks in generalized complex Ginzburg-Landau equations. Phys. D: Nonlinear Phenom. **56**(4), 303–367 (1992)
16. N.N. Rosanov, I transverse patterns in wide-aperture nonlinear optical systems. Prog. Opt. **35**, 1–60 (1996)
17. U. Bortolozzo, S. Residori, Storage of localized structure matrices in nematic liquid crystals. Phys. Rev. Lett. **96**(3), 037801 (2006)

18. A. Jacobo, D. Gomila, M.A. Matías, P. Colet, Logical operations with localized structures. New J. Phys. **14**(1), 13040 (2012)
19. F. Pedaci, G. Tissoni, S. Barland, M. Giudici, J. Tredicce, Mapping local defects of extended media using localized structures. Appl. Phys. Lett. **93**(11), 111104–111104-3 (2008)
20. C. Mou, S.V. Sergeyev, A.G. Rozhin, S.K. Turitsyn, Bound state vector solitons with locked and precessing states of polarization. Opt. Express **21**(22), 26868–26875 (2013)
21. X. Hu, J. Guo, G.D. Shao, Y.F. Song, L.M. Zhao, L. Li, D.Y. Tang, Dissipative dark-bright vector solitons in fiber lasers. Phys. Rev. A **101**, 063807 (2013)
22. V. E. Lobanov, A. E. Shitikov, R. R. Galiev, K. N. Min'kov, O. V. Borovkova, N. M. Kondratiev, Generation of vector flat-top solitons and hybrid bright-flat-top soliton complexes in optical microresonators via modulated pump. Phys. Rev. A **104**, 063511 (2021). (Dec)
23. S.V. Sergeyev, C. Mou, E.G. Turitsyna, A. Rozhin, S.K. Turitsyn, K. Blow, Spiral attractor created by vector solitons. Light.: Sci. & Appl. **3**(1), e131 (2014)
24. C. Mou, S. Sergeyev, A. Rozhin, S. Turistyn, All-fiber polarization locked vector soliton laser using carbon nanotubes. Opt. Lett. **36**(19), 3831–3833 (2011)
25. S.T. Cundiff, B.C. Collings, N.N. Akhmediev, J.M. Soto-Crespo, K. Bergman, W.H. Knox, Observation of polarization-locked vector solitons in an optical fiber. Phys. Rev. Lett. **82**(20), 3988–3991 (1999)
26. V.J. Sánchez-Morcillo, I. Pérez-Arjona, G.J. de Valcárcel F. Silva, E. Roldán, Vectorial kerr-cavity solitons. Opt. Lett. **25**(13), 957–959 (2000)
27. S.V. Sergeyev, C. Mou, A. Rozhin, S.K. Turitsyn, Vector solitons with locked and precessing states of polarization. Opt. Express **20**(24), 27434–27440 (2012)
28. V. Tsatourian, S.V. Sergeyev, C. Mou, A. Rozhin, V. Mikhailov, B. Rabin, P. S. Westbrook, S.K. Turitsyn, Polarisation dynamics of vector soliton molecules in mode locked fibre laser. Sci. Rep. **3**, 3154 (2013)
29. F. Leo, S. Coen, P. Kockaert, S.-P. Gorza, P. Emplit, M. Haelterman, Temporal cavity solitons in one-dimensional kerr media as bits in an all-optical buffer. Nat. Photonics **4**(7), 471–476 (2010)
30. A.B. Matsko, A.A. Savchenkov, W. Liang, V.S. Ilchenko, D. Seidel, L. Maleki, Mode-locked kerr frequency combs. Opt. Lett. **36**(15), 2845–2847 (2011)
31. L.A. Lugiato, F. Prati, M.L. Gorodetsky, T.J. Kippenberg, From the Lugiato–Lefever equation to microresonator-based soliton kerr frequency combs. Philos. Trans. R. Soc. A: Math., Phys. Eng. Sci. **376**(2135), 20180113 (2018)
32. P.J. Delfyett, S. Gee, M-T. Choi, H. Izadpanah, W. Lee, S. Ozharar, F. Quinlan, T. Yilmaz, Optical frequency combs from semiconductor lasers and applications in ultrawideband signal processing and communications. J. Lightwave Technol. **24**(7), 2701 (2006). (Jul)
33. Y.K. Chembo, Kerr optical frequency combs: theory, applications and perspectives. Nanophotonics **5**(2), 214–230 (2016)
34. T.J. Kippenberg, A.L. Gaeta, M. Lipson, M.L. Gorodetsky, Dissipative kerr solitons in optical microresonators. Science **361**(6402), eaan8083 (2018)
35. T. Fortier, E. Baumann, 20 years of developments in optical frequency comb technology and applications. Commun. Phys. **2**(1) (2019)
36. M. Lipson, J.S. Levy, A. Gondarenko, M.A. Foster, A.C. Turner-Foster, A.L. Gaeta, Cmos-compatible multiple-wavelength oscillator for on-chip optical interconnects. Nat. Photonics **4**(1), 37–40 (2010)
37. P. Del'Haye, T. Herr, E. Gavartin, M.L. Gorodetsky, R. Holzwarth, T.J. Kippenberg, Octave spanning tunable frequency comb from a microresonator. Phys. Rev. Lett. **107**, 063901 (2011). (Aug)
38. A.A. Savchenkov, A.B. Matsko, W. Liang, V.S. Ilchenko, D. Seidel, L. Maleki, Kerr combs with selectable central frequency. Nat. Photonics **5**(5), 293–296 (2011)
39. R.A. Chipman, W.-S.T. Lam, G. Young. *Polarized Light and Optical Systems* (CRC Press, 2018)
40. G.P. Agrawal, *Nonlinear Fiber Optics*, 6th Edn. (Academic Press, 2019)

41. Kostet, Y. Soupart, K. Panajotov, M. Tlidi, Coexistence of dark vector soliton Kerr combs in normal dispersion resonators. Phys. Rev. A **104**, 053530 (2021)
42. M. Haelterman, S. Trillo, S. Wabnitz, Polarization multistability and instability in a nonlinear dispersive ring cavity. J. Opt. Soc. Am. B **11**, 446–456 (1994)
43. T. Hansson, M. Bernard, S. Wabnitz, Modulational instability of nonlinear polarization mode coupling in microresonators. J. Opt. Soc. Am. B **35**, 835–841 (2018)
44. A.U. Nielsen, B. Garbin, S. Coen, S.G. Murdoch, M. Erkintalo, Coexistence and interactions between nonlinear states with different polarizations in a monochromatically driven passive kerr resonator. Phys. Rev. Lett. **123**, 013902 (2019)
45. S. Coen, M. Tlidi, Ph. Emplit, and M. Haelterman, Convection versus Dispersion in Optical Bistability. Phys. Rev. Lett. **83**, 2328 (1999)
46. E. Averlant, M. Tlidi, K. Panajotov, L. Weicker, Coexistence of cavity solitons with different polarization states and different power peaks in all-fiber resonators. Opt. Lett. **42**, 2750–2753 (2017)
47. J. Burke, E. Knobloch, Homoclinic snaking: structure and stability. Chaos: Interdiscip. J. Nonlinear Sci. **17**(3), 037102 (2007)
48. P. Parra-Rivas, E. Knobloch, D. Gomila, L. Gelens, Dark solitons in the Lugiato-Lefever equation with normal dispersion. Phys. Rev. A **93**, 063839 (2016). (Jun)
49. B. Kostet, S. S. Gopalakrishnan, E. Averlant, Y. Soupart, K. Panajotov, and M. Tlidi, Vectorial dark dissipative solitons in Kerr resonators. OSA Continuum **4**, 1564–1570 (2021)
50. P. Coullet, L. Gil, D. Repaux, Defects and subcritical bifurcations. Phys. Rev. Lett. **62**(25), 2957 (1989)
51. J. Knobloch, T. Wagenknecht, Homoclinic snaking near a heteroclinic cycle in reversible systems. Phys. D: Nonlinear Phenom. **206**(1–2), 82–93 (2005)
52. Y.-P. Ma, J. Burke, E. Knobloch, Defect-mediated snaking: a new growth mechanism for localized structures. Phys. D: Nonlinear Phenom. **239**(19), 1867–1883 (2010)
53. A. Dhooge, W. Govaerts, Yu.A. Kuznetsov, Matcont: a matlab package for numerical bifurcation analysis of odes. ACM Trans. Math. Softw. (TOMS) **29**(2), 141–164 (2003)